Revise A2 Chemistry For AQA

Paddy Gannon

Heinemann

Heinemann Educational Publishers

Halley Court, Jordan Hill, Oxford, OX2 8EJ

a division of Reed Educational & Professional Publishing Ltd

Heinemann is a registered trademark of Reed Educational & Professional Publishing Ltd

OXFORD MELBOURNE AUCKLAND
JOHANNESBURG BLANTYRE GABORONE
IBADAN PORTSMOUTH NH (USA) CHICAGO

First published 2002

ISBN 0 435 58309 3

05 04 03 02
10 9 8 7 6 5 4 3 2

Commissioning/development editor: Paddy Gannon

Edited by Mary Korndorffer

Index compiled by Jean Macqueen

Typeset by Saxon Graphics Ltd, Derby

Printed and bound in Great Britain by Thomson Litho Ltd, Glasgow

Acknowledgements

The author wishes to thank Roy Farrow for his contribution to this book.

The publishers have made every effort to trace the copyright holders, but if they have inadvertently overlooked any, they will be pleased to make the necessary arrangements at the first opportunity.

Tel: 01865 888058 www.heinemann.co.uk

Contents

Introduction

This book is designed to help you study the **AQA Advanced level General Certificate of Education in Chemistry** (Specification 6421). It concentrates on the facts you need to know to understand the concepts and provides you with opportunities practice answering questions.

The content of each module follows the specification and divides it up into manageable chunks. Each chunk is covered by a number of pages in which you'll see:

- The specification material **condensed** and **summarised** into the most important points and squeezed unbelievably into 112 pages.
- **Tips boxes**: these point out things you need to watch out for or give you general tips about how to get the Chemistry right.
- **Diagrams**: these are drawn simply, so you can reproduce them in exams, if needed.
- **Quick check questions**: these check that you're understanding the material covered on a particular double page spread.
- **Exam *style* questions**: these dig a little deeper than the quick check questions and give you an idea of the things AQA may ask you in exams.
- **Answers**: it's always nice to double check you are heading in the right direction, so brief answers are given at the back of the book.
- **Key words**: highlighted in **bold**, so you can immediately see that a new or important word has been introduced or used. Make sure that you learn the meaning of these words, as you are going to need them when it comes to those horrid module test and exams.

Advanced level Chemistry

Well if you're reading this book, the chances are that you have done AS chemistry, so well done for that. I hope you found it both **essential** and **fascinating**. I'm sure that there were times when you thought AS chemistry was more like 'confusing and challenging'...well if you didn't, either you weren't paying attention or you're very bright. Hopefully you realised that a knowledge of the basic facts and concepts helps you build up a sound understanding of the subject.

At A2 you build on AS. The chemical skills, knowledge and understanding you gained are taken a bit further and you are expected to apply your chemistry to slightly unfamiliar situations. Mmmm....this is where it gets a bit tricky, but nothing to lose sleep over. When looking at A2 questions, the skill is to extract the familiar chemistry from the unfamiliar context. The synoptic questions are questions which make you link different modules and indeed the practical work you've carried out. This is where practice comes in — the more you practise what they ask, the easier it gets. Honest!

This all sounds more difficult than it really is, with hard and continuous work, you should reach your true potential. Good luck.

Just remember...

...Chemistry is pHun... ☺

Hope you continue to have pHun....

Paddy Gannon
The Lakes

Chemistry for AQA — Assessment

A2 Chemistry

As you probably know if you're taking the second part the full A level, the scheme of assessment involves carrying over the marks you gained from doing **AQA AS GCE in Chemistry (5421)** which involved three exams covering the three modules and the lovely practical assessment.

For the full **Advanced level** (AS + A2) qualification you study the **A2 modules, 4** and **5.** You then sit the appropriate assessment units tests.

The assessment for **AQA A2 Chemistry (6421)** involves more exams and some more lovely practical assessment. The marks from these units of assessment are added to the AS assessment units with the appropriate weighting.

Synoptic assessment

Some of the questions AQA will ask in the A2 exams will be on more than one module. These questions can overlap with AS topics as A2 builds on AS material.

These will draw together knowledge, understanding and skills learned in the whole of the AS and A2 modules and remember that longer answers will be marked for quality of written language used.

Re-sits

You can re-sit each assessment unit once only, where the better result will count.

Simple rate equations

All those eons ago, when you studied, AS level, you'll have learnt there are two ways of increasing the rate of a chemical reaction:

1 Increase the frequency of collisions. How?
- increase the **concentration** of the reactants present
- increase the **pressure** of gaseous reactants
- increase the **surface area** of solid reactants
- increase the **temperature** of the reactions

2 Increase the percentage of particles with energy $\geq$ the activation energy, (E_a) How?
- and/or add a **catalyst** (which includes **light**)

Facts you should know about rates of reaction

- **rate of reaction =**
  $$\frac{\textbf{change in concentration}}{\textbf{time}}$$

- the units of rate are usually **mol dm^{-3} s^{-1}**

- the rate of reaction can be determined experimentally and graphically

- the rate of a reaction changes with time, getting slower towards the end of the reaction.

- The initial rate is often used when investigating rates so there is zero concentration of products, which may cause a reverse reaction.

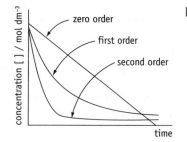

The rate equation

✓ Quick check 1, 2

The rate equation tells you the major factors which govern the speed of the reaction

$$A + B \rightarrow \textbf{products}$$

- The overall **order** = m + n
- A **rate equation** is for a certain reaction and can only be worked out experimentally.
- The rate equation is not the same as the balanced (stoichiometric) equation.
- The order of a reactant is the power to which its concentration is raised in the rate equation.
- The larger the value of k, the faster the reaction goes.
- m and n are usually 0, 1 or 2, (but could be 3 as given in the example below).

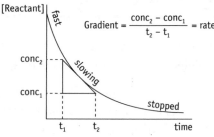

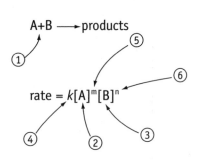

① A and B are the reactants
② [A] is the initial conc of A (i.e. at the beginning) in mol dm^{-3}
③ [B] is the initial conc of B in mol dm^{-3}
④ k = rate constant (relates the rate of a chemical reaction to the reactant concentrations)
⑤ m = order of the reaction w.r.t. reactant A
⑥ n = order of the reaction w.r.t. reactant B

Order of reaction

The order of a reaction with respect to (w.r.t.) a reactant, tells us something of the effect that changing the reactant concentration has on the rate of reaction. It kind of indicates how important a substance is, as far as altering the rate goes.

$$2NO + O_2 \rightarrow 2NO_2$$

$$\text{rate} = k\,[NO]^2[O_2]$$

The rate is

- first-order with respect to O_2
- second-order with respect to NO
- third-order overall

NO is more influential in altering the rate than oxygen, the other reactant.

AQA state that individual orders will be 0, 1, 2 only

✓ Quick check 3, 4

Examination of simple data

The order of reaction and the rate equation can be derived from experimental data relating the initial rate of reaction to the concentrations of the reactants.

The following data was found for the reaction: A + B → C

Experiment run	[A] mol dm^{-3}	[B] mol dm^{-3}	Rate (mol dm^{-3} s^{-1})
1	4.0×10^{-3}	3.0×10^{-3}	5.2×10^{-4}
2	4.0×10^{-3}	6.0×10^{-3}	41.6×10^{-4}
3	8.0×10^{-3}	3.0×10^{-3}	20.8×10^{-4}

The idea is to look at what reactant is changing and see what effect it has on the rate.

Looking at experiment 1 and 2: Double [B] and you increase the rate by × 8 (2^3)

$$\text{Rate } \alpha \text{ [B]}^3$$

Look at experiment 1 and 3: double [A] and you quadruple the rate (i.e. rate × 4 or 2^2).

$$\text{Rate } \alpha \text{ [A]}^2$$

$$\textbf{Rate} = k[A]^2\,[B]^3$$

✓ Quick check 5

Order w.r.t. B is 3
Order w.r.t. A is 2
Overall order is 5

Quick check questions

In the equation: Rate = $k[A]^x[B]^y$

1 What do the square brackets, [ ] represent?
2 What does k stand for?
3 What is the order w.r.t. A and B?
4 What is the overall order of the reaction?
5 If x = 1 and y = 2 what will happen to the rate if both concentrations are doubled?

Determining the rate equation

You must be able to work out a **rate equation** from given data. The order of reaction is determined by investigating the effects of altering the initial concentration of one reactant on the reaction rate, whilst keeping constant other factors which affect the rate (like temperature). Interpret the data to identify the rate equation.

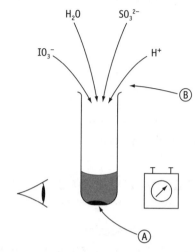

Worked example

Determine the rate equation for the iodate(V) and sulphite ions in the presence of an acid by examining the data in the table.

$$2IO_3^-(aq) + 5SO_3^{2-}(aq) + 2H^+(aq) \rightarrow I_2(s) + 5SO_4^{2-}(aq) + H_2O(l)$$

Run	$IO_3^-(aq)/cm^3$	$SO_3^{2-}(aq)/cm^3$	$H^+(aq)/cm^3$	water/cm^3	time/s
1	20	20	20	40	26
2	20	40	20	20	26
3	20	20	40	20	13
4	40	20	20	20	6.5

(A) starch solution indicates the presence of iodine by turning blue-black

(B) the concentration of one reactant is changed while the others remain constant

Compare the results

Compare runs 1 and 3: [Acid] doubles ($\times$ 2)

- Doubling the concentration of the acid halves the time taken for the reaction.
- Since the rate doubles as the [H$^+$] doubles, the [H$^+$] is proportional to the rate. ([H$^+$] $\times$ 2 means rate $\times$ 2)
- The reaction is therefore **first order** with respect to the acid.
- [any number]^1 = [the number] e.g. 157 ^{1}= 157

Compare runs 1 and 4: the [iodate] ions doubles ($\times$ 2).

- Doubling the iodate concentration increases the rate of reaction by a factor of four. (26/6.5 = 4)
- Since the change in rate is proportional to the square of the change in concentration (i.e. $\times 2^2$).
- The reaction is **second order** with respect to the iodate.

Compare runs 1 and 2: the [sulphite] doubles (i.e. $\times$ 2)

- Doubling the concentration of sulphite does **not** effect the rate of reaction.
- The rate of reaction is **zero order** with respect to the sulphite.
- [any number]^0 = 1, e.g. 157 ^{0}= 1

Overall **rate $\propto$ [H$^+$] [IO$_3^-$]^2**

So the rate equation is **rate = k [H$^+$] [IO$_3^-$]^2**

- Where k is a proportionality constant known as the **rate constant**
- overall order = 1 + 2 = 3. It is a third order reaction.

They sometimes give more awkward numbers, just look what's doubling or tripling and what effect this has on the rate. Also note rate α 1/time

rate $\propto$ [H$^+$(aq)]^1

rate $\propto$ [IO$_3^-$(aq)]^2

rate $\propto$ [SO$_3^{2-}$(aq)]^0

✓ *Quick check 1,2*

▶ This is the famous iodine 'clock' experiment. Great for A level questions, lousy for telling the time.

Using graphs to determine order

You might be asked to calculate the initial rates from experimental data with various concentrations of a reactant.

If results are plotted, the shape of the graph indicates the order with respect to that reactant for that reaction.

zero order
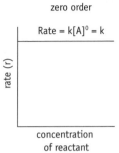
Rate = $k[A]^0$ = k
rate (r) — concentration of reactant

first order
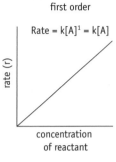
Rate = $k[A]^1$ = $k[A]$
rate (r) — concentration of reactant

second order
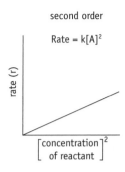
Rate = $k[A]^2$
rate (r) — [concentration of reactant]^2

The rate constant k

- The rate constant is **temperature-dependent.**
- Its value for a specific reaction, at a particular temperature is constant, but the value increases as temperature increases.
- The rate of reaction increases with temperature, as increasing the temperature increases the proportion of molecules having the minimum energy needed to react (i.e. ≥ the **activation energy**). A higher temperature *increases* the value of k, so the reaction goes faster.
- For a given rate equation e.g.   rate = $k[A][B]^2$
  If the concentrations of the reactants are kept constant and the rate increases, the value of k must increase as the temperature rises.
- The lower the temperature the lower the value of k, (i.e. the reaction slows down).

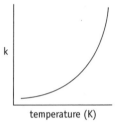
k — temperature (K)

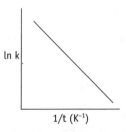
ln k — 1/t (K^{-1})

▶ Remember large k = fast reaction; small k = slow reaction

Calculations with k

You are regularly asked to calculate the units of the rate constant.

- These vary depending on the overall order of the reaction. For example if.... **rate = k [A][B]**

Rearranging this gives:

$$k = \frac{rate}{[A][B]} \quad \text{where the units are} \quad \frac{mol\ dm^{-3}s^{-1}}{(mol\ dm^{-3})^2}$$

Overall order	rate equation	Units of k
First	rate = $k[A]^0$	s^{-1}
Second	rate = $k[A]^1$	$dm^3\ mol^{-1}\ s^{-1}$
Third	rate = $k[A]^2$	$dm^6\ mol^{-2}\ s^{-1}$

Cancelling this gives the units as **$mol^{-1}\ dm^3\ s^{-1}$**

- You often are asked to calculate the value of k from given data. Simply rearrange the data to give k as the subject and place the data in the equation.

✓ *Quick check 3,4*

? Quick check questions

Consider the reaction rate = $k[V][W]^3$.

1 What will happen if a) [V] is doubled, b) [W] is tripled, c) both [V][W] are tripled and d) what are the units of k?

2 Which has the greater effect on rate: trebling [V] or doubling [W]?

3 What happens to the rate if [V] and [W] are kept constant and the temperature is increased?

4 What is changing in the rate equation to make this happen?

The equilibrium constant K_c

You should recall all the work you did on Le Chatelier's principle, which helps us understand the qualitative effects of changes in temperature, pressure and concentration on the position of **dynamic chemical equilibrium**. In A2 you examine quantitative aspects of equilibria for **homogeneous** systems.

Expressing K_c

Consider the reaction which has reached the steady state of dynamic equilibrium

$$a\text{A} + b\text{B} \rightleftharpoons c\text{C} + d\text{D}$$

An equation involving the equilibrium constant, K_c, known as the equilibrium law can be written by taking the concentration of the reactant or product and raising it to the power of the number of moles shown in the reaction equation.

$$K_c = \frac{[\text{C}]^c\,[\text{D}]^d}{[\text{A}]^a\,[\text{B}]^b}$$

- K_c is the ratio of the concentration of the products to reactants.
- Every reversible reaction has its own K_c value.
- K_c values can be used to predict the equilibrium position and whether a reaction is exothermic or endothermic (see below).
- The units of K_c vary depending on the reaction.
- If the value of the equilibrium constant is very large (say 100) then the equilibrium position lies very much to the right.
- If the value of the equilibrium constant is very small (say 10^{-2}) then the position of the equilibrium lies very much to the left.

Example:

$$\text{CH}_3\text{COOH(aq)} + \text{CH}_3\text{CH}_2\text{OH(aq)} \rightleftharpoons \text{CH}_3\text{COOCH}_2\text{CH}_3\text{(aq)} + \text{H}_2\text{O(l)}$$

$$K_c = \frac{[\text{CH}_3\text{COOCH}_2\text{CH}_3]\,[\text{H}_2\text{O}]}{[\text{CH}_3\text{COOH}]\,[\text{CH}_3\text{CH}_2\text{OH}]} = 3.95 \quad \frac{\text{mol dm}^{-3}\ \text{mol dm}^{-3}}{\text{mol dm}^{-3}\ \text{mol dm}^{-3}} \therefore \text{ no units}$$

The value of K_c for the reverse reaction is $1/3.95 = 0.253$. This means that the equilibrium lies to the right hand side.

More facts you should know about K_c

Temperature

K_c is changed when the temperature is changed. (unless $\Delta H^{\ominus} = 0$)

- exothermic reactions are favoured by a decrease in temperature.
  A temperature rise in an exothermic reaction will increase the rate of the reverse reaction more than the forward. The equilibrium will favour the reverse reaction — the opposite happens for a decrease in temperature.

> Read the section on Kinetics in the AS book and remember LCP — If a factor which affects the position of equilibrium is altered, the equilibrium changes and shifts in the direction which tends to reduce (and oppose) the change

> Don't mix up K_c with the rate constant k. K_c tells us nothing of how fast a reaction is.

> Where [ ] represents the equilibrium concentration of the reactant in mol dm^{-3}; (aq) is omitted for simplicity and is not in the equation; a, b, c and d are the number of moles reacting.

> You should only be asked about homogenous systems. If a solid or liquid appears omit it from the K_c expression.

✓ **Quick check 1**

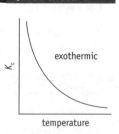

exothermic

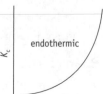

endothermic

- Endothermic reactions are favoured by an increase in temperature.
  A temperature rise in an endothermic reaction will increase the rate of reaction and the equilibrium constant, favouring the forward reaction. The opposite happens for a decrease in temperature.

	exothermic reaction	endothermic reaction
temperature ↑	K_c ↓	K_c ↑
temperature ↓	K_c ↑	K_c ↓

Concentration and pressure

✓ *Quick check 3*

K_c is not changed when the concentration or pressure changes. Pressure only affects gaseous equilibria where the reaction involves a change in the number of gas particles. Increasing the pressure shifts the equilibrium in the direction of the reaction with fewer gaseous moles, but K_c does not change.

▶ $2SO_2(g) + O_2(g)$ $\rightleftharpoons 2SO_3(g)$

$K_c = \dfrac{[SO_3]^2}{[SO_2]^2[O_2]}$

If $[SO_3]$ increases then $[SO_2]$ decreases to keep K_c constant

Catalysts

Whilst catalysts decrease the time taken to reach equilibrium, they have no effect on the equilibrium position and so K_c is not changed when a catalyst is added.

Nasty K_c calculations

If you are given the *equilibrium* concentrations (molarities)

The equation for K_c uses the concentrations at equilibrium. If you are given these, simply substitute into the K_c equation:

Worked example

Calculate K_c for: $3H_2(g) + N_2(g) \rightleftharpoons 2NH_3(g)$

Given that at equilibrium $[H_2] = 0.763$ mol dm^{-3}
                                                 $[N_2] = 0.922$ mol dm^{-3}
                                                 $[NH_3] = 0.157$ mol dm^{-3}

Step 1 Write out the expression for K_c from the stoichiometric equation.

$$K_c = \frac{[NH_3]^2}{[H_2]^3[N_2]}$$

Step 2 Substitute with the equilibrium concentrations you are given.

✓ *Quick check 4*

$$K_c = \frac{[NH_3]^2}{[H_2]^3[N_2]} = \frac{(0.157)^2}{(0.763)^3 \times 0.922} = 6.02 \times 10^{-2} \text{ dm}^6 \text{ mol}^{-2}$$

? Quick check questions

1 Give the K_c expression for the reaction: $I_2(aq) + I^-(aq) \rightleftharpoons I_3^-(aq)$ and state its units.

  Use the equation $H^+(aq) + OH^-(aq) \rightleftharpoons H_2O(l)$ $\Delta H = -57.3$ kJ mol^{-1} to answer the next questions:

2 Which direction will the equilibrium move if the temperature is decreased?

3 What will happen to the equilibrium if a small amount of sodium hydroxide is added?

4 What effect will these two changes have on the value of the equilibrium constant?

More K_c calculations

Calculate a value for K_c at a certain temperature from given molar amounts

You can calculate concentrations from initial amounts and volume of the reactants and the equilibrium concentration of the product. If you know how many moles of product are present at equilibrium, the stoichiometric equation will tell you how many moles of reactant are present at equilibrium.

Worked example 1

Calculate K_c for the equilibrium $2SO_3(g) \rightleftharpoons 2SO_2(g) + O_2(g)$

When 0.0200 mole of SO_3 was introduced into the equilibrium vessel (volume 1.48 dm^3) 0.0139 mole of SO_2 was found at equilibrium.

> ▶ Making up a table helps you to think through these calculations in a stepwise manner. Use the equation as headings.

Step 1 Construct the equation for K_c

Step 2 Put the values you know under the appropriate species in the table under the equation.

Step 3 Calculate the number of moles of the other species.

Step 4 Work out the concentration by dividing the number of moles by the volume.

Step 5 Substitute these values into the expression for K_c.

Step 1
$$\frac{[SO_2]^2[O_2]}{[SO_3]^2}$$

Step		$2SO_3$	$\rightleftharpoons$	$2SO_2$ +	O_2
2	initial amounts/mol	0.0200 mol		0 mol	0 mol
3	equilibrium amounts/mol	0.0200 - 0.0139 = 0.0061		0.0139	$\dfrac{0.0139}{2} = 0.00695$ (½ the moles of SO_2)
4	equilibrium concentration/ mol dm^{-3}	$\dfrac{0.0061}{1.48} = 0.00412$ mol dm^{-3}		$\dfrac{0.0139}{1.48} = 0.00939$ mol dm^{-3}	$\dfrac{0.00695}{1.48} = 0.00470$ mol dm^{-3}
5	Substitute these values into the expression for K_c	$K_c = \dfrac{[SO_2]^2[O_2]}{[SO_3]^2} = \dfrac{(0.00939)^2(0.00470)}{(0.00412)^2}$ $= 0.0244 = 2.44 \times 10^{-2}$ mol dm^{-3}			

Worked example 2

If a mol of acid react with b mol of alcohol and is allowed to reach equilibrium at constant temperature so that x mol of ester are formed in volume, V, then:

Step 1 $K_c = \dfrac{[CH_3COOCH_2CH_3][H_2O]}{[CH_3COOH][CH_3CH_2OH]}$

Step		$CH_3COOH(aq)$	$CH_3COOH(aq)$	$\rightleftharpoons$	$CH_3COOCH_2CH_3(aq)$	$H_2O(l)$
2	initial amounts/mol	a	b		0	0
3	equilibrium amounts/mol	$a - x$	$b - x$		x	x
4	equilibrium concentration/ mol dm^{-3}	$a - x/V$	$b - x/V$		x/V	x/V
5	Substitute these values into the expression for K_c cancel V	$$K_c = \frac{[CH_3COOCH_2CH_3][H_2O]}{[CH_3COOH][CH_3CH_2OH]}$$ $$K_c = \frac{(x/V) \times (x/V)}{((a-x)/V \times ((b-x)/V)}$$ $$K_c = \frac{x^2}{(a-x)(b-x)}$$			cancel V	

Worked example 3

1.00 mole of acid is reacted with 0.326 moles of alcohol. At equilibrium 0.297 moles of ester is present. Calculate K_c.

$$CH_3COOH(aq) + CH_3CH_2OH(aq) \rightleftharpoons CH_3COOCH_2CH_3(aq) + H_2O(l)$$

Step		$CH_3COOH(aq)$	$CH_3CH_2OH(aq)$	$\rightleftharpoons$	$CH_3COOCH_2CH_3(aq)$	$H_2O(l)$
2	Initial amounts/mol	1.00	0.326		0	0
3	Equilibrium amounts/mol	1-0.297	0.326-0.297		0.297	0.297
4	Equilibrium concentration/mol dm^{-3}	0.703/V	0.0290/V		0.297/V	0.297/V
5	Substitute these values into the expression for K_c	$$K_c = \frac{[CH_3COOCH_2CH_3][H_2O]}{[CH_3COOH][CH_3CH_2OH]} \text{ cancel V}$$ $$K_c = \frac{0.297 \times 0.297}{0.703 \times 0.0290}$$ $$K_c = 4.33$$				

Alternatively the values can be substituted into the expression:

$$K_c = x^2/(a - x)(b - x)$$

> Chances are you'll not remember this, so make sure you are able to use the method above.

? Quick check questions

Use this data to answer the questions that follow. When 0.0228 mole hydrogen and 0.0228 mole iodine were mixed at 760K in a 2 dm^3 vessel 0.0353 mole hydrogen iodide were produced at equilibrium.

1 Give the K_c expression for the reaction.

2 What was the initial amount of each of the reactants?

3 What was the amount of H_2 and I_2 at equilibrium?

4 What were the equilibrium concentrations of all reactants?

5 Calculate a value for K_c at 760K and give its units, (if it has no units say so).

> Some questions give you a % dissociation of a compound. This just means how much of it has turned into the product.

K_p – using partial pressures

Equilibria involving gases uses **partial pressures** rather than concentrations. In these homogeneous equilibria reactions, we get an equilibrium constant, K_p, in terms of the partial pressures of the reactants and products.

> ▶ Homogeneous = same phase

Writing an expression for K_p

> **Partial pressure, p, of a gas in a mixture of gases, is the pressure that the gas would exert if it alone occupied the available volume.**

The **total pressure P_{total}** of a mixture of gases = **sum of the partial pressures** exerted by each of the gases in the mixture. $P_{total} = p_1 + p_2 + p_3 + \ldots\ldots\ldots$

where p_1, p_2 etc are the partial pressures of the individual gases.

For example: if there are two gases, A and B

The partial pressure of each gas = mole fraction × total pressure

$$p_A = \text{mole fraction of A} \times P_{total}$$

> ✓ Quick check 1, 2

> ▶ n_B is the number of moles of B.

The mole fraction of A $= X_A = \dfrac{\text{No moles of A}}{\text{Total number of moles of gas in mixture}} = \dfrac{n_A}{n_A + n_B}$

Therefore $p_A = \dfrac{n_A}{n_A + n_B} \times P_{total}$ and so $p_B = \dfrac{n_B}{n_A + n_B} \times P_{total}$

For $aA + bB \rightleftharpoons cC + dD$ ∴ $K_p = \dfrac{pC^c \times pD^d}{pA^a \times pB^b}$

> ▶ If you know the values of the partial pressures of the gases at equilibrium, you can calculate the value of the equilibrium constant, K_p by putting the values in this expression.

- Partial pressures are measured in pascals, Pa, kilopascals, kPa, or even megapascals, mPa. Concentration is measured in mol dm^{-3}.
- An older unit was the **atmosphere**, where one atmosphere is 101,325 Pa.
- K_p is a constant at constant temperature. Like K_c its value will alter if the temperature changes, depending on whether the forward reaction is exothermic or endothermic.
- It does *not* change when the *pressure* is changed.
- As with K_c — a large value of K_p means a high yield of products.
- A small value of K_p means that a high proportion of reactants are present.

Units of K_p

Consider the Haber Process:

> ▶ Never use [ ] in this expression and only include gases in the K_p expression. The spec. says only homogeneous systems will be asked about.

$$N_2(g) + 3H_2(g) \rightleftharpoons 2NH_3(g)$$

$$K_p = \frac{(pNH_3)^2}{(pN_2)(pH_2)^3}$$

where pNH_3, pN_2 and pH_2 are the partial pressures of the three gases at equilibrium.

The units of K_p for this reaction: $\dfrac{Pa^2}{Pa \times Pa^3} = \dfrac{1}{Pa^2} = Pa^{-2}$

▶ Take care with units! Gas pressure is usually given in kPa (10^3Pa).

Nasty calculations involving K_p

To calculate K_p, use the same method as you did for K_c. But life is not that simple — before you can substitute into the K_p equation there are often two extra steps for gaseous reactions.

1 You have to work out the **mole fractions**

2 Then calculate the **partial pressures**

Worked example

In the equilibrium $N_2(g) + 3H_2(g) \rightleftharpoons 2NH_3(g)$ there were 6.1 mole of H_2, 1.5 mole of NH_3, and 2.2 mole of N_2. The total pressure was 100 kPa. Calculate the value of K_p for this equilibrium.

✓ Quick check 3,4

Step 1 Write the expression for K_p. $\quad K_p = \dfrac{(p\mathrm{NH_3})^2}{(p\mathrm{N_2})(p\mathrm{H_2})^3}$

Step 2 Calculate the mole fraction of each gas.

mole fraction of H_2 $\quad = \dfrac{\text{no. moles } H_2}{\text{total no. of moles}} \quad = \dfrac{6.1}{6.1+1.5+2.2}$

$= 0.622$ (no units)

mole fraction of NH_3 $\quad = \dfrac{\text{no. moles } NH_3}{\text{total no. of moles}} \quad = \dfrac{1.5}{6.1+1.5+2.2}$

$= 0.153$

▶ Check that the mole fractions add up to 1.

mole fraction of N_2 $\quad = \dfrac{\text{no. moles } N_2}{\text{total no. of moles}} \quad = \dfrac{2.2}{6.1+1.5+2.2}$

$= 0.224$

Step 3 Calculate the partial pressure of each gas.

$p(H_2) = 0.622 \times 100$ kPa, $p(NH_3) = 0.153 \times 100$ kPa and $p(N_2) = 0.224 \times 100$ kPa

Step 4 Put the values of the partial pressures into the expression for K_p, and calculate:

$$K_p = \frac{(p\mathrm{NH_3})^2}{(p\mathrm{N_2})(p\mathrm{H_2})^3} = \frac{(15.3)^2}{(22.4) \times (62.2)^3} = 4.34 \times 10^{-5}$$

Step 5 Add the units: $\quad \dfrac{kPa^2}{kPa \times kPa^3} \quad = \dfrac{1}{kPa^2}$

$\therefore K_p = 4.34 \times 10^{-5} \, kPa^{-2}$

? Quick check questions

1 In a mixture of 1 mole SO_2, 2 mole O_2 and 3 mole SO_3 what is the mole fraction of a) SO_2 and b) SO_3?

2 If the total pressure of the mixture in 1) was 150 kPa what was the partial pressure of SO_2?

3 Give the K_p expression for the reaction $PCl_5(g) \rightleftharpoons PCl_3(g) + Cl_2(g)$ and give its units.

4 What change in conditions can cause K_p to change?

Acids and bases

- **Brønsted and Lowry** noticed that acids ionise or **dissociate** in water releasing $H^+(aq)$ ions. They defined an acid (HA) as a **proton donor**.

$$HA(aq) + H_2O(l) \rightarrow H_3O^+(aq) + A^-(aq)$$

- Brønsted–Lowry theory defines a base as a **proton acceptor**. In solution a base (B) accept protons from water molecules.

$$B(aq) + H_2O(l) \rightarrow BH^+(aq) + OH^-(aq)$$

> $H_3O^+(aq)$ is the oxonium ion, sometimes to simplify equations it is left as $H^+(aq)$

Acids and bases

- Acids and bases are classified as **strong** or **weak** depending upon the degree to which the acid ionises in solution.

- Strong acids *fully* ionise in solution. For example sulphuric and nitric acid.

$$H_2SO_4 + water \rightarrow H^+(aq) + HSO_4^-(aq)$$

$$HNO_3 + water \rightarrow H^+(aq) + NO_3^-(aq)$$

- Weak acids only *partially* ionise in solution. For example with organic acids like ethanoic acid, CH_3COOH the equilibrium lies to the LHS.

$$CH_3COOH(aq) \rightleftharpoons CH_3COO^-(aq) + H^+(aq)$$

- Strong bases fully ionise in solution, for example ionic compounds like potassium and sodium hydroxide.

$$Na^+OH^-(s) + water \rightarrow Na^+(aq) + OH^-(aq)$$

$$K^+OH^-(s) + water \rightarrow K^+(aq) + OH^-(aq)$$

- Weak bases only partially ionise in solution. For example with ammonia and amines (see page 38) the equilibrium lies to the LHS.

$$NH_3(aq) + H_2O(l) \rightleftharpoons NH_4^+(aq) + OH^-(aq)$$

$$RNH_2(aq) + H_2O(l) \rightleftharpoons RNH_3^+(aq) + OH^-(aq)$$

- A **base** is a substance that neutralises acids to form salts and water.

$$Neutralisation = H^+(aq) + OH^-(aq) \rightarrow H_2O(l)$$

- pH tells us how acidic a substances is. pH = $-\log_{10}[H^+]$, (where [ ] = concentration of hydrogen ions in mol dm^{-3}. (See page 14 for more details).

> Strong and weak refer to the degree of dissociation of an acid or base they don't mean concentrated or dilute which refer to the amount of acid or base in a certain volume of water.

> ✓ *Quick check 1, 2*

> A soluble base is called an **alkali**; these form $OH^-(aq)$ in water.

Conjugate pairs

In the equation opposite, HA is acting as an acid as it donates H^+ and A^- is acting as a base in the reverse reaction. They are known as a **conjugate pair**:

- A^- is the called the **conjugate base** of the acid HA.

- $H_3O^+(aq)$ is the conjugate acid of the base H_2O.

- Note that water can act as an acid and a base so it can be described as **amphoteric.**

- Acids in the presence of stronger acids can act as bases.

conjugate pair

$$HA(aq) + H_2O(l) \rightleftharpoons H_3O^+(aq) + A^-(aq)$$
acid   base   acid   base

conjugate pair

The acid dissociation constant — K_a

Where there's an equilibrium there's a equilibrium constant. For a weak acid HA...

$$HA(aq) + H_2O(l) \rightleftharpoons H_3O^+ (aq) + A^- (aq)$$

$$\therefore K_c = \frac{[H_3O^+(aq)][A^-(aq)]}{[HA(aq)][H_2O(l)]}$$

> where [ ] = the equilibrium concentrations (mol dm^{-3}) and $[H_3O^+(aq)] = [H^+(aq)]$

- For a dilute weak acid $[H_2O(l)]$ is pretty much constant and in excess. You combine it with K_c to give the equilibrium expression...

$$K_a = \frac{[H^+(aq)][A^-(aq)]}{[HA(aq)]} \quad Units = \frac{mol\ dm^{-3}\ mol\ dm^{-3}}{mol\ dm^{-3}} = mol\ dm^{-3}$$

> ▶ This means that $K_a = K_c[H_2O(l)]$.

Example

For ethanoic acid and water

$$CH_3COOH(aq) + H_2O(l) \rightleftharpoons CH_3COO^-(aq) + H_3O^+(aq)$$

The dissociation constant K_a, for ethanoic acid is...

$$K_a = \frac{[H^+(aq)][CH_3COO^-(aq)]}{[CH_3COOH(aq)]} = 1.78 \times 10^{-5}\ mol\ dm^{-3}$$

> ▶ You need to be able to construct these K_a expressions for weak acids.
> Note:
> $K_a < 1$ = weak acid
> $K_a > 1$ = strong acid

- K_a indicates the extent of acid dissociation — ethanoic acid is *weaker* than methanoic acid its K_a value is *smaller*.
- Most values of K_a are quoted at a certain temperature, i.e. 298K.
- The dissociation of an acid is an **endothermic** process so as the temperature increases the amount of dissociation will increase.
- Increasing the temperature increases the value of the acid dissociation constant.

Formula	Name	K_a / mol dm^{-3}
HNO_3	Nitric acid	10
$(COOH)_2$	Ethanedioic acid	5.9×10^{-2}
$CH_2ClCOOH$	Chloroethanoic acid	1.3×10^{-3}
HCOOH	Methanoic acid	1.6×10^{-4}
C_6H_5COOH	Benzene carboxylic acid	6.3×10^{-5}
CH_3COOH	Ethanoic acid	1.7×10^{-5}

pK_a for weak acids

$$pK_a = -\log_{10} K_a$$

- The scale makes the small values of K_a just a bit more manageable.
- pK_a is often more useful than pH since it is independent of concentration.
- The stronger the acid, the lower its pK_a value.

> ▶ Example: For ethanedioic acid $K_a = 5.9 \times 10^{-2}$ mol dm^{-3}

You need to be able to calculate the pH of weak acid from K_a

Calculate the pH of a 0.10 M solution of a weak acid given that it has a pK_a value of 5.20 at 298K.

Using $pK_a = -\log_{10}K_a$ i.e. $5.20 = -\log_{10}K_a$ so $K_a = 6.30 \times 10^{-6}$ mol dm^{-3}

$[H^+]^2 = K_a c = 6.30 \times 10^{-6} \times 0.10 = 6.30 \times 10^{-7}$ (approx. see page 15)

$[H^+] = \sqrt{6.30 \times 10^{-7}} = 7.90 \times 10^{-4}$ mol dm^{-3}

now pH $= -\log_{10}[H^+(aq)] \therefore$ pH $= -\log_{10}(7.90 \times 10^{-4}) = 3.10$

Formula	pK_a
HNO_3	−1.0
$(COOH)_2$	1.23
$CH_2ClCOOH$	2.8
HCOOH	3.8
C_6H_5COOH	4.2
CH_3COOH	4.7

❓ Quick check questions

1 Give the Brønsted–Lowry definition of a base and define the term pH.

2 What is the difference between a strong and a weak acid?

pH and K_w

$$pH = -\log_{10}[H^+]$$

- pH tells us how acidic or alkaline a substance is.

- You must be able to convert a concentration of hydrogen ions to a pH value and vice versa. If $-\log_{10}[H^+(aq)] = pH$ then $[H^+(aq)] = $ antilog pH or 10^{-pH}

✓ Quick check 1, 2

- Where [ ] = concentration of hydrogen ions in mol dm^{-3}

- pH varies with $[H^+]$ and temperature

- Indicators like **Universal Indicator** and **pH meters** are used to measure pH (see page 20). Meters are first calibrated with buffers of known pH.

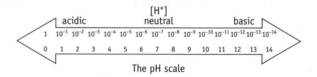
The pH scale

The ionic product of water — K_w

Water weakly dissociates as shown in the equation below...

$$H_2O(l) + H_2O(l) \rightleftharpoons H_3O^+(aq) + OH^-(aq)$$

Simplified to
$$H_2O(l) \rightleftharpoons H^+(aq) + OH^-(aq)$$

$$\therefore K_c = \frac{[H^+(aq)][OH^-(aq)]}{[H_2O(l)]}$$

but $[H_2O(l)]$ can be considered constant as the equilibrium lies to LHS

$$\therefore K_w = [H^+(aq)][OH^-(aq)] = 10^{-14} \text{ mol}^2 \text{ dm}^{-6} \text{ at 298K (25°C)}$$

▶ $K_w = K_c[H_2O]$

In pure water $[H^+(aq)] = [OH^-(aq)]$

So $K_w = [H^+(aq)]^2$

Hence $[H^+(aq)] = \sqrt{K_w}$

At 298K $[H^+(aq)] = \sqrt{(1 \times 10^{-14})} = 1 \times 10^{-7} \text{ mol dm}^{-3}$

So in pure water, at 298K, the pH value is 7.

▶ For acid solutions
$[H^+] > [OH^-]$
For alkaline solutions
$[OH^-] > [H^+]$
For a neutral solution
$[H^+] = [OH^-]$

- This dissociation is endothermic, therefore an increase in temperature *increases* the value of K_w.

- At higher temperatures the concentration of $H^+(aq)$ in pure water increases — this means it has a lower pH.

- This means that hot water is neutral, since $[H^+(aq)] = [OH^-(aq)]$, but it will not have a pH 7.

Temp/°C	K_w
18	0.61×10^{-14}
25	1×10^{-14}
40	2.92×10^{-14}

Calculating the pH of a strong acid — use $[H^+]$

- 1 mole of strong **monoprotic** acid fully dissociates in water to give 1 mole of H^+ ions $\therefore [H^+(aq)] = $ concentration of the acid.

$$HA(aq) + water \rightarrow H^+(aq) + A^-(aq) \quad \text{where } [HA(aq)] = [H^+(aq)]$$

▶ H_2SO_4 can provide 2 H^+ ions.

So for 0.1 M HNO_3 the $[H^+(aq)] = 0.1$ ∴ $pH = -\log_{10}[0.1] = 1$

So for 0.2 M HCl the $[H^+(aq)] = 0.2$ ∴ $pH = -\log_{10}[0.2] = 0.7$

In exams, you will often be asked questions about HNO_3, H_2SO_4, HCl, HBr, HI and $HClO_4$

Calculating the pH of a weak acid — use K_a

For a weak acid we have a problem...

- They are only **partially ionised** in solution.
- The $[H^+(aq)]$ is *not* the same as the concentration of the acid.
- [HA] is much greater than $[H^+]$.

The pH is therefore calculated using K_a

$$HA(aq) \rightleftharpoons H^+(aq) + A^-(aq) \quad \text{where HA is any weak acid}$$

$$K_a \frac{[H^+(aq)][A^-(aq)]}{[HA(aq)]}$$

For a weak acid
$$[H^+(aq)] = [A^-(aq)]$$

$$\therefore K_a = \frac{[H^+(aq)]^2}{[HA(aq)]} \text{ mol dm}^{-3}$$

If the concentration of the weak acid (HA), is c mol dm^{-3} then

$$[H^+(aq)]^2 = K_a c \quad \therefore [H^+(aq)] = \sqrt{K_a c}$$

The questions usually give you two out of K_a, c and $[H^+(aq)]$ or pH, so with a bit of thinking and jigging about of equations, you can work out the unknown.

Worked example

A 0.2 M solution of a weak acid, HA, has a pH value of 3.50. Calculate the value for the acid dissociation constant, K_a.

Step 1 Use $pH = -\log_{10}[H^+(aq)]$ to calculate the concentration of $H^+(aq)$.

$$3.50 = -\log_{10}[H^+(aq)] \quad \therefore [H^+(aq)] = 3.16 \times 10^{-4} \text{ mol dm}^{-3}$$

Step 2 Use $K_a = \frac{[H^+(aq)]^2}{[HA]}$ mol dm^{-3} to calculate K_a

$$\therefore K_a = \frac{(3.16 \times 10^{-4})}{0.2} = 5.0 \times 10^{-7} \text{ mol dm}^{-3}.$$

> Negative pH values are possible. e.g. If $[H^+(aq)] = 2$ pH = – 0.3 Weird eh? — Very negative pHs are unlikely.

> Any acid over 50% ionised is considered strong.

> You have to assume that the quoted concentration of the acid, c, is the equilibrium concentration. Really, c = [HA(aq)] – [H⁺(aq)], but we assume [HA(aq)] – [H⁺(aq)] ≅ [HA(aq)] since [H⁺(aq)] is so very small.

✓ Quick check 3, 4

Quick check questions

1 What is the pH of the following?
   a) $[H^+] = 2.0$ mol dm^{-3} b) $[H^+] = 1.0$ mol dm^{-3}
2 What is the hydrogen ion concentration of the following?
   a) pH = 2;   b) pH = 9;   c) pH = 7
3 By what name is the expression $[H^+][OH^-]$ known? Give its value at 298K.
4 Calculate the pH of water at 18°C and 40°C (see box on page 14).

More pHun calculations

These calculations carry on from those on the last page...

The pH of a strong base — use K_w

We use the ionic product of water to calculate the pH for a solution of a strong base...

✓ Quick check 1, 2

$$K_w = [H^+(aq)][OH^-(aq)]$$

$$\text{Hence} \quad [H^+(aq)] = \frac{K_w}{[OH^-(aq)]}$$

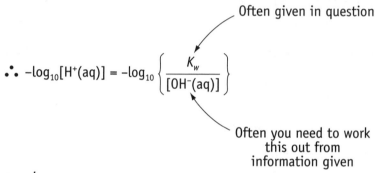

Often given in question

$$\therefore -\log_{10}[H^+(aq)] = -\log_{10}\left\{\frac{K_w}{[OH^-(aq)]}\right\}$$

Often you need to work this out from information given

Worked example

Calculate the pH of 0.2 M NaOH.
(NaOH is a strong base so is 100% or fully ionised in water.)

Step 1 Work out concentration of OH^-

0.2 M solution of NaOH

$$\therefore \text{the } [OH^-(aq)] = 0.2 \text{ mol dm}^{-3}$$

Step 2 Rearrange K_w expression to get $[H^+(aq)]$ and so pH.

At 298K then $K_w = [H^+(aq)][OH^-(aq)] = 1 \times 10^{-14} \text{ mol}^2 \text{ dm}^{-6}$

$$\therefore [H^+(aq)] = \frac{1 \times 10^{-14}}{0.2} = 5 \times 10^{-14} \text{ mol dm}^{-3}$$

Using pH $= -\log_{10}[H^+(aq)] = -\log_{10}(5 \times 10^{-14}) = 13.3$

You could use this expression to calculate the pH of an alkali **but only at 298K**

$$\text{pH} + \text{pOH} = 14$$

$$\text{where pOH} = -\log_{10}[OH^-(aq)]$$

pH of partially neutralised solutions

✓ Quick check 3

Worked example

Calculate the pH of a solution formed when 30 cm^3 of 0.2 M HCl and 30 cm^3 of 0.1 M NaOH are mixed together

Step 1 Work out the initial moles of acid and alkali:

$$\text{No. moles} = cV/1000$$

Some might use
No. moles = $MV/1000$

Initial moles of HCl $= (0.2 \times 30)/1000 = 6 \times 10^{-3}$ moles

Initial moles of NaOH $= (0.1 \times 30)/1000 = 3 \times 10^{-3}$ moles

Step 2 Calculate the number of moles of acid in excess (unreacted).

Therefore moles of acid unreacted is
$(6 \times 10^{-3} - 3 \times 10^{-3}) = 3 \times 10^{-3}$ moles in a new volume of 60 cm^3

Step 3 Calculate the concentration of [H$^+$] and so the pH.

Hence the new concentration of the acid is $(3 \times 10^{-3} \times 1000)/60 = 5.0 \times 10^{-2}$ M

The acid is a strong acid $\therefore$ [H$^+$] $= 5.0 \times 10^{-2}$ mol dm^{-3}

$$\text{pH} = -\log_{10}[\text{H}^+(\text{aq})] = -\log_{10} 5.0 \times 10^{-2} = 1.30$$

pH of diluted solutions

Worked example

Calculate the change in pH of the acid when 15 cm^3 of 0.1 M HCl is added to 200 cm^3 of water.

✓ Quick check 4

Step 1 Calculate the initial concentration of [H$^+$(aq)]

The acid is a strong monoprotic acid so [H$^+$(aq)] = [HCl] = 0.1 mol dm^{-3}

Initial pH = $\log_{10}[\text{H}^+(\text{aq})]$ = $\log_{10}(0.1)$ = 1

Step 2 Calculate the new pH after the addition of water.

Initial moles of H$^+$(aq) = $cV/1000$ = $0.1 \times 15/1000$ = 0.0015 mol

But the H$^+$(aq) ions are in a new volume of 215 cm^3 (200 + 15)

$$n = cV/1000 \therefore c = 1000 \times n/V$$

$$c = 1000 \times 0.0015/215 = 6.98 \times 10^{-3} \text{ mol dm}^{-3}$$

The final **pH = $\log_{10}[\text{H}^+(\text{aq})]$ = $\log_{10}(6.98 \times 10^{-3})$ = 2.16**

The **change** in pH is from 1 to 2.16

Quick check questions

1 What is the pH of a 0.010 M solution of hydrochloric acid?

2 What is the pH of a 0.020 M solution of potassium hydroxide?

3 What is the pH of a solution made by mixing equal volumes of a 0.010 M hydrochloric acid and 0.020 M potassium hydroxide?

4 If the pH of a solution at 298K is 5.73 what is the concentration of OH$^-$?

pH curves

Acid/base titration reactions can be followed and pH curves produced. pH values can be monitored using a **pH meter** or a **pH probe** and **data logger**. The shape of the curve depends on the combination of acid and base used. There are four typical curves.

pH curves — monoprotic

a) strong acid/strong base e.g. 0.1M HCl/NaOH

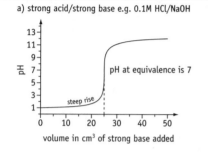

b) strong acid/weak base e.g. 0.1M HCl/NH₃

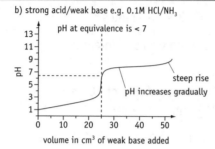

✓ Quick check 1,2

c) weak acid/strong base e.g. 0.1M CH₃COOH/NaOH

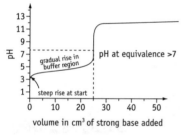

d) weak acid/weak base e.g. 0.1M CH₃COOH/NH₃

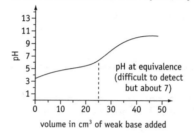

▶ Mirror image curves can easily be obtained.

	SA/SB	SA/WB	WA/SB	WA/WB
The initial pH	1.0	1.0	3.0	3.0
Equivalence pH	7	< 7	> 8	7
The pH range at equivalence	3–11	3–6.5	7.5–11	Not a clear change
The final pH	12	9	12	9

pH curves — diprotic

✓ Quick check 3

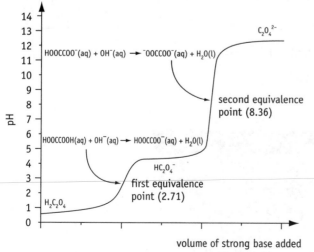

Ethanedioic acid HOOCCOOH can lose two protons and so it has two **equivalence points** and two pK_a values.

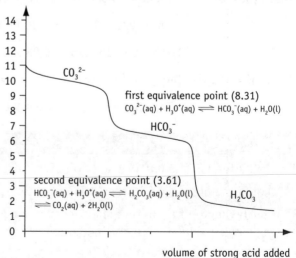

An almost mirror image curve is seen with sodium carbonate and hydrochloric acid as the carbonate ion can accept two protons.

pH at a certain point in a titration

Worked example 1 (SA/SB)

Calculate the pH of the resulting solution when 20 cm^3 of 0.2 mol dm^{-3} HCl is mixed with 20 cm^3 of 0.1 mol dm^{-3} solution of NaOH.

Step 1 Calculate the no. moles present of excess substance i.e. H$^+$(aq)

$n = cV/1000$

No. moles of H$^+$(aq) (initially) = 0.2 × 20/1000 = 4 × 10^{-3} mol

No. moles of OH$^-$(aq) (initially) = 0.1 × 20/1000 = 2 × 10^{-3} mol

No. moles of H$^+$(aq) (in excess) = (4 × 10^{-3} mol) – (2 × 10^{-3} mol) = 2 × 10^{-3} mol

Step 2 Divide the excess moles of substance (H$^+$) by total volume of solution to get [H$^+$(aq)] and so pH

H$^+$(aq) = 2 × 10^{-3} mol in volume = 20 + 20 = 40 cm^3 (or 0.04 dm^3)

$$\therefore [H^+(aq)] \text{ in 1 dm}^3 = \frac{2 \times 10^{-3}}{40} \times 1000 = 5 \times 10^{-2} \text{ mol dm}^{-3}$$

$$\therefore \textbf{pH} = \textbf{--log}_{10} \textbf{(5} \times \textbf{10}^{-2}\textbf{)} = \textbf{1.3}$$

Worked example 2 (WA/SB)

What is the pH of a solution in a titration where 20 cm^3 of 0.2 mol dm^{-3} NaOH is mixed with 20 cm^3 of 0.5 mol dm^{-3} solution of ethanoic acid?
($K_a = 1.76 \times 10^{-5}$ mol dm^{-3})

Step 1 Calculate the no. moles present of excess substance i.e. CH$_3$COOH

$n = cV/1000$

No. moles of CH$_3$COOH (initially) = 0.5 × 20/1000 = 1 × 10^{-2} mol

No. moles of OH$^-$(aq) (added) = 0.2 × 20/1000 = 4 × 10^{-3} mol

No. moles CH$_3$COO$^-$ produced = 4 × 10^{-3} mol

No. moles of CH$_3$COOH (present) = 1 × 10^{-2} mol – 4 × 10^{-3} mol = 6 × 10^{-3} mol

Step 2 *Rearrange the K$_a$ expression to get [H$^+$(aq)] and thus pH.*

$$K_a = \frac{[H^+(aq)][CH_3COO^-(aq)]}{[CH_3COOH(aq)]} \qquad [H^+(aq)] = K_a \times \frac{[CH_3COOH(aq)]}{[CH_3COO^-(aq)]}$$

$$[H^+(aq)] = K_a \times \text{mole ratio} = 1.76 \times 10^{-5} \times \frac{6 \times 10^{-3}}{4 \times 10^{-3}} = 2.64 \times 10^{-5} \text{ mol dm}^{-3}$$

$$\therefore \textbf{pH} = \textbf{-log}_{10} \textbf{2.64} \times \textbf{10}^{-5} = \textbf{4.58}$$

$\dfrac{[CH_3COOH(aq)]}{[CH_3COO^-(aq)]} = \dfrac{n(CH_3COOH(aq))}{n(CH_3COO^-(aq))}$

conc ratio = mole ratio

▶ WB/SA calculations are not covered.

❓ Quick check questions

1 Give two differences and two similarities between the pH curves for 0.10 M HCl + 0.10 M NaOH and 0.10 M CH$_3$COOH + 0.10 M NaOH.

2 Which combination of acid plus base give a pH curve on which the addition of one drop of acid or base does not give a significant change in pH at the equivalence point?

3 Why does the pH curve for ethanedioic acid and sodium hydroxide have two equivalence points?

Indicators and buffers

Using pH curves to work out K_a

For a weak acid and strong base titration, we see the curve opposite. K_a may be calculated from the curve by finding the pH at the half-equivalence point.

We know that...

$$K_a = \frac{[H^+(aq)][A^-(aq)]}{[HA(aq)]} \text{ mol dm}^{-3}$$

But $[HA(aq)] = [A^-(aq)]$ when half of the acid has been neutralised.

- This point is known as the half-equivalence point

$$K_a = [H^+(aq)] \text{ so pH at this point} = pK_a$$

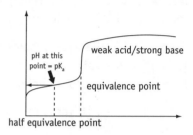

Indicators

✓ Quick check 1, 2

- **Indicators** are usually weak organic acids
- The undissociated and dissociated forms have different colours

$$HIn(aq) \rightleftharpoons H^+(aq) + In^-(aq)$$

	colour 1	colour 2
	Add acid $[H^+(aq)]$	Add alkali $[OH^-(aq)]$
	Equilibrium shifts to LHS to remove $[H^+(aq)]$	Equilibrium shifts to RHS to remove $[OH^-(aq)]$
	To give colour 1	To give colour 2

e.g. methyl orange: red $\rightleftharpoons$ yellow

Choosing an indicator

A good indicator will...

- change colour sharply on addition of no more than 1 drop of acid/base.

- have at least two distinct colours.

- have an **end point** (colour change point) which coincides with the equivalence point or pH range of the acid/base reaction. For this reason not all indicators are suitable for all reactions.

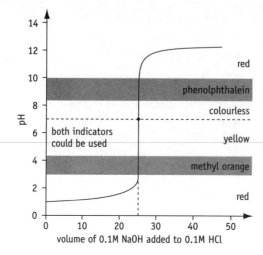

- In methyl orange when HIn(aq) = In⁻(aq) the colour is orange (half way between red and yellow). The pH of the end point is called the pK_{In} of the indicator.

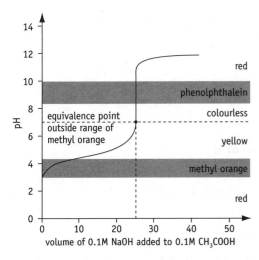

volume of 0.1M NaOH added to 0.1M CH₃COOH

Indicator	pK_a value (298K)	Useful pH range	Colour change (acid → alkali)
Methyl orange	3.7	3.0 — 4.4	red → yellow
Methyl red	5.1	4.1 — 6.2	red → yellow
Bromothymol blue	7.0	6.0 — 7.6	yellow → blue
Phenol red	7.9	6.8 — 8.5	yellow → blue
Phenolphthalein	9.3	8.4 — 10.0	colourless → pink

Buffers — p(ic)K up a proton

A buffer solution will retain its pH despite the addition of small amount of acid or alkali or on dilution.

Acidic buffers

- Contain a weak acid and one of its salts
- The pH < 7

e.g. ethanoic acid (CH₃COOH) and sodium ethanoate (CH₃COONa)

- The acid is slightly dissociated

$$CH_3COOH \rightleftharpoons CH_3COO^- + H^+$$

- The salt is fully dissociated

$$CH_3COONa \rightarrow CH_3COO^- + Na^+$$

The solution has a large concentration of CH₃COOH and CH₃COO⁻

Basic buffers

- Contain a weak base and one of its salts
- The pH > 7

e.g. ammonium hydroxide (NH₄OH) and ammonium chloride NH₄Cl

- The base is slightly dissociated

$$NH_3 + H_2O \rightleftharpoons NH_4^+ + OH^-$$

- The salt is fully dissociated

$$NH_4^+ \rightarrow NH_3 + H^+$$

The solution has a large concentration of NH₃ and NH₄⁺

cont.

Buffers —p(ic)K up a proton continued

How they work	**How they work**

How they work

Since $K_a = \dfrac{[CH_3COO^-][H^+]}{[CH_3COOH]}$

$[H^+] = K_a \dfrac{[CH_3COOH]}{[CH_3COO^-]}$

so pH depends on K_a and ratio of acid to anion

1) Addition of H⁺

The H⁺ combines with the CH_3COO^- to form CH_3COOH. As there is a large supply of CH_3COO^- the acid is removed.

2) Addition of OH⁻

The OH⁻ combines with H⁺ to form water. More acid dissociates to restore the equilibrium. If the CH_3COOH is of a high enough concentration the acid will restore the H⁺ concentration.

In summary

- High CH_3COO^- from salt removes H⁺
- High CH_3COOH from acid removes OH⁻
- Acid/anion ratio is constant on dilution
- So pH remains stable.

How they work

Since $[H^+] = K_a \dfrac{[NH_4^+]}{[NH_3]}$

so pH depends on K_a and the ratio $\dfrac{[NH_4^+]}{[NH_3]}$

1) Addition of H⁺

The H⁺ combines with the OH⁻ to form water. More NH_3 reacts with H_2O to restore the OH⁻ concentration.

2) Addition of OH⁻

The OH⁻ combines with NH_4^+ to form NH_3 and H_2O. The high concentrations of NH_4^+ ensures that the OH⁻ is removed.

In summary

- NH_3 and H_2O from base removes H⁺
- High NH_4^+ from salt removes OH⁻
- Dilution will not affect the above ratio
- So pH remain stable.

pH of an acidic buffer

✓ *Quick check 3–6*

$$\text{Since } K_a = \dfrac{[A^-][H^+]}{[HA]} \quad [H^+] = K_a \dfrac{[acid]}{[salt]}$$

K_a is often given, so simply calculate the concentration of the acid and the salt and put them into the expression above.

If the question is about the change in pH of a buffer caused by adding H⁺(aq), you need to work out by how many moles the acid (HA) has increased, and how much the anion (A⁻) has decreased. The pH is then calculated from the above expression.

Uses of buffers
- Industrial processes which are pH sensitive e.g. fermentation.
- Biological systems e.g. blood.
- Synthetic food and medicine production.
- Natural buffers include H_2CO_3/HCO_3^- and proteins.

? *Quick check questions*

1 Define the term pK_a.

2 What is an indicator and what is a buffer solution?

3 What is needed to make a basic buffer?

4 Explain how an acidic buffer works.

5 If an acid has a pK_a value of 2.54 at 298K, calculate the pH of a 0.5 M solution.

6 a) Calculate the pH of a buffer solution produced by adding 4.00 g of sodium ethanoate to 1 dm^3 of 0.01 M ethanoic acid. The K_a of ethanoic acid is 1.84×10^{-5} mol dm^{-3} at 300K.
   b) Calculate the pH of this buffer if 5 cm^3 of 0.1 M HCl is added.

Naming organic compounds

Most of the rules for naming organic compounds are covered in Revise AS chemistry for AQA, so are not repeated here. In addition to the seven **homologous series** that you have come across, there are three others you need to know about along with aromatic compounds.

Anoraks see www.acdlabs.com/iupac/nomenclature for a full and exciting explanation of nomenclature. ☺

Aliphatic compounds

Homologous series	Functional group	Name (suffix)	Example
8 esters*	$-\overset{\displaystyle O}{\underset{\displaystyle O-R}{C}}$	-oate	$CH_3COOCH_2CH_3$ ethyl ethanoate
9 acyl halide (or acid halide)	$-\overset{\displaystyle O}{\underset{\displaystyle X}{C}}$	-oyl halide	CH_3COCl ethanoyl chloride
10 acid anhydrides	$-\overset{\displaystyle O}{C}$ $-\overset{}{C}\overset{\displaystyle O}{}$ with O	-oic anhydride	$(CH_3CO)_2O$ ethanoic anhydride
11 benzene derivatives	C_6H_5-	Phenyl- or -benzene	See below

* Numbering carries on from AS book

Ethanoyl chloride and ethanoic anhydride are usually the only ones asked about.

Aromatic compounds

- Contain **phenyl** group C_6H_5-; the hydrocarbons are known as **arenes**.
- They are named by giving the position and the name of the **substituent group**.
- If two or more substituent groups are present they are numbered with lowest possible and listed in alphabetical order.

✓ **Quick check 1,2**

A benzene ring without a circle in it is cyclohexane **not** benzene.

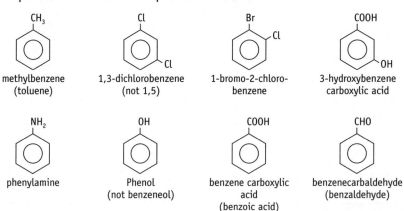

methylbenzene (toluene)   1,3-dichlorobenzene (not 1,5)   1-bromo-2-chloro-benzene   3-hydroxybenzene carboxylic acid

phenylamine   Phenol (not benzeneol)   benzene carboxylic acid (benzoic acid)   benzenecarbaldehyde (benzaldehyde)

Isomerism

Details and examples can be found in the AS book. Learn the definitions below.

Isomerism
Isomers are compounds with the *same molecular* formulae but *different arrangement* of atoms in space.

Structural Isomerism
The same molecular formula but the structural formula is different.

Stereoisomerism
The same molecular and structural formula, but different orientation of bonds in space.

Positional Isomerism
The molecular formula and the functional group are the same, but the position of the functional group is different.

Geometrical Isomerism
The molecular formula and the structural formula are the same, but the displayed formula is different. Sometimes called **cis–trans** isomerism. See AS book.

Chain Isomerism
The molecular formula and the functional group are the same, but the arrangement of the carbon atoms is different.

Optical Isomerism
Contain an **asymmetric carbon** so have no plane of symmetry. Isomers are **non-superimposable** on each other.

Functional Group Isomerism
The molecular formula is the same, but the functional group is different.

✓ *Quick check 3,4,5*

Optical isomerism

When four different atoms or groups are joined to a carbon, the carbon is said to be asymmetric, as it has no centre, plane or axis of symmetry. The molecule is described as **chiral**, the asymmetric carbon being the **chiral** carbon.

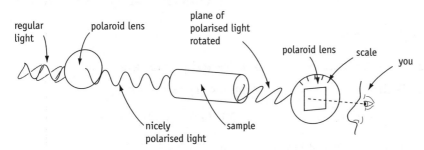

e.g.

2-hydroxypropanoic acid: (lactic acid) found in muscles after activity

- **Optical isomers** are known as **enantiomers**.
- Enantiomers cannot be superimposed on one another.
- They have the same chemical and physical properties, but will rotate the plane of polarised light either clockwise or anti-clockwise.

- The **dextroenantiomer** (+) rotates it clockwise (to the right).
- The **laevoenantiomer** (−) rotates it anti-clockwise (to the left).
- If equal quantities of + and − enantiomer are present the mixture appears optically inactive and is called a **racemic** mixture or a racemate.
- Chemical reactions tend to produce **racemate** while biological reactions produce single enantiomers.

Quick check questions

1 Give the structural formula of a) methylpropene, b) 2-methylbutan-2-ol, c) 3-ethylpentanone and d) butylmethanoate.

2 Name the following a) $CH_3CH_2CH=CHCH_3$, b) $(CH_3)_2CHCH(OH)$ CH_3, c) $CH_3CH_2CH_2CHClCHBrCH_3$, d) $CH_3CH_2COOCH_2CH_2CH_3$.

3 Give a definition of stereoisomerism.

4 What must organic molecules have if they are to exhibit optical isomerism?

5 What is a racemic mixture?

Aldehydes and ketones

You have come across **aldehydes** and **ketones** in AS chemistry. Here you look more at the redox reactions they undergo and the mechanism of their reactions.

$$CH_3$$
$$|$$
$$CH_3(CH_2)_7CH — CHO$$

An aldehyde found in Chanel No 5

Oxidation of aldehydes

Aldehydes are **carbonyl compounds** which are readily oxidised to carboxylic acids. Warm acidified potassium dichromate(VI) solution or mild oxidising agents like **Fehling's** or **Tollen's** reagent distinguishes between aldehydes and ketones. A positive test indicates the presence of an aldehyde.

$RCHO + [O] \rightarrow$
$RCOOH$

✓ *Quick check 1,2,3*

▶ The salt of the acid is formed as alkali is used.

Tollen's Reagent with aldehydes:

$$\textbf{Colourless} \quad \rightarrow \quad \textbf{gently warm} \quad \rightarrow \quad \textbf{silver mirror}$$
$$\mathbf{[Ag(NH_3)_2]^+} \qquad\qquad \rightarrow \qquad\qquad \textbf{Ag}$$
$$\textbf{(Silver nitrate in ammonia solution)} \qquad \textbf{(metal)}$$

$$RCHO(aq) + [Ag(NH_3)_2]^+(aq) + H_2O(l) \rightarrow \qquad RCOOH(aq) + Ag(s) + 2NH_4^+(aq)$$

Fehling's solution with aldehydes:

$$\textbf{Blue} \quad \rightarrow \quad \textbf{gently warm} \quad \rightarrow \quad \textbf{brick red precipitate}$$
$$\textbf{(Cu}^{2+}\textbf{ ions in alkali)} \qquad \rightarrow \qquad \textbf{(copper (I) oxide, Cu}_2\textbf{O)}$$

$$RCHO(aq) + 2Cu^{2+}(aq) + 4OH^-(aq) \rightarrow \qquad RCOOH(aq) + Cu_2O(s) + 2H_2O(l)$$

Note: Ketones are resistant to oxidation so show a negative result with both Tollen's reagent and Fehling's solution.

Reduction of aldehydes and ketones

✓ *Quick check 4*

1 Using NaBH$_4$ — addition of hydride ion (:H$^-$)

Aldehydes and ketones are reduced to alcohols by sodium tetrahydridoborate(III) (NaBH$_4$) in methanol (aqueous) (see AS Module 3) or LiAlH$_4$ in ethanol.

Equations: **Aldehyde:** $RCHO + 2[H] \rightarrow$ RCH_2OH
Primary alcohol

Ketone: $RCOR' + 2[H] \rightarrow$ $RCH(OH)R'$
Secondary alcohol

- NaBH$_4$ is given the symbol [H] for simplicity.
- NaBH$_4$ or LiAlH$_4$ do not reduce alkenes so compounds with C=C and C=O can be selectively reduced.
- :H$^-$ is unlikely to be free in solution.

The mechanism — nucleophilic addition

- NaBH$_4$ or LiAlH$_4$ donate the hydride ion (:H$^-$) — a **nucleophile**.
- The nucleophile attacks the δ+ carbonyl carbon atom and a pair of electrons in the carbon-oxygen double bond are transferred to the oxygen atom.
- The **intermediate** produced then gains a proton from water (or a weak acid), which is used as a solvent in the reaction: (but not with LiAlH$_4$)

> The 'curly' arrow must start on the pair of electrons or on the bond from which a pair is moving.

Visualising the reaction

$$H: \frown C = O \longrightarrow H - C - O:\frown H^+ \longrightarrow RCH_2OH$$

(with R above and H below on carbon)

2 Catalytic hydrogenation — addition of hydrogen using a catalyst

A nickel or platinum catalyst will add hydrogen across the C=O bond, completely saturating all double bonds in aldehydes and ketones to produce alcohols.

$$RCHO + H_2 \rightarrow RCH_2OH$$
$$RCOR + H_2 \rightarrow RCH(OH)R$$

> H$_2$ adds across C=C ∴ is not selective like NaBH$_4$

3 Additon of hydrogen cyanide (:CN$^-$)

Aldehydes and ketones react with HCN to produce hydroxynitriles. HCN is often formed in situ (in the reaction), as HCN is a bit nasty.

e.g. CH$_3$CHO + HCN $\rightarrow$ **CH$_3$CH(OH)CN**

ethanal   2-hydroxypropanenitrile

> In situ formation of HCN
> NaCN + HCl → HCN + NaCl

The mechanism — nucleophilic addition

> ✓ *Quick Check 5,6*

- The lone pair of electrons of the :CN$^-$ attacks the δ+ carbon atom of the C=O of the aldehyde or ketone.
- The intermediate gains a proton from the solvent to form the product

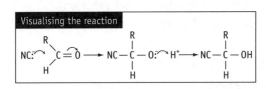

$$NC:\frown C = O \longrightarrow NC - C - O:\frown H^+ \longrightarrow NC - C - OH$$

> Make sure the lone pair goes on the carbon :CN$^-$, NOT CN:$^-$.

The nitrile group reacts further and undergoes hydrolysis in acidic conditions to form a carboxylic acid group.

CH$_3$CH(OH)CN + 2H$_2$O + HCl $\rightarrow$ **CH$_3$CH(OH)COOH + NH$_4$Cl**

2-hydroxypropanenitrile   2-hydroxypropanoic acid (lactic acid)

> 2-hydroxypropanoic acid is optically active, but in the lab a racemic mixture forms in the reaction.

Catalytic hydrogenation reduces the cyanides to the amine group

CH$_3$CH(OH)CN + 2H$_2$ $\xrightarrow{\text{Ni}}$ **CH$_3$CH(OH)CH$_2$NH$_2$**

2-hydroxypropanenitrile   1-aminopropan-2-ol

❓ Quick check questions

1. How could you distinguish between an aldehyde and a ketone by means of a chemical test?
2. Which type of alcohol can be oxidized to an aldehyde?
3. What precaution must be taken if an aldehyde is required from an alcohol?
4. Give the name or formula of a reagent that can be used to reduce a ketone.
5. Name the type of mechanism involved in the reaction between an aldehyde and hydrogen cyanide.
6. Why is it common to use hydrochloric acid and potassium cyanide rather than hydrogen cyanide in the type of reaction in question 5?

Carboxylic acids

Carboxylic acids (R – COOH) have two groups, the **carbonyl** (C=O) and the **hydroxyl** (– OH) so the – COOH is known as the **carboxyl group.**

$$R-C \begin{matrix} \diagup\diagup O \\ \diagdown O-H \end{matrix}$$

Production of carboxylic acids

Oxidation of a primary alcohol or an aldehyde:

$$RCH_2OH + [O] \rightarrow RCHO + H_2O$$

$$\text{e.g. } CH_3CH_2CH_2CH_2OH + [O] \rightarrow CH_3CH_2CH_2CHO + H_2O$$
$$\textbf{butan-1-ol} \qquad\qquad \textbf{butanal}$$

Name	Formula
Methanoic acid	$HCOOH$
Ethanoic acid	CH_3COOH
Propanoic acid	CH_3CH_2COOH
Butanoic acid	$CH_3CH_2CH_2COOH$

The aldehyde can be further oxidised to carboxylic acid by refluxing with excess oxidising agent:

$$RCHO + [O] \rightarrow RCOOH$$

$$\text{e.g. } CH_3CH_2CHO + [O] \rightarrow CH_3CH_2COOH$$
$$\textbf{propanal} \qquad\qquad \textbf{propanoic acid}$$

▶ [O] = oxidising agent e.g. warm acidified potassium dichromate(VI) solution.

✓ *Quick check 4*

Carboxylic acid reactions

1 Acid reactions

● Carboxylic acids are weak acids but they will release carbon dioxide on reaction with carbonates or hydrogencarbonates. This can be used to test for carboxylic acids.

$$2RCOOH + CO_3^{2-} \rightarrow 2RCOO^- + CO_2 + H_2O$$

$$\text{e.g. } 2CH_3CH_2COOH + Na_2CO_3 \rightarrow 2CH_3CH_2COO^-Na^+ + CO_2 + H_2O$$
$$\textbf{sodium propanoate}$$

✓ *Quick check 1,2,3*

● Carboxylic acids are weak acids so will neutralise bases like NaOH.

$$RCOOH + OH^- \rightarrow RCOO^- + H_2O$$

$$\text{e.g. } CH_3COOH + NaOH \rightarrow CH_3COO^-Na^+ + H_2O$$
$$\textbf{sodium ethanoate}$$

2 Esterification

● Carboxylic acids react with alcohols (ROH) in the presence of a strong acid catalyst, (such as concentrated sulphuric acid) to make esters.

✓ *Quick check 5,6*

$$RCOOH + R'OH \overset{H^+}{\rightleftharpoons} RCOOR' + H_2O$$

$$\textbf{Acid + alcohol} \rightleftharpoons \textbf{ester + water}$$

$$CH_3COOH + CH_3OH \rightleftharpoons CH_3COOCH_3 + H_2O$$

$$\textbf{ethanoic acid + methanol} \rightleftharpoons \textbf{methyl ethanoate + water}$$

$$HCOOH + CH_3CH_2OH \rightleftharpoons HCOOCH_2CH_3 + H_2O$$

$$\textbf{methanoic acid + ethanol} \rightleftharpoons \textbf{ethyl methanoate + water}$$

▶ R' means another carbon chain which could be the same or different from R.

- **Esterification** can be thought of as a condensation reaction since it involves the elimination of water. More accurately, it is an **addition–elimination** reaction
- It is the oxygen from the alcohol that forms the 'bridging' oxygen, not the oxygen on the acid.

$$R_1-C\overset{O}{\underset{O-H}{\Big\backslash}} \cdots \overset{+}{H} \cdots \overset{*}{O}-R_2 \longrightarrow R_1-C\overset{O}{\underset{\overset{*}{O}-R_2}{\Big\backslash}} + H\overset{O}{\diagdown}H$$

- If the alcohol's oxygen contains the ^{18}O isotope, it can be traced to the ester produced in the reaction.
- This is **isotopic labelling**, — interesting but rarely asked about.

The carboxylate ion

When the acid loses a proton and becomes dissociated, we often write the structure as having a minus charge on the oxygen which was attached to the hydrogen.

$$R-C\overset{O}{\underset{OH}{\Big\backslash}} \longrightarrow R-C\overset{O}{\underset{O^{\text{:}-}}{\Big\backslash}} + H^+$$

The point charge is stabilised by being spread evenly onto the two oxygen atoms. The resonance arrow denotes this is happening.

$$-C\overset{O^{\text{:}-}}{\underset{O}{\Big\backslash}} \longleftrightarrow -C\overset{O}{\underset{O^{\text{:}-}}{\Big\backslash}} = -C\overset{O}{\underset{O^{\text{:}}}{\Big\backslash}}{}^-$$

Esters

Esters have the general structure

$$R-C{\overset{\displaystyle O}{\diagup}}{\underset{\displaystyle O-R'}{\diagdown}}$$

- They are formed by reacting a carboxylic acid with an alcohol in the presence of concentrated sulphuric acid (see page 27).
- The alcohol used determines the start of the ester's name and the acid determines the ending of the ester.
- Esters are usually volatile.
- However animals fats contain solid esters while vegetable oils are liquid esters.

ethyl methanoate

From the alcohol – ethanol   From carboxylic acid – methanoic acid

Uses of esters

✓ *Quick check 1*

They are used as **food flavourings**, **solvents** and **plasticisers**.

Food flavourings
Esters have strong sweet smells and are naturally found in foods such as fruits. For this reason esters are used as **artificial flavourings**, e.g. ethyl methanoate smells of raspberries.

Solvents
Esters are almost insoluble in water due to a lack of an OH group in their structure, but their molecules are slightly polar, due to the carbon–oxygen bonds. For this reason, they make good solvents for polar organic compounds, e.g. glues.

Plasticisers
These are added to plastics to make them softer and more flexible e.g. in car interior (leather-like) upholstery and children's toys. They allow the chains of polymers to slide past each other, increasing flexibility.

There is concern that **esters** added to cling film may diffuse into fatty foods causing harm.

Reactions of esters

- Esters undergo **hydrolysis** (in acid or alkaline conditions).
- Alkaline hydrolysis using NaOH produces an alcohol and the sodium salt of the carboxylic acid

$$RCOOR' + NaOH \rightarrow RCOO^-Na^+ + R'OH$$

For example

$$CH_3CH_2COOCH_3 \ + \ NaOH \ \rightarrow \ CH_3CH_2COO^-Na^+ \ + \ CH_3OH$$
methyl propanoate   **sodium propanoate**   **methanol**

- Alkaline hydrolysis is known as **saponification**.
- When excess of dilute sulphuric or hydrochloric acid is added to the sodium salt of a carboxylic acid then the carboxylic acid forms, e.g. sodium propanoate would form propanoic acid.

The **hydrolysis** will take place with just water present, but in acidic or even better hot alkaline solution, it is much quicker.

Triesters or triglycerides

✓ Quick check 2,4,5,6

- Natural fats and oils are **triesters**.
- They are esters of long-chained carboxylic acids and propane-1,2,3-triol.
- Fats contain mainly the **saturated fatty acids**, e.g. stearic acid and oils contain fatty acids with some degree of **unsaturation**, e.g. linoleic acid.

propane – 1,2,3-triol (glycerol) + palmitic acid (a fatty acid) → a triester (a triglyceride)

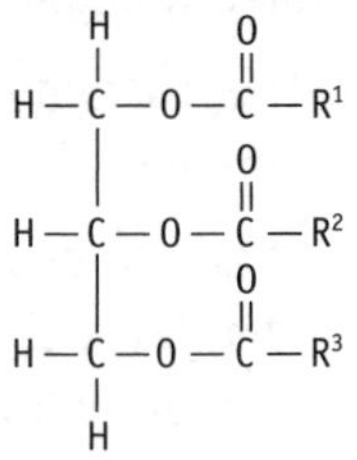

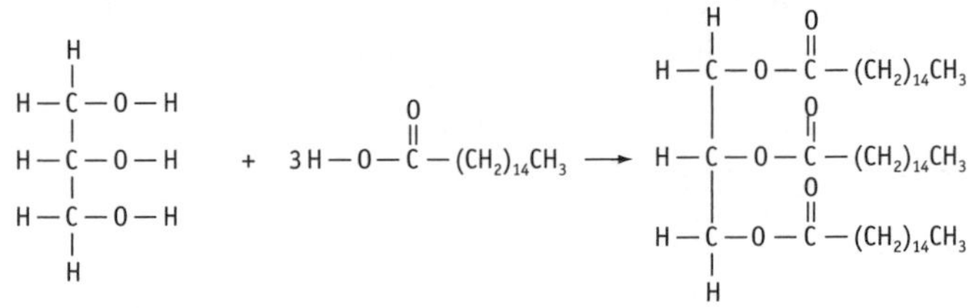

- Since **glycerol** contains three alcohol groups, the **triesters** it forms have three carboxylic acid attached. The 'R' groups of the triester may be the same or different but are often straight-chained hydrocarbons like those acids shown in the table below. Triesters therefore have the general structure shown opposite.

Long-chained carboxylic acids are known as fatty acids. Propane-1,2,3-triol is known as **glycerol** which is also a useful solvent.

a general triester

Fatty acid	Structure	Saturated or unsaturated
Stearic acid	$CH_3(CH_2)_{16}COOH$	Saturated
Oleic acid	$CH_3(CH_2)_7CH{=}CH(CH_2)_7COOH$	Unsaturated
Palmitic acid	$CH_3(CH_2)_{14}COOH$	Saturated
Linoleic acid	$CH_3(CH_2)_4CH{=}CHCH_2CH{=}CH(CH_2)_7COOH$	Unsaturated

✓ Quick check 3

- Doctors tell us that fats and oils with a higher degree of unsaturation are better than those that are saturated, as they don't 'clog' blood vessels as much.

Formation of soap

- Like all esters, triesters can be **hydrolysed** by boiling with sodium hydroxide solution.
- They form glycerol and the sodium salts of the long-chain carboxylic acids.
- These sodium salts are used in the manufacture of soap e.g. sodium stearate.

propane – 1,2,3-triol (glycerol)

sodium octadecanoate (stearate) – a soap

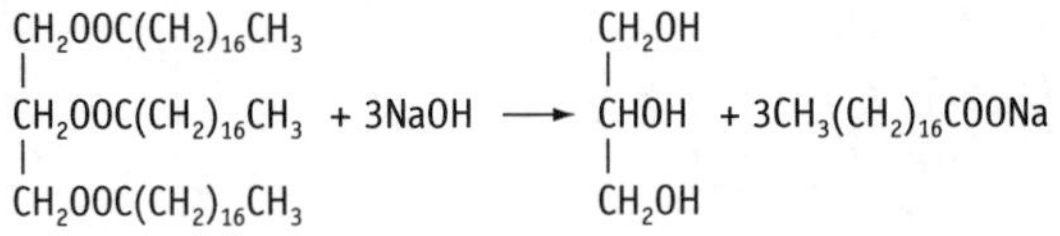

Palmolive® soap (sodium palmitate) is made by reacting NaOH with the ester formed from glycerol and palmitic acid.

? Quick check questions

1 Give two uses of esters.

2 Name the type of reaction that occurs when a long-chained ester is hydrolysed in the presence of sodium hydroxide. Give a use for the product of this type of reaction.

3 Which tri-ol is present in natural fats and oils?

4 Name the ester $CH_3CH_2COOCH_2CH_2CH_2CH_3$ and its hydrolysis products with NaOH.

5 Why is hydrolysis almost complete when sodium hydroxide is used as the catalyst but only partial when sulphuric acid is used?

6 Which ester is isomeric with ethanoic acid?

Acylation

- **Acyl chlorides** and **acid anhydrides** are derived from carboxylic acids.

  e.g. $CH_3COOH + SOCl_2 \rightarrow CH_3COCl + SO_2 + HCl$

- The carbonyl carbon has a relatively large $\delta+$ due to the presence of electronegative groups like oxygen and chlorine, so is attacked by nucleophiles.

- The reactivity of the derivative depends on the size of the carbon's $\delta+$ which is determined by the electron releasing or attracting ability of Z — the group attached.

- It is also determined by how easily Z will leave — a good leaving group facilitates the reaction (helps it happen). The reactivity towards nucleophiles can be seen below.

Visualising the reaction

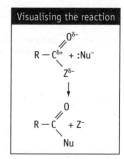

✓ **Quick check 1**

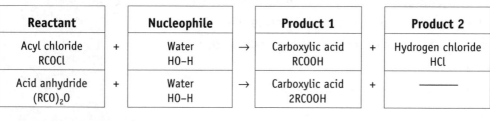

| acyl chlorides | acid anhydrides | acids | esters | amides |

$\longleftarrow$ Z group e⁻ withdrawing   Z group e⁻ releasing $\longrightarrow$

- The mechanism involves nucleophilic addition followed by elimination, so funnily enough it is called **nucleophilic addition–elimination**, though for short, they are known as acylation reactions.

Acylation reactions you should know and understand

✓ **Quick check 2**

Reactant		Nucleophile		Product 1		Product 2
Acyl chloride RCOCl	+	Water HO–H	→	Carboxylic acid RCOOH	+	Hydrogen chloride HCl
Acid anhydride $(RCO)_2O$	+	Water HO–H	→	Carboxylic acid 2RCOOH	+	———

▶ The mechanism for acid anhydrides isn't required for the AQA spec.

Visualising the reaction with water

step 1

$H_2O: \rightarrow \overset{CH_3}{\underset{Cl}{C}} = O \longrightarrow H - \overset{+}{O} - \overset{CH_3}{\underset{\underset{Cl}{|}}{C}} - O:\overset{-}{}$

The nucleophile, here H_2O attacks the $C^{\delta+}$ and a pair of electrons in the carbon-oxygen double bond are transferred to the oxygen atom. This is addition.

step 2

The pair of electrons on the oxygen atom reform the double bond and the chlorine atom is eliminated as a chloride ion. (This bonds to a hydrogen atom on the protonated hydroxyl group to form HCl.)

Acyl chloride RCOCl	+	Alcohol R'O–H	→	Ester RCOOR'	+	Hydrogen chloride HCl
Acid anhydride $(RCO)_2O$	+	Alcohol R'O–H	→	Ester RCOOR'	+	Carboxylic acid RCOOH

Visualising the reaction with alcohol

$$CH_3CH_2O: \overset{CH_3}{\underset{\underset{Cl}{|}}{C}} = O \longrightarrow CH_3CH_2\overset{+}{O} - \overset{CH_3}{\underset{\underset{Cl}{|}}{C}} - \overset{..}{O}:^- \xrightarrow{-Cl:^-} CH_3CH_2\overset{+}{O} - \overset{CH_3}{\underset{\underset{H}{|}}{C}} = O \xrightarrow{-H^+} CH_3CH_2OCCH_3$$

Acyl chloride RCOCl	+	Ammonia H–NH$_2$	→	Acid amide RCONH$_2$	+	Hydrogen chloride HCl
Acid anhydride (RCO)$_2$O	+	Ammonia H–NH$_2$	→	Acid amide RCONH$_2$	+	Carboxylic acid RCOOH

▸ HCl is produced but it reacts with NH$_3$ to form NH$_4^+$Cl$^-$.

✓ *Quick check 3, 4*

▸ HCl formed reacts with RNH$_2$ to make RNH$_3^+$Cl$^-$.

Visualising the reaction with ammonia

$$H_3N: \overset{CH_3}{\underset{\underset{Cl}{|}}{C}} = O \longrightarrow H_2\overset{+}{N} - \overset{CH_3}{\underset{\underset{Cl}{|}}{C}} - \overset{..}{O}:^- \xrightarrow{-Cl^-} H_2\overset{+}{N} - \overset{CH_3}{\underset{\underset{H}{|}}{C}} = O \longrightarrow CH_3CONH_2 + NH_4^+$$

Further acylation is unlikely as amides are weakened as nucleophiles because the carbonyl withdraws e$^-$ from the nitrogen.

Acyl chloride RCOCl	+	Primary amine R'NH-H	→	N-substituted acid amide RCONHR'	+	Hydrogen chloride HCl
Acid anhydride (RCO)$_2$O	+	Primary amine R'NH-H	→	N-substituted acid amide RCONHR'	+	Carboxylic acid RCOOH

Visualising the reaction with amine

$$CH_3\overset{..}{N}H_2 \overset{CH_3}{\underset{\underset{Cl}{|}}{C}} = O \longrightarrow CH_3 - \overset{H}{\underset{\underset{Cl}{|}}{\overset{+}{N}}} - \overset{CH_3}{\underset{\underset{Cl}{|}}{C}} - \overset{..}{O}:^- \xrightarrow{-Cl:^-} CH_3 - \overset{H}{\underset{\underset{H}{|}}{\overset{+}{N}}} - \overset{CH_3}{C} = O \xrightarrow{-H^+} CH_3NHCOCH_3$$

"N" indicates that the R' group is attached to the nitrogen atom. Acid amides do not react further (like NH$_3$ and RX). The O withdraws e$^-$ from the –NH$_2$ making it slightly positive.

Making aspirin

- Aspirin is an **analgesic** (or painkiller) and an **anti-pyretic** (or a reducer of body temperature).
- It is made by acylating 2-hydroxybenzoic acid (AKA 2-hydroxybenzenecarboxylic acid or salicylic acid).
- Ethanoic anhydride is used as the acylating agent in preference to ethanoyl chloride, as it is 1) cheaper 2) less corrosive 3) less susceptible to hydrolysis 4) safer 5) easier to say... !

2-hydroxybenzene carboxylic acid + ethanoic anhydride → 2-ethanoyl oxybenzene carboxylic acid (aspirin) + ethanoic acid

✓ *Quick check 5*

▸ Paracetamol is manufactured by acylating 4-aminophenol. The preparation is no headache...

? Quick check questions

1 Give the general formula for an acyl group.

2 Name the product/s of the reaction between ethanoic anhydride and water.

3 What must be reacted with an acyl chloride to produce an acid amide?

4 Give an equation for the reaction between ethanoyl chloride and methanol and name the products.

5 Why are acid anhydrides used in industry in preference to acyl chlorides?

Aromatic chemistry

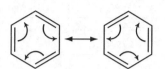

Benzene has an **empirical formula** CH and a M_r of 78, which suggests a **molecular formula** C_6H_6. You are unlikely to use benzene in the laboratory as it is **carcinogenic**, but you will come across derivatives like methylbenzene. Most arenes (derivatives of benzene) have similar relative numbers of carbons and hydrogen atoms, so are easy to spot.

In 1865 **Kekulé** suggested that benzene was planar and had an alternative double/ single bond structure. He suggested that the two structures opposite rapidly switch between each other. Four bits of evidence now suggest that benzene does not have the three-conjugated double bond structure.

1 **X-ray diffraction data**

✓ Quick check 1

2 **Chlorination of benzene** produces only one 1,2-disubstituted structure

3 **Thermodynamic data** suggest it is more stable than expected

4 Benzene does not undergo **addition reactions**, like alkenes, it performs **electrophilic substitution**

1 X-ray diffraction data

Modern X-ray diffraction shows benzene to have planar molecules with electron density around the carbon atoms equal, i.e. they have equivalent bonds. These are part way between single and double bonds. The notation used to indicate a **delocalised structure** is shown by a circle within a hexagon.

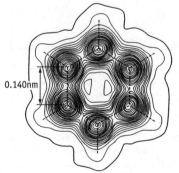

Electron density map of benzene

	Bond	Bond length/nm
Alkane	C — C	0.154
Benzene	C — C	0.140
Alkene	C = C	0.134

2 Chlorination of benzene

✓ Quick check 2

When benzene is chlorinated, (with $AlCl_3$ catalyst) it produces 1,2-dichlorobenzene, $C_6H_4Cl_2$, which has only one structure formed. This indicates that all the hydrogen atoms in benzene are equivalent. If Kekulé's structure was correct we would see the two isomers shown opposite.

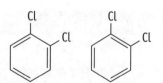

Kekulé's structure would give two positional isomers.

3 Thermodynamic data

If benzene had three double bonds, the theoretical enthalpy of hydrogenation should be three times that of cyclohexene.

cyclohexene + H_2(g) $\longrightarrow$ cyclohexane $\quad \Delta H^{\ominus} = -121$ kJ mol^{-1}

∴ theoretically

Kekule's benzene (cyclohexa-1,3,5,-triene) + $3H_2$(g) $\longrightarrow$ cyclohexane $\quad \Delta H^{\ominus} = -363$ kJ mol^{-1} which is 3 x 121 kJ mol^{-1}

Difference = 363 – 208 = 155 kJ mol^{-1}

but experimentally

+ $3H_2$(g) $\longrightarrow$ $\quad \Delta H^{\ominus} = -208$ kJ mol^{-1}

✓ Quick check 3

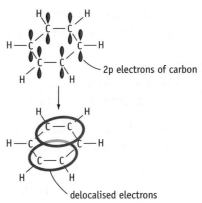

2p electrons of carbon

delocalised electrons

π electrons spread out around the ring in benzene

The 155 kJ mol^{-1} difference is a measure of the stability of benzene's delocalised structure. In other words benzene's structure is more stable than is predicted. It is thought that the spare p electrons form π electrons and delocalise or are shared between the carbon atoms in the ring. This means that there is no great concentration of electrons as in an alkene but there are regions of electron density above and below the ring; this produces a stable structure.

	Alkenes	Arenes
Attacking species	electrophile	electrophile
Type of mechanism	addition	substitution
Ease of reaction	readily react	requires vigorous conditions or a catalyst

Electrophilic substitution

✓ Quick check 4

The electrons in the benzene ring attract electrophiles but substitution allows the ring to reform. Since this is the most stable route, electrophilic substitution is favoured over electrophilic addition, which you would expect from a triene.

▶ E^+ = an **electrophile** — an atom or ion that accepts a pair of electrons.

Step 1 Addition of the electrophile

A pair of electrons from the delocalised π electron system form a bond with the electrophile. A highly unstable cationic intermediate is produced.

Step 2 Removal of a hydrogen

A hydrogen atom is then lost from the intermediate ion to reform the delocalised (and stable) π electron system.

Visualising the reaction

E^+ → step 1 → (+) → – H^+ step 2 →

? Quick check questions

1. Draw a diagram of the benzene molecule and give the bond angles.

2. Explain why the bond length is neither 0.154 nm (C–C single) nor 0.134 nm (C=C double) in length.

3. The enthalpy of hydrogenation of cyclohexene is –121 kJmol^{-1} and the corresponding value for benzene is –208 kJ mol^{-1}. Explain why the value for benzene is not three times that for cyclohexene.

4. Name the reaction mechanism, typical for the reactions of benzene.

Reactions of benzene

In many **electrophilic substitution** reactions the electrophile must be generated before the reaction can proceed. You need to know the details for the **nitration** and **Friedel–Crafts alkylation** and **acylation** reactions for benzene.

Nitration

Conditions: Conc. nitric and sulphuric acid, **warm**

Generating the electrophile: Concentrated sulphuric acid acts in this reaction as a catalyst which, when added to nitric acid, produces the **nitronium ion** — the electrophile.

$$H_2SO_4 + HNO_3 \rightleftharpoons HSO_4^- + H_2NO_3^+$$

The protonated nitric acid breaks down forming the **nitronium ion** NO_2^+

$$H_2NO_3^+ \rightarrow H_2O + NO_2^+$$

$$O = \overset{+}{N} = O$$
Nitronium ion

Overall forming E$^+$: $2H_2SO_4 + HNO_3 \rightarrow 2HSO_4^- + H_3O^+ + NO_2^+$

Overall equation: $C_6H_6 + HNO_3 \rightarrow C_6H_5NO_2 + H_2O$
nitrobenzene

✓ *Quick check 1,2*

🕭 This equation is the one to remember as is most often asked about.

Mechanism: Electrophilic substitution

Step 1 Electrophilic attack

The NO_2^+ ion is a powerful electrophile and attracts electrons out of the benzene ring to form a positively charged intermediate.

Step 2 Removal of hydrogen

The electrons in the positively charged intermediate are only partially delocalised making it unstable. To regain the stable delocalised system, a hydrogen ion is lost to give nitrobenzene.

Visualising the reaction

The importance of nitration of benzene

Nitration of benzene is an important synthetic process as production of many organic compounds goes via nitrobenzene and its derivatives.

- **TNT** (trinitrotoluene or methyl-2,4,6-trinitrobenzene) an explosive, is formed from methylbenzene by nitration.

- The nitro– group in benzene derivatives can be reduced by using tin and conc. hydrochloric acid to produce **aromatic amines**. These aromatic amines are used to make **synthetic dyes** (see page 38).

methyl-2,4,6-trinitrobenzene (TNT)

Friedel–Crafts reactions

A Friedel–Crafts reaction is one where electrophilic substitution occurs between an aromatic compound and an electrophile, which is a carbocation or a highly polarised complex with a positive carbon. Sounds difficult, but it isn't, there are only two you need to worry about.

1 Alkylation

Alkylation, as the name suggests, is the introduction of one or more alkyl groups such as CH_3— . This requires a halogen carrier or catalyst to get things going, which is often aluminium chloride.

$$C_6H_6 + RCl \xrightarrow{AlCl_3} C_6H_5R + HCl$$

Conditions: When chloroethane, benzene and aluminium chloride are warmed, ethylbenzene forms, (all under anhydrous conditions).

Generating the electrophile:

$$CH_3CH_2\!-\!Cl \quad AlCl_3 \rightarrow CH_3CH_2{}^+[AlCl_4]^-$$

R^+ (the carbocation) is the electrophile

> Aluminium chloride has a vacant orbital so acts as a **Lewis acid** and accepts a lone pair of electrons from the carbon-chlorine bond in the chloroalkane.

Overall Equation: $\quad C_6H_6 + CH_3CH_2Cl \rightarrow C_6H_5CH_2CH_3 + HCl$

chloroethane   ethylbenzene

The catalyst, $AlCl_3$, is regenerated

$$H^+ + [AlCl_4]^- \rightarrow AlCl_3 + HCl$$

Mechanism:

The reaction may not stop here, further alkylation may occur.

Ethylbenzene is used to make phenylethene, the monomer of polystyrene.

Visualising the reaction

2 Acylation

Acylation produces an aromatic ketone from an acyl chloride (or acid anhydride) with an aluminium chloride catalyst.

$$C_6H_6 + RCOCl \xrightarrow{AlCl_3} C_6H_5COR + HCl$$

Conditions: Ethanoyl chloride, benzene and aluminium chloride warmed (all under anhydrous conditions).

Generating the electrophile:

$$CH_3CO\!-\!Cl \quad AlCl_3 \rightarrow CH_3C^+O[AlCl_4]^-$$

The electrophile RCO^+ is an acylium cation

> Industrially ethylbenzene is made from benzene, HCl, $AlCl_3$ and ethene to make the R^+ (at 90°C)
> $CH_2{=}CH_2 + HCl$
> $\rightarrow CH_3CH_2{}^+ + Cl^-$

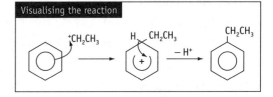

RCO – is the acyl group

Overall equation:

$$C_6H_6 + CH_3COCl \rightarrow C_6H_5COCH_3 + HCl$$

ethanoyl chloride phenylethanone

- The catalyst, $AlCl_3$, is regenerated as with alkylation
- The reaction can be stopped when one RCO^+ has reacted, as this group withdraws electrons from the ring, reducing the chance of electrophilic attack.

Mechanism:

Visualising the reaction

phenylethanone

Quick check questions

1   Give an equation to show the formation of the reactive species required to nitrate benzene.
2   Give the mechanism for the nitration of benzene.
3   Name the catalyst used in the alkylation of benzene.
4   Why is the catalyst needed?

Amines

Facts you should know about amines

- They are derivatives of ammonia (NH_3).
- The functional group is the **amino** $-NH_2$ group.
- They have a general formula $C_nH_{2n+1}NH_2$ (1° only).
- They are classified as **primary**, **secondary** or **tertiary.**

general formula   R — NH_2
amines

| | primary | secondary | tertiary |
| no. of alkyl groups | 1 | 2 | 3 |

e.g. ethylamine CH_3CH_2 NH_2 (1°)

Preparation

- They are produced from haloalkanes and ammonia by nucleophilic substitution (as opposite, not fantastic as it gives mixed products).

$$CH_3CH_2Br + 2NH_3 \rightarrow CH_3CH_2NH_2 + NH_4Br$$
bromoethane   ethylamine

- They are produced by reduction of nitriles with a nickel catalyst.

$$CH_3CN + 2H_2 \rightarrow CH_3CH_2NH_2 \quad \text{(a better method)}$$
ethanenitrile   ethylamine

- Aryl amines can be prepared by reduction of **nitro derivatives** using a mixture of tin or iron and concentrated hydrochloric acid (or $H_2(g)$ + Ni).

$$C_6H_5NO_2 + 6[H] \rightarrow C_6H_5NH_2 + 2H_2O$$
nitrobenzene   phenylamine

Visualising the reaction

This reaction occurs via nucleophilic substitution.

Properties of amines

The nitrogen atom in the amine molecule can donate its lone pair of electrons to another species. This means that amines behave as bases, nucleophiles and ligands.

- as bases: they donate a lone pair to a proton.
- as nucleophiles: they donate a lone pair to a δ+ carbon atom.
- as ligands: they donate a lone pair to a transition metal ion (see page 72).

Name	Formula	pK_a
ammonia	NH_3	9.25
methylamine	CH_3NH_2	10.64
ethylamine	$CH_3CH_2NH_2$	10.73
propylamine	$CH_3CH_2CH_2NH_2$	10.84
phenylamine	$C_6H_5NH_2$	4.62
(phenylmethyl)amine*	$C_6H_5CH_2NH_2$	9.37

*Brackets in the name, e.g. (phenylmethyl)amine, means that the phenyl group isn't attached to the nitrogen atom.

Basic properties of primary amines

- Nitrogen's lone pair can form a co-ordinate bond with a proton.

$$CH_3CH_2NH_2 + H^+ \rightarrow CH_3CH_2NH_3^+$$
ethylamine   methylammonium ion

- As they are proton acceptors, primary amines are Brønsted–Lowry bases.
- In water, they produce OH⁻ ions so are weak alkalis.

$$CH_3CH_2NH_2 + H_2O \rightleftharpoons CH_3CH_2NH_3^+ + OH^-$$

The basicity of the amino ($-NH_2$) group depends on the availability of the lone pair, anything that increases the electron density of the lone pair on the nitrogen atom, increases the basic strength.

The stronger the base the *higher* the pK_a value.

RNH$_2$	<	NH$_3$	<	ArNH$_2$

- Primary **aliphatic amines**
- These are stronger bases than ammonia
- The alkyl groups are electron releasing (relative to hydrogen)
- They "push" electrons onto the nitrogen and so increase the availability of the lone pair
- This effect increases with increasing the length of the carbon chain (up to 2/3 carbons long)

- Ammonia
- The lone pair on nitrogen is donated to a proton

- Aromatic **aryl amines**
- These are weaker bases than ammonia
- The lone pair on the nitrogen atom become delocalised with electrons in the benzene ring
- The lone pair is therefore less available for bonding with a H$^+$ ion

If the N of the amine group is not directly attached to the benzene ring (like in (phenylmethyl)amine, $C_6H_5CH_2NH_2$) the molecule is an aliphatic primary amine as the lone pair does not become delocalised into the ring.

Nucleophilic properties of amines

1 With haloalkanes (and ammonia)

- Propylamine is produced from bromoethane according to the equation

$$CH_3CH_2CH_2Br + 2NH_3 \rightarrow CH_3CH_2CH_2NH_2 + NH_4Br$$
a primary amine

- The product further reacts with bromoethane in successive substitution reactions to form dipropylamine, a **secondary amine**

$$CH_3CH_2CH_2Br + CH_3 CH_2CH_2NH_2 \rightarrow (CH_3 CH_2CH_2)_2NH + HBr$$
a secondary amine

- The product of this reaction can further react with bromoethane to form tripropylamine, a **tertiary amine**, and a **quaternary ammonium salt**

$$CH_3CH_2CH_2Br + (CH_3CH_2CH_2)_2NH \rightarrow (CH_3 CH_2CH_2)_3N + HBr$$
tripropylamine

$$CH_3CH_2CH_2Br + (CH_3CH_2CH_2)_3N \rightarrow (CH_3 CH_2CH_2)_4N^+Br^-$$
a quarternary ammonium salt

No more protons can be removed from the nitrogen and the positive charge resides on the nitrogen atom, so the quaternary ammonium ion cannot continue to behave as a nucleophile.

2 With acyl chlorides and acid anhydrides

Primary amines react with acyl chlorides and acid anhydrides to form substituted amides

$$CH_3NH_2 + CH_3COCl \rightarrow CH_3CONHCH_3 + HCl$$
ethanoyl chloride   N-methylethanamide

$$CH_3CH_2NH_2 + (CH_3CO)_2O \rightarrow CH_3CONHCH_2CH_3 + CH_3COOH$$
ethanoic anhydride   N-ethylethanamide

✓ *Quick check 2,3,4*

▶ The primary amine has a lone pair of electrons so it can further act as a nucleophile.

▶ Mixed products is not great for synthesis — but excess NH$_3$ favours production of the 1° amine, while excess haloalkane favours the quartenary ammonium salt.

✓ *Quick check 5,6*

▶ Quaternary ammonium salts are **cationic surfactants** used in **fabric conditioners** and hair products. They reduce the static by coating the surface of the cloth or hair with charge.

? *Quick check questions*

1 Which feature of the amine molecule allows it to act as a base?
2 Give an equation for the reaction between bromoethane and ethylamine.
3 What is likely to happen if there is an excess of bromoethane?
4 Name the type of mechanism involved.
5 Give two methods of preparing aliphatic amines.
6 Which of the two methods is likely to give the best yield?

Amino acids

- Their general formula is RC*H(NH₂)COOH and generally, their displayed formula can be represented as shown opposite.

- They contain an amine group –NH₂ and an acid group –COOH.

- Amino acids can exhibit **optical isomerism** if they have a carbon atom* attached to four different groups (a **chiral** carbon*)

- When the amino group is attached to the carbon atom next to the carboxyl group, the amino acids are classified as α-**amino acids**.

- There are 23 naturally occurring amino acids which are often known by their trivial names.

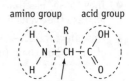

General structure of an amino acid.

✓ *Quick check 1*

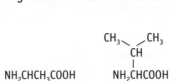

NH₂CH₂COOH   NH₂CHCH₃COOH   NH₂CHCOOH

glycine   alanine   valine

The acidic and basic properties

✓ *Quick check 2,3*

Amino acids have both an acidic and basic nature and react with both acids and bases. They are therefore amphoteric.

- As acids: The acidic carboxyl (–COOH) group donates a proton, here reacting with a base to form a salt plus water.

$$R\text{–}CH(NH_2)COOH + NaOH \rightarrow R\text{–}CH(NH_2)COO^-Na^+ + H_2O$$

- As bases: The basic amine (–NH₂) group accepts a proton from an acid to form a salt.

$$R\text{–}CH(COOH)\text{–}NH_2 + HCl \rightarrow R\text{–}CH(COOH)\text{–}NH_3^+Cl^-$$

- Pure amino acids are white crystalline solids which easily dissolve in water and have unusually high melting points. This is because in the pure state amino acids exist as **zwitterions**.

- At high pH the proton on the –NH₃⁺ is removed leaving an overall negative charge.

- At low pH the –COO⁻ group accepts a proton giving an overall positive charge.

- At intermediate pH (called the isoelectric point), the negative and positive charges are equal and the amino acid will have no overall charge.

- The positive and negative charges in the zwitterion cause ionic bonds which are stronger than the expected hydrogen bonding, so they have high melting points.

RCHCOOH
|
⁺NH₃   acidic pH

⇅

RCHCOO⁻
|
⁺NH₃   neutral pH

⇅

RCHCOO⁻
|
NH₂   basic pH

General amino acid at low, medium and high pH

Proteins

- Proteins are natural sequences of amino acids joined together by an amide bond called a peptide link.

- Many amino acids join up to form a polymer known as a **polypeptide**.

2 amino acid units = dipeptide
3 amino acid units = tripeptide
Loads of amino acid units = polypeptide

- They join up by addition–elimination reactions which involve the loss of water. (a kind of condensation reaction).

Condensation polymerisation reactions in the body are catalysed by enzymes.

$$H_2N-\underset{\underset{CH_3}{|}}{\overset{\overset{H}{|}}{C}}-COOH \; + \; H_2N-\underset{\underset{H}{|}}{\overset{\overset{H}{|}}{C}}-COOH \longrightarrow \; -N-CH-\overset{\overset{O}{\|}}{C}-N-CH-\overset{\overset{O}{\|}}{C}-$$

peptide link

e.g alanine + glycine forming a protein with a peptide link.

✓ Quick check 6

- Proteins are fibrous (long chained) or globular (spherical).

Hydrolysis of proteins

hydrolysis = splitting of water hydrolysis of food = digestion

- Hydrolysis is kind of the reverse of peptide formation or polymerisation. Water is effectively added which breaks the peptide bonds. This produces the individual amino acids from the polypeptide.

- The protein is refluxed with concentrated hydrochloric acid (~ 6 M) for several hours (or a specific enzyme may catalyse the reaction).

- If this is done from one end of the polymer, one amino acid at a time, it reveals the amino acids present and can give a clue as to the primary structure of the protein i.e. what amino acids were present and in what order.

Hydrolysis of a peptide bond

Hydrogen bonding in proteins

Hydrogen bonding in proteins determines the secondary structure of the polymer chains. They usually produce an α helix shape or a β pleated sheet.

The presence of many hydrogen bonds makes the protein structures very stable.

You do not need to remember secondary structures, but you must know that hydrogen bonding, ionic bonding and disulphide bridges are important in forming them.

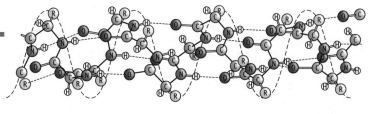

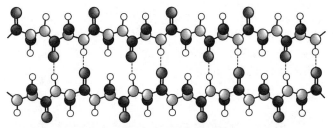

Helix shape and beta pleated sheet showing hydrogen bonding.

? Quick check questions

1 Give the general structure of an α-amino acid.

2 Give an equation showing 2-aminopropanoic acid acting as both an acid and a base.

3 Which biologically important group of compounds is produced by combining amino acids to form large linear molecules?

4 What is the name given to the linkages between the amino acid sub-units in these molecules?

5 The properties of these molecules are determined by their three dimensional shape. Which type of intermolecular forces are important in maintaining their shape?

6 Which type of reaction is involved in the formation of individual amino acids from these polymeric molecules?

Polymers

Polymers are made from smaller repeating units called monomers.

The process of polymer formation is called polymerisation. There are two types of polymers you need to know about — **addition** and **condensation**.

monomer
+
monomer
+   = polymer
monomer
+
monomer

1 Addition polymers

✓ Quick check 1

- These form directly from compounds containing a C=C bond.
- They are referred to as chain-growth polymers as they form by addition of a monomer to the end of the chain, usually by a free radical mechanism.
- Reaction conditions involve high pressure and an initiator, which provides the free radicals needed to start the reaction.
- The length of the polymer chain can vary depending on the properties required, but typically is between 100 and 100 000 monomers long.
- You must be able to draw polymers from monomers and vice versa.

Example	Monomer		Polymer	Use	
Poly(ethene)	n $CH_2=CH_2$	→	$-[CH_2-CH_2]_n$	Shopping bags, cling-film	
Poly(styrene)	n $CH_2=CHC_6H_5$	→	$-[CH_2-CH]_n$ C_6H_5	Protection in packaging, thermal insulation	
PVC	n $CH_2=CHCl$	→	$-[CH_2-CH]_n$ Cl	Fake leather, pipes, flooring	
Poly(propene)	n $CH_2=CHCH_3$	→	$-[CH_2-CH]_n$ CH_3	Washing up bowls, ropes	
PTFE	n $CF_2=CF_2$	→	$-[\overset{F}{\underset{F}{C}}-\overset{F}{\underset{F}{C}}]_n$	Coat electric irons and frying pan for non-stick properties	
Acrylic (Perspex®) [poly(methyl-propenoate)]	n $CH_2=CCOOCH_3$ CH_3	→	$COOCH_3$ $-[CH_2-C]_n$ CH_3	Fake/safety glass	

Poly(alkene)s are chemically inert (like alkanes) due to saturation and lack of polar groups bonded to the carbon chain...which has its good and bad points.

✓ Quick check 2

Good points
- They can be used to make many products that do not affect the substances they are in contact with, like food or chemicals

Bad points
- Non-biodegradable so difficult to dispose of (e.g. in landfill sites)
- Some are flammable
- Some give off toxic fumes on incineration

2 Condensation polymers

When two monomers with functional groups at *both* ends join, and a small molecule like water is eliminated — you have a condensation polymer.

These are known as **step growth** polymers and there are two main types of polymers you need to know about, **polyesters** and **polyamides**.

Polyesters

As you know, esters are formed by the reaction of a carboxylic acid and an alcohol, (see page 30).

✓ *Quick check 3*

$$RCOOH + R'OH \rightarrow RCOOOR' + H_2O$$

But if a **di**carboxylic acid and a **di**ol react, long chain polyesters form.

e.g. Terylene® (manufactured from benzene-1,4-dicarboxylic acid and ethane-1,2-diol)

Terylene

Polyamides

When a carboxylic acid and an amine react, a substituted amide is produced.

✓ *Quick check 4*

$$RCOOH + R'NH_2 \rightarrow RCONHR' + H_2O$$

So when a **di**carboxylic acid and a **di**amine react a chain reaction occurs, and a long chain condensation polyamide or amino acid forms.

Nylon-6,6 is formed from the condensation reaction between hexanedioic acid and 1,6-diaminohexane:

nylon-6,6

Other polyamides include Kevlar® (a polymer of benzene-1,4-dicarboxylic acid and benzene-1,4-diamine) and nylon-6.

> There are many polyamides known as nylon. The name nylon-6,6 shows that the two monomers each contain six carbon atoms. They're used for things like textiles, ropes and fishing lines.

Hydrolysis of polyesters and polyamides

Both polyesters and polyamides undergo hydrolysis in acid or alkaline conditions so are biodegradable.

e.g. Hydrolysis of Terylene

✓ *Quick check 5*

> You don't need to know the mechanism for these hydrolysis reactions....phew...

? Quick check questions

1 What feature is essential in a monomer for it to be able to form an addition polymer?

2 Why is poly(ethene) chemically inert and what are the ecological implications of this?

3 Draw the repeating unit when propanedioic acid forms a polymer with ethane 1,2-diol.

4 What are the monomers used to form nylon-6,6?

5 What happens when a polyamide is added to a solution of sodium hydroxide?

Organic synthesis and analysis

Synthesis is all about tailor-making molecules for a particular job. In exams you are often given a molecule that you have to synthesise; the idea is that you learn the flow charts below so that you can work back from the target molecule and propose a synthetic pathway, which may take a few steps.

Reactions of the main functional groups

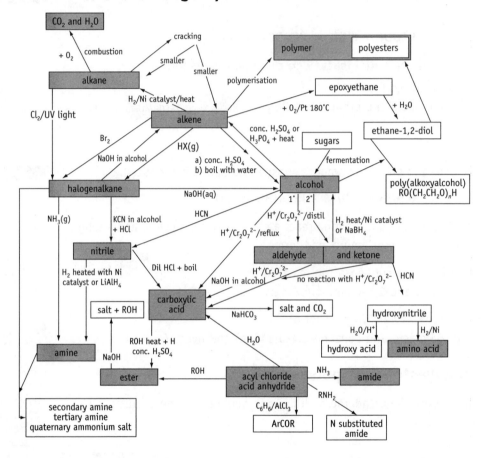

Reactions of benzene

Analysis

To analyse an unknown compound the following techniques are usually used to determine its structure...

- **Empirical formulae** determination from % by mass analysis data (see AS book).
- Relative molecular mass determination from **mass spectra** (see AS book).
- **Functional group tests** (see below).
- Functional group analysis from **i.r. spectral data** (see page 48).
- Structural determination from fragmentation patterns on **mass spectra** (see page 46).
- Structural determination from **n.m.r. spectra** (see page 50).

Infra-red spectra are good at confirming the results of functional group tests and **mass spectra** with **empirical formula data** are useful for finding the molecular formula of a compound. Lastly the **n.m.r. spectra** help to confirm the position of hydrogen atoms and so alkyl groups in a molecule.

Functional group test

Functional group	Group tested for	Test	Positive result
Alkenes	C=C	Add bromine water	Orange-brown decolorised
Haloalkanes	R—X	NaOH(aq) + warm /add HNO_3(aq) + $AgNO_3$(aq)	ppt of AgX: white = chloride; cream = bromide; yellow = iodide; no ppt = fluoride
Alcohols	R—OH	Acidified $K_2Cr_2O_7$, CH_3COOH +warm + cH_2SO_4	Orange to green if 1° or 2° alcohol; sweet smelling ester
Aldehydes	R—CHO	Acidified $K_2Cr_2O_7$, warm Fehling's solution, warm Tollen's solution	Orange → green; Red ppt of Cu_2O; Silver ppt (silver mirror)
Carboxylic acids	R—COOH	Add $NaHCO_3$(aq)	CO_2(g) gas evolves
Acyl chlorides	R—COCl	Add $AgNO_3$(aq)	ppt of white AgCl (quite a violent reaction)

? Quick check questions

1 Name the three different types of reagent that attack organic molecules.

2 Give an example of each type.

3 Name the six reaction mechanisms you have come across.

4 Which type of reaction do the following undergo a) alkanes, b) alkenes and c) acyl chlorides?

5 How could you make a carboxylic acid from a haloalkane in two steps?

6 How could you make the alcohol ⬡– CH(OH)CH$_3$ from benzene plus one other organic compound plus any inorganic reagents you wish?

Mass spectrometry

Mass spectrometry uses the fact that heavier ions are deflected less than lighter ones in a magnetic field. A molecule is vaporised and ionised to form a **molecular ion:**

$$M(g) \rightarrow M^{+\bullet}(g)$$

- The molecular ion produces a **molecular ion peak (the M⁺ peak).** This peak is the last but one of all the peaks on the mass spectrum.
- The very last one is a small peak caused by the presence of carbon-13 (^{13}C) — the **M+1 peak.** About 1% of all naturally occurring carbon is ^{13}C.
- The *m/z* **value** for the molecular ion peak = the M_r of the compound if the molecular ion is unipositive.
- The more stable ions produce the larger peaks. The most stable fragment forms the **base peak.** Other peak heights are given as a percentage of the base peak and represent the relative abundance.

Fragmentation

Fragmentation is when a covalent bond in the molecular ion breaks due to electron impact and produces an ion and a free radical.

$$M^{+\bullet}(g) \rightarrow X^{+}(g) + Y^{\bullet}(g) \text{ Or } M^{+\bullet}(g) \rightarrow X^{\bullet}(g) + Y^{+}(g)$$

The *ions* are detected in the mass spectrometer.

- Stable ions like the carbocations CH_3^{+}, $CH_3CH_2^{+}$ and acylium ions CH_3CO^{+} and $CH_3CH_2CO^{+}$ are common. These are able to hold a +ve charge.
- Identifying the fragments helps to determine the molecular structure of a compound and to distinguish between isomers.

Example

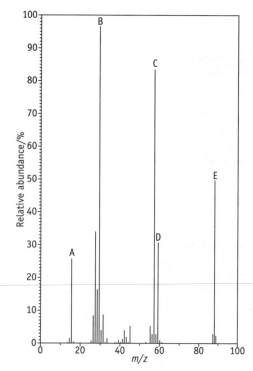

m/z
A = 15 = CH_3^{+}
B = 29 = $CH_3CH_2^{+}$
C = 57 = $CH_3CH_2CO^{+}$
D = 59 = $COOCH_3^{+}$
E = 88 = (molecular ion)

Mass spectrum of methyl propanoate

m/z value	Possible +ve ions
15	CH_3^{+}
29	$CH_3CH_2^{+}$
31	CH_2OH^{+}
43	$(CH_3)_2CH^{+}$ or CH_3CO^{+}
57	$CH_3CH_2CO^{+}$
77	$C_6H_5^{+}$
91	$C_6H_5CH_2^{+}$

▶ Read the section in the AS book for background information on workings of the mass spectrometer. Make sure you can explain the five steps involved.

▶ The molecular ion is a free radical (i.e. it has an unpaired electron) and is also called a **radical cation.**

▶ As you'd expect, the weaker bonds break easiest. Remember the radicals are not detected.

▶ Always look out for methyl (15) and ethyl (29) these nearly always show up.

Presence of isotopes

The interesting thing to notice in examples which contain chlorine and bromine (e.g. in haloalkanes which are commonly asked about), is that you get a peak for each isotope in the ratio that they usually exist.

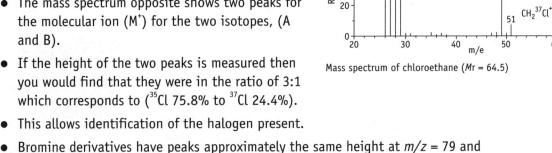

Mass spectrum of chloroethane (Mr = 64.5)

- Chlorine has two common isotopes ^{35}Cl and ^{37}Cl.

- The mass spectrum opposite shows two peaks for the molecular ion (M^+) for the two isotopes, (A and B).

- If the height of the two peaks is measured then you would find that they were in the ratio of 3:1 which corresponds to (^{35}Cl 75.8% to ^{37}Cl 24.4%).

- This allows identification of the halogen present.

- Bromine derivatives have peaks approximately the same height at m/z = 79 and 81, corresponding to ^{79}Br 50.7% and ^{81}Br 49.3%, so here M^+ has three peaks.

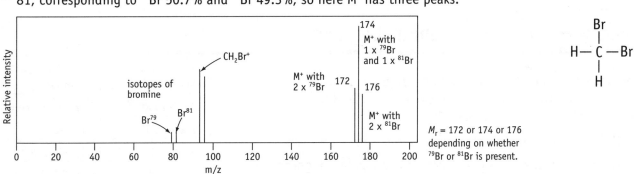

Mass spectrum of dibromomethane

M_r = 172 or 174 or 176 depending on whether ^{79}Br or ^{81}Br is present.

Exact molecular mass values

- Many molecules have what seems to be the same relative molecular mass (same integral formula mass).

- But if high resolution mass spectra are used, precise atomic masses can be calculated to four or five decimal places and so a difference between compounds of similar mass can be distinguished.

Examples

Formula	M_r	Exact M_r
$C_6H_5NO_2$	123	123.0320
$C_8H_{13}N$	123	123.1047

? Quick check questions

1 How can the molecular mass of a compound be found from its mass spectrum?

2 Give an equation showing the fragmentation of a molecular ion $M^{+\bullet}$

3 The mass spectrum of butanone has prominent peaks at m/z = 15, 29, 43 and 57. Give the formulae for the species that give rise to these peaks.

Infra-red spectroscopy

Infra-red spectroscopy is a technique that is mainly used in analytical chemistry to determine the functional groups present and to determine the degree of purity. Certain bonds have a natural tendency to vibrate and do so at certain frequencies. If radiation across the full frequency range is passed through a substance, the bonds absorb i.r. energy corresponding to their natural frequency of vibration.

The frequency they absorb depends on:

● The bond strength

● The mass of the atoms in the bond

Weak bonds and high masses absorb at low frequencies

Strong bonds and low masses absorb at high frequencies

> ▶ If two samples have identical i.r. spectra, they must be the same compound.

> ✓ **Quick check 1**

Facts you should know about i.r. spectra

The frequencies of vibration tend to be large numbers, e.g. 1.2×10^{14} Hz, so the unit of wavenumber is used for simplicity, which is $1/\lambda$ (units cm^{-1}).

● There are four regions of the i.r. spectrum shown opposite.

● The region between 1500–400 cm^{-1} is know as the **fingerprint region**. Every compound produces its own i.r. fingerprint pattern which can be used to identify compounds by comparison of an unknown with the spectra of known compounds.

● Since impurities produce additional peaks, the fingerprint region can be used to check the purity of a compound.

● Detailed data tables are given in questions. They usually break down the same bond for specific homologous series. For example C—H bond will vibrate slightly differently in alkanes, alkenes, aldehydes and arenes, because of the effect of adjacent groups.

Region/cm^{-1}	Absorption
4000–2500	C—H, O—H, N—H
2500–2000	C≡N, C≡C
2000–1500	C=C, C=O, C=N
1500–400	C—C, C—O C—N, C—X

Bond	Bond in	range/cm^{-1}
C=O	aldehydes, ketones, carboxylic acids, esters	1680–1750
C—H	aliphatic organic molecules	2850–3300
O—H	alcohols	3230–3550
O—H	carboxylic acids	2500–3000
C—C	alkanes	750–1100
C=C	alkenes	1620–1680

> ▶ Get an idea of where to look for peaks, but it is pointless learning exact ranges, unless you really want to.

> ✓ **Quick check 2**

● The intensity of absorbencies varies depending on the polarity of the bonds. Polar bonds like C=O produce large absorbency peaks while C=C does not.

Examples:

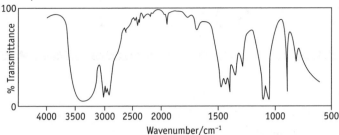

Infra-red spectrum of an alcohol (ethanol CH$_3$CH$_2$OH)

Comments:
The absorption around 3350 cm^{-1} is due to the OH group.
C—O absorption 1000–1300 cm^{-1}

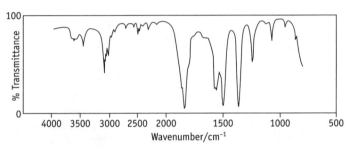

Infra-red spectrum of a ketone (propanone CH$_3$COCH$_3$)

Comments:
C–H absorption at around 2950 cm^{-1}
C=O absorption 1680–1750 cm^{-1}

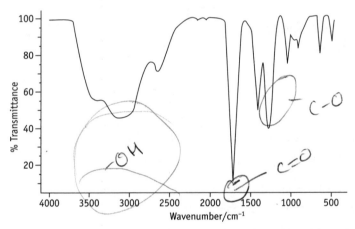

Infra-red spectrum of a carboxylic acid (ethanoic acid CH$_3$COOH)

Comments:
O–H of acid gives the very broad absorption between 2500 and 3500 cm^{-1}
C=O of carbonyl is seen around 1680 cm^{-1}
C—O between 1200 and 1030 cm^{-1}

✓ *Quick check 3, 4*

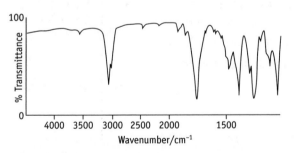

Infra-red spectrum of an ester (ethyl ethanoate CH$_3$COOCH$_2$CH$_3$)

Comments:
Absorption around 1750 cm^{-1} is due to C=O and that around 1200 cm^{-1} C—O. No O—H present so no large fat absorption peak around 3300 cm^{-1}

? *Quick check questions*

1 What is the section of the infra-red spectrum with wavenumbers greater than 1500 cm^{-1} used to identify?

2 Other than using the fingerprint region, how could you use i.r. spectroscopy distinguish between
a) but-2-ene and butanone and b) ethanol and ethanoic acid?

3 How would you use i.r. spectrometry to prove that a compound was propanone?

4 Why don't the i.r. breathalysers used by the police determine the alcohol content of the breath by measuring the height of the O—H peak?

Nuclear magnetic resonance

Proton nuclear magnetic resonance spectroscopy (or n.m.r. for short), gives information about the arrangement of atoms in a molecule by detailing the relative number and positions of hydrogen atoms, allowing the chemical structure of an organic compound to be determined. It is commonly used to identify the positions of alkyl groups in a molecule.

The Theory — the nasty bit

- Nuclei with odd mass numbers have spin which induces a weak magnetic field. The sample is placed in a strong magnetic field and its absorption of electromagnetic radiation measured.

- This can interact with an applied magnetic field and either align with it or against it. With the correct frequency of radiation the spin of the nucleus may resonate between these two states.

- To complicate things, the electrons around the protons tend to **shield** them from the external magnetic field. How much **shielding**, depends on the position of the proton in the molecule and what it is near to, for example electron donating/withdrawing species or polar bonds.

> **Deshielding = electron density reduced (larger δ value)**
> **Shielded = electron density increased (lower δ value)**

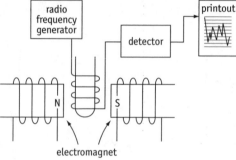

- **Chemical shift** δ is a measure of how much a proton is shielded and is measured in parts per million (ppm) compared to a standard or reference compound called tetramethylsilane (TMS, $(CH_3)_4Si$)

- TMS is used as its twelve methyl protons are well shielded so resonate up field. They are equivalent, so produce one large peak away from most others and TMS can easily be extracted from a sample if need be, (because it has a low b.p.).

- TMS is given $\delta = 0$ and all samples are compared to it dissolved in a proton free solution (like $CDCl_3$ a deuterated solvent or CCl_4)

> The Theory (the nasty bit) necessary, but thankfully rarely asked about.

> TMS is non-toxic and inert which is good to know.

> ✓ *Quick check 1,2*

The spectra — the nice bit

When you look at an n.m.r. trace, look at the following:

1 **The number of absorption peaks** — this tells you the number of non-equivalent protons.

2 **Positions of absorption peaks** — tells you what the proton may be near to (its environment).

3 **Intensities of the absorption** (the integration trace) — tells you the relative heights of the peak and so tells you the relative numbers of protons in a certain environment generating that peak.

4 **The splitting** — tells you about the neighbouring protons.

For example, a low-resolution spectrum (with no splitting) (left) and high resolution spectrum (right) of ethanol (CH_3CH_2OH) is shown below.

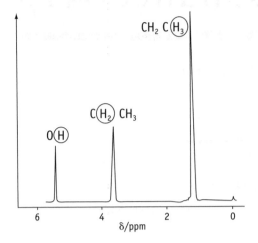

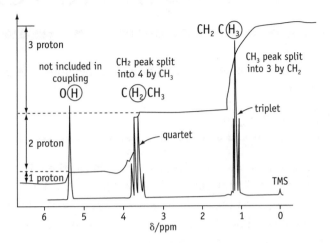

Low resolution

1   The number of absorption peaks — the spectrum has 3 peaks as there are 3 different proton environments present CH_3CH_2OH.

2   Positions of absorption peaks — their positions indicate how close they are to the oxygen of the OH group.

3   Area under the peaks gives us ratio of peak 1:2:3 which tells us the number of protons in that environment, i.e. $1 \times OH$, $2 \times CH_2$, and $3 \times CH_3$.

High resolution

4   Intensities of the absorption (the integration trace) — the height of the three peaks show the relative numbers of each type of proton, 3 for the CH_3 protons, 2 for the CH_2 protons and 1 for the OH proton. 1:2:3 corresponds to $OH:CH_2:CH_3$.

5   The splitting — tells you about the neighbouring protons:

—CH_2 peak is split into a quartet by —CH_3. —CH_3 peak is split into a triplet by adjacent —CH_2 (see next page for more details).

CH_3CH_2OH

3 peaks means 3 proton environments

? *Quick check questions*

1 What is the standard used in nuclear magnetic resonance?

2 Why is this compound used as the standard?

3 What are the four pieces of information that can be obtained from an absorbance in a n.m.r. spectrum?

Nuclear magnetic resonance (II)

Coupling with the spin of a neighbouring proton, often splits the peak associated with another proton.

Using the n + 1 rule

Sounds hard to follow I know, but all it means is that if there is 1 adjacent proton, then the peak splits into 2 peaks (n + 1 = 1 + 1 = 2), if it is next to 2 protons then there will be 3 peaks (2 + 1 = 3). This is known as spin–spin coupling. The number and intensities of these peaks depends on the possible ways the spins can align.

1 With one neighbouring non-equivalent hydrogen, the magnetic field from this hydrogen can be aligned with (↑) or against (↓) the external field, producing two different magnetic fields around the proton. The peak splits into a pair of peaks called a **doublet**:

2 With two neighbouring non-equivalent hydrogen atoms, then the magnetic field aligns as in the diagram opposite producing a **triplet**:

3 With three neighbouring non-equivalent hydrogen atoms then there are four different ways that the coupling can occur giving a **quartet**:

Therefore

> **The number of peaks formed by splitting = (n + 1)**
> (*where n is the number of non-equivalent neighbouring hydrogen atoms*)

1)
↑↓ and ↓↑   2
One adjacent proton produces a doublet

2)
↑↑   1
↑↓ and ↓↑   2
↓↓   1
Two adjacent protons cause the peak to split into three smaller peaks (**a triplet**) with a ratio of 1:2:1

3)
↑↑↑   1
↑↑↓ ↑↓↑ ↓↑↑   3
↓↓↑ ↓↑↓ ↑↓↓   3
↓↓↓   1
Three adjacent protons causes **a quartet** of peaks in the ratio 1:3:3:1

Pascal's triangle

✓ *Quick check 1,2*

Pascal's triangle allows you to predict the relative intensities (the ratios) of the split peaks at high resolution. Each number is the sum of the two above it. A triplet therefore has three peaks in a ratio of 1:2:1.

1 ← singlet
1   1 ← doublet
1   2   1 ← triplet
1   3   3   1 ← etc.

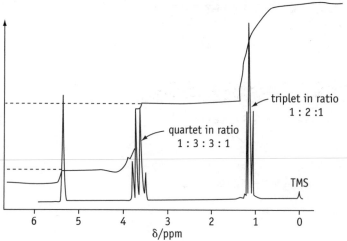

triplet in ratio
1 : 2 :1

quartet in ratio
1 : 3 : 3 : 1

TMS

δ/ppm

High resolution n.m.r. of ethanol

▶ Questions about the n+1 rule usually ask you to USE it, not to explain it.

Worked example 1: $C_2H_3Cl_3$

1 No. of peaks = 2 ∴ Two environments for the proton.

2 Positions of peaks: One environment is slightly more deshielded than the other, i.e. nearer to an electron-withdrawing group.

3 Ratio of peaks = 1:2. One proton on a carbon and two on the next carbon.

4 Splitting of peaks: triplet and doublet, suggests a single proton next to two protons.

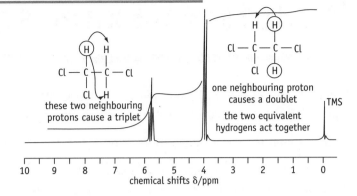

Two equivalent hydrogens in ratio 1:2. One neighbouring proton gives a doublet and two neighbouring protons gives a triplet.

Therefore this is a high resolution n.m.r. spectrum of 1,1,2-trichloroethane.

example 1

```
     H   H
     |   |
Cl — C — C — Cl
     |   |
     Cl  H
```

Worked example 2: $C_5H_{12}O$

1 No. of peaks = 4 so four non-equivalent proton environments

2 Positions of peaks: The proton environments indicate how near they are to the oxygen.

3 Ratio of peaks (integration) = 2:1:6:3 (4:2:12:6)

4 Splitting of peaks:

- large single peak so no hydrogens on group adjacent to that group (six).

- quartet due to neighbouring $-CH_3$

- triplet due to neighbouring $-CH_2$

Therefore this is a high resolution n.m.r. spectrum of $C_5H_{12}O$.

example 2

```
     H   H   CH_3
     |   |   |
H — C — C — C — O — H
     |   |   |
     H   H   CH_3
```

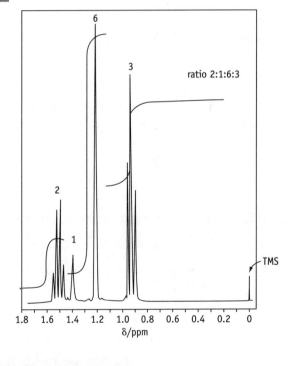

? Quick check questions

1 What information can be obtained from the number of absorbances in the n.m.r. of a compound?

2 Explain the n + 1 rule.

3 Why is it necessary to use deuterated solvents when preparing samples for n.m.r. spectroscopy?

Module 4: Exam style questions

Module 4...tricky stuff eh? I'm afraid so...it makes AS look like a walk in the park. But don't get too worried about it, there is a solution.

Quality revision? Eh? Yes, quality revision is the secret to success.

You are going to have to decide what 'quality revision' is for you, and no doubt it will incorporate many different methods like note taking, making summary charts etc, but one thing it isn't, is just reading a revision guide — that'll not get you a decent grade for sure. Quality revision is **active**, which means using your brain. One of the best ways of learning the stuff in this book is by trying A level type questions, which is exactly what you see below. Try these questions, they should make you think and apply the knowledge you have. Mark them, (see page 102) then try them again, at a later stage. If you get stuck try reading the section in the book again and if you are still having problems, see Teach (they don't bite you know). Enjoy.

1 The reaction between potassium manganate(VII) and sodium ethanedioate can be represented by the ionic equation:

$$2MnO_4^- + 5C_2O_4^{2-} + 16H^+ \rightarrow 2Mn^{2+} + 10CO_2 + 8H_2O$$

An experiment was carried out to determine the rate equation. The results in the table were obtained when the rate of the reaction was measured at constant temperature in solution buffered at pH = 0 but at various concentrations of MnO_4^- and $C_2O_4^{2-}$.

Experiment	Initial concentration of MnO_4^- / mol dm^{-3}	Initial concentration of $C_2O_4^{2-}$ / mol dm^{-3}	Initial rate/ mol dm^{-3} s^{-1}
1	0.00200	0.0100	0.000172
2	0.00500	0.0100	0.000430
3	0.0150	0.0300	0.00387

It was assumed that the reaction was first order for H^+ ions.

a Why was it not possible to find the order with respect H^+ in this experiment? (1)

b Why would you expect this ionic reaction to be slow when most other ionic reactions are fast? (2)

c Use the data to determine the order with respect to MnO_4^- ions. (2)

d Find the order with respect to $C_2O_4^{2-}$ ions. (3)

e Assuming that the reaction was first order for H^+ ions write out the rate equation. (1)

f What is the overall order for the reaction? (1)

When the reaction was carried out at the same temperature and pH in the presence of iron(II) sulphate the results in the table below were obtained. It was found that changing the pH had no effect as long as the pH remained below 1.

Initial concentration of MnO_4^-/ mol dm^{-3}	Initial concentration of $C_2O_4^{2-}$/ mol dm^{-3}	Initial concentration of Fe^{2+}/ mol dm^{-3}	Initial rate/mol dm^{-3} s^{-1}
0.00100	0.00500	0.000100	0.00658
0.00200	0.0100	0.000200	0.0132
0.00200	0.0100	0.000100	0.00658

g Use the data to find the rate equation under these conditions. (3)

h What is the role of the Fe^{2+} ions in this reaction? What has led you to your deduction? (2)   Total (15)

2 The reaction between sulphur(IV) oxide and oxygen to form sulphur(VI) oxide is important in the production of sulphuric acid. It can be represented by the equation:

$$2SO_2(g) + O_2(g) \rightleftharpoons 2SO_3(g).$$

Three moles of sulphur(IV) oxide and three moles of oxygen were mixed and allowed to react at a constant pressure of 1000 kPa and a temperature of 810K. At equilibrium there were equal numbers of moles of oxygen and sulphur(VI) oxide.

a Calculate the mole fraction of sulphur(IV) oxide and oxygen at equilibrium. (7)

b Give an equation that shows the relationship between mole fraction and partial pressure of a gas. (1)

c Calculate the partial pressure of sulphur(IV) oxide and oxygen. (2)

d Write an expression for K_p for the reaction. (1)

e Calculate the value of K_p at 810K and give its units. (3)   Total (14)

3 In a titration, 0.750 M sodium hydroxide was added to 25.0 cm^3 of a solution of propanoic acid at 298K. The pH was noted after adding each 5.00 cm^3 portion of NaOH. The results are given in the table.

Volume of NaOH/cm^3	0.00	5.00	10.00	15.00	20.00	25.00	30.00
pH	2.35	3.92	4.26	4.50	4.69	4.87	5.05
Volume of NaOH/cm^3 (cont.)	35.00	40.00	45.00	50.00	55.00	60.00	65.00
pH	5.24	5.47	5.82	9.51	12.27	12.57	12.75

a Use the results to plot a titration curve or pH curve. (4)

b Use your plot to determine the volume of sodium hydroxide needed to neutralize the acid solution. (1)

c Calculate the molarity of the acid. (2)

d Which indicator could have been used to determine the end-point of the titration? (1)

e Explain your answer to d. (2)

f Use an appropriate value to calculate the K_a of propanoic acid at 298K. (4)

g How else might a value for K_a have been found? (4)

h Sketch a graph to show how the pH would have changed if 65.00 cm^3 of the propanoic acid had been added to 25.00 cm^3 of the 0.750 M sodium hydroxide. (4)   Total (22)

Enthalpy change — $\Delta H^{\ominus}$

The first thing you need to learn in this section is a bunch of definitions and then you need to know how to apply them.

Standard enthalpy change ($\Delta H^{\ominus}$)

Standard enthalpy changes are those which occur under **standard conditions**

Standard conditions = 100 kPa, 298 K and 1 mol dm^{-3}

- the **standard state** is the physical state of something under standard conditions
- $\Delta H^{\ominus}$ is the heat change at constant pressure, i.e. the **enthalpy change**.

1 Standard enthalpy change of formation ($\Delta H_f^{\ominus}$)

The enthalpy change when 1 mole of a compound is formed from its elements in their standard states, under standard conditions.

For example: $C(gr) + O_2(g) \rightarrow CO_2(g)$ $\Delta H_f^{\ominus} = -394$ kJ mol^{-1}

The standard enthalpy change of formation, $\Delta H_f^{\ominus}$ ...

- for elements is zero
- can be positive or negative
- is illustrated by equations which must be balanced to produce 1 mole of the substance.

> **Remember Hess's Law** states that the total energy change in a reaction is independent of the route.

> You should know this definition already. See AS book for definition of enthalpy of combustion.

> ✓ *Quick check 1,2*

2 Standard enthalpy of atomisation ($\Delta H_{at}^{\ominus}$)

The enthalpy of atomisation of an element is the enthalpy change when one mole of gaseous atoms is formed from the element in its standard state.

e.g. $Na(s) \rightarrow Na(g)$ $\Delta H_{at}^{\ominus} = +107$ kJ mol^{-1}

- is sometimes $= \Delta H_{sub}^{\ominus}$ where sub = **sublimation**
- the reaction is always endothermic
- is from elements in their standard states so could be from (s) or (l) to (g) or (g) to (g) as below.
- for diatomic molecules $\Delta H_{at}^{\ominus} = \frac{1}{2} \Delta H_{diss}^{\ominus}$ where $\frac{1}{2} Cl_2(g) \rightarrow Cl(g)$

3 Ionisation enthalpy ($\Delta H_i^{\ominus}$)

The first ionisation enthalpy of an element is the energy required to form one mole of gaseous unipositive ions from one mole of gaseous atoms.

e.g. $Na(g) \rightarrow Na^+(g) + e^-$ $\Delta H_i^{\ominus} = +494$ kJ mol^{-1}

- This process involves the formation of positive ions.
- It is usually associated with metal ions.
- The reaction is endothermic.
- The second ionisation would be represented by $M^+(g) \rightarrow M^{2+}(g) + e^-$

> The second ionisation enthalpy > first ionisation enthalpy.

Electron affinity ($\Delta H_{ea}^{\ominus}$)

> The first electron affinity of an element is the standard enthalpy change when one mole of gaseous atoms each gain an electron to form one mole of gaseous negative ions.

$$\text{e.g. } Cl(g) + e^- \rightarrow Cl^-(g) \quad \Delta H^{\ominus} = -364 \text{ kJ mol}^{-1}$$

- involves the formation of negatively charged ions
- is associated with non-metal ions.
- the reaction is exothermic (second electron affinities are endothermic).

Lattice enthalpy of formation ($\Delta H_L^{\ominus}$)

> Lattice enthalpy of formation is the standard enthalpy change when one mole of an ionic solid is formed from its gaseous ions.

$$Na^+(g) + Cl^-(g) \rightarrow Na^+Cl^-(s) \quad \Delta H_L^{\ominus} = -771 \text{ kJ mol}^{-1}$$

- is also called the lattice formation enthalpy
- denotes an exothermic reaction, as energy is released when bonds are formed.

Bond dissociation enthalpy ($\Delta H_{diss}^{\ominus}$)

> Bond dissociation enthalpy is the standard enthalpy required when a particular covalent bond in a gaseous molecule is broken to form gaseous atoms.

$$\text{e.g. } Cl_2(g) \rightarrow 2Cl(g) \quad \Delta H_{diss}^{\ominus} = +242 \text{ kJ mol}^{-1}$$

- the reaction is endothermic
- involves **homolytic fission** of the bond and forms free radicals
- its equation usually omits the free radical dot (Cl•) unlike the mechanism for the chlorination of methane.

The enthalpy of hydration ($\Delta H_{hyd}^{\ominus}$)

> The enthalpy of hydration is the enthalpy change when one mole of isolated gaseous ions is dissolved in water forming one mole of aqueous ions under standard conditions.

e.g. $Na^+(g) \rightarrow Na^+(aq) \Delta H^{\ominus} = -406 \text{ kJ mol}^{-1}$ or $Cl^-(g) \rightarrow Cl^-(aq) \Delta H_{hyd}^{\ominus} = -364 \text{ kJ mol}^{-1}$

Enthalpies of solution ($\Delta H_{sol}^{\ominus}$)

> The enthalpy of solution of a compound is the standard enthalpy change when one mole of a compound dissolves completely in water at constant pressure.

$$\text{e.g. } NaCl(s) + water \rightarrow Na^+(aq) + Cl^-(aq) \Delta H_{sol}^{\ominus} = +1 \text{ kJ mol}^{-1}$$

- the reaction really occurs in two steps, so can either be exothermic or endothermic depending on the lattice enthalpy and the enthalpy of hydration:

Enthalpy of solution = lattice enthalpy + enthalpy of hydration
$$= +771 + (-406 + -364) = +1 \text{ kJ mol}^{-1}$$

Just to confuse you, some books define this the other way round...as the enthalpy of lattice dissociation $Na^+Cl^-(s) \rightarrow Na^+(g) + Cl^-(g)$ where $\Delta H^{\ominus} = +771 \text{ kJ mol}^{-1}$. The thing here is to always show an equation, then you won't mess up the definition.

Free radicals... sounds like some freedom fighting organisation but they're not. See the AS book (Organic section) for more exciting information.

✓ *Quick check 3,4*

? Quick check questions

1 Define enthalpy of formation.
2 Give an equation, plus state symbols, showing the enthalpy of formation of propanal $CH_3CH_2CHO(l)$.
3 Give a definition of the enthalpy of hydration.
4 Give an equation showing the enthalpy of hydration of calcium chloride.

Born–Haber cycles

Born–Haber cycles are really just fancy energy diagrams. They show the theoretical energy changes associated with the formation of an ionic compound. You need to be able to draw out Born–Haber cycles for simple ionic compounds and apply Hess's law to calculate such things as lattice enthalpy.

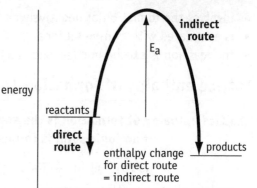

✓ **Quick check 1**

Questions involving Born–Haber cycles

The questions normally give you a shed full of numbers and you have to rearrange them to find a missing one — usually the lattice enthalpy (as you can't calculate this by experiment).

The two largest enthalpy changes are the ionisation enthalpy and the lattice enthalpy. It is the balance between these two enthalpies that determines whether or not an ionic compound can be formed.

Use Hess's law and follow the cycle round to find the unknown.

To find the unknown lattice formation enthalpy

Step 1 Write down what you know

Route 1 = Route 2

Enthalpy of formation = Enthalpy of sublimation + Enthalpy of atomisation + Enthalpy of ionisation + Electron affinity + Lattice formation enthalpy

$$\Delta H_f^\ominus = \Delta H_{sub}^\ominus + \tfrac{1}{2} \Delta H_{diss}^\ominus + \Delta H_i^\ominus + \Delta H_{ea}^\ominus + \Delta H_L^\ominus$$

Step 2 Place the figures in the equations and rearrange to get lattice enthalpy as the subject of the equation

$-411 = 109 + 121 + 496 + (-364) + (\Delta H_L^\ominus)$

$\Delta H_L^\ominus = -411 - 109 - 121 - 496 - (-364) = -773$ kJ mol^{-1}

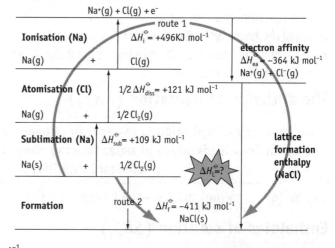

Step 3 Bingo ☺ you have the lattice enthalpy.

- These theoretical lattice enthalpies assume that the solid is 100% ionic. But when lattice enthalpies are calculated using experimental values, they differ from the theoretical values.

- Since bonding is not 100% ionic, the two values differ.

- When doing calculations with divalent cations, like in $MgCl_2$, remember that magnesium needs its first and second ionisation enthalpies in the cycle. If a divalent anion is used like O^{2-} the first and second electron affinities must be used.

- Charge and size of the ion affect lattice enthalpy.

NaCl −773 kJ mol^{-1}
NaBr −742 kJ mol^{-1}
NaI − 705 kJ mol^{-1}
- anion size increases
- larger distance between ions
- attraction less
- so LE less negative

Enthalpy of solution calculations

✓ *Quick check 2*

Enthalpy of solution involves ionic solids becoming totally dissolved in water. To do this

1 Energy is needed to break down the crystal lattice (i.e. opposite to the lattice formation enthalpy). This takes energy so it is endothermic.

2 Water molecules will then bond or associate with the ions from the substance...this releases energy so is exothermic.

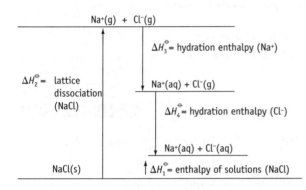

Step 1 the **lattice dissociation**

$NaCl(s) \rightarrow Na^+(g) + Cl^-(g)$ $\Delta H^\ominus = +771$ kJ mol^{-1}

Step 2 the **hydration** of the ions: see above

By Hess's Law: $\Delta H_1 = \Delta H_2 + \Delta H_3 + \Delta H_4$

Where $\Delta H_1 = ?$

$\Delta H_2 = + 771$ kJ mol^{-1}

$\Delta H_3 = - 406$ kJ mol^{-1}

$\Delta H_4 = - 364$ kJ mol^{-1}

Enthalpy of solution for sodium chloride
= 771 − (406 + 364) = +1 kJ mol^{-1}

Using mean bond enthalpies

The **mean bond** enthalpy is the average value of the bond dissociation energy of a particular type of covalent bond in a range of different gaseous compounds. Enthalpy values calculated from mean bond enthalpies will vary from those calculated experimentally as the mean value rarely represents the true bond enthalpy.

❶ The bond enthalpy for a C—H bond in one molecule is different from that in another. Even within a molecule they are different — if a C—H bond is broken the environment of the other bonds changes and so does their strength.

$$C_5H_{12} + 8O_2 \rightarrow 6H_2O + 5CO_2$$

4 × (C—C) = 4 × 347 = 1388	12 × (O—H) = 12 × 464 = 5568
12 × (C—H) = 12 × 413 = 4956	10 × (C=O) = 10 × 743 = 7430
8 × O=O = 8 × 498 = 3984	
Energy in = 10328	Energy out = 12998

Energy change = Σ bonds broken − Σ bonds formed
= 10328 −12998 = −2670 kJ mol^{-1}

✓ *Quick check 3,4*

❓ *Quick check questions*

1 What is a Born–Haber cycle?

2 Name the three enthalpy changes needed to calculate the enthalpy of solution of potassium bromide.

3 Define the term mean bond enthalpy.

4 Why are the values obtained for enthalpy changes using mean bond enthalpies different from those found using energy cycles?

Entropy

Stuff happens in this world of ours. Chemists use **enthalpy** to try to explain why this is. There is another factor called **entropy** which has a lot to do with it as well, and we think it is the balance between enthalpy and entropy which determines whether something happens or not.

Entropy is a measure of disorder

▶ The second law of thermodynamics says the total entropy of the universe is on the increase.

You usually talk about entropy of a **system** – a system being a certain definable area, or reaction. Reactions tend to increase the **disorder** of a system, (rather like bedrooms — as soon as you tidy they become disordered).

Facts you should know about entropy

✓ *Quick check 1*

- Gases have higher entropy than liquids and liquids higher than solids.
- When a highly ordered solid melts, its entropy increases, so entropy is temperature dependent.
- When a highly ordered solid dissolves, its entropy increases.
- Simple substances have low entropy values, whereas larger and more complex substances have much higher entropy values.
- S is the symbol for entropy and the units of entropy are J K^{-1}mol^{-1}.
- The entropy values are given in J K^{-1} mol^{-1} (that's joules (J) and not kilojoules (kJ) got it...)
- $S^{\ominus}$ represents standard entropy at 298K and 100 kPa pressure.
- At 0K there is no entropy (zero entropy).
- $\Delta S^{\ominus}$ is the change in entropy and is very useful. $\Delta S^{\ominus}$ is calculated from the entropy values of the reactants and products. (Bit like $\Delta H^{\ominus}$ values).

$$\Delta S^{\ominus} = \Sigma S^{\ominus}_{(products)} - \Sigma S^{\ominus}_{(reactants)}$$

Substance	Entropy, $S^{\ominus}$/ J K^{-1} mol^{-1}
$H_2O(s)$	62.1
$H_2O(l)$	70.0
$H_2O(g)$	188.7

Worked Example

What is the entropy change which accompanies the reaction given below:

Use the data in the table opposite.

$$2KHCO_3 \rightarrow K_2CO_3(s) + H_2O(g) + CO_2(g)$$

$$\Delta S^{\ominus} = \Sigma S^{\ominus}_{(products)} - \Sigma S^{\ominus}_{(reactants)}$$

$$= (156 + 189 + 214) - (116 \times 2)$$

$$= 559 - 232$$

$$= +327 \text{ J K}^{-1} \text{ mol}^{-1}$$

	$S^{\ominus}$/JK^{-1} mol^{-1}
KHCO$_3$	116
K$_2$CO$_3$	156
H$_2$O	189
CO$_2$	214

▶ Always show values as + or – to indicate increase or decrease in entropy.

Feasible or not feasible that is the question...

✓ Quick check 2,3,6

A **feasible** or **spontaneous** reaction occurs without the help of anything (like heat). The relationship between enthalpy and entropy is given by the expression:

$\Delta G^{\ominus} = \Delta H^{\ominus} - T\Delta S^{\ominus}$ where $\Delta G^{\ominus}$ is the free energy change (Gibbs free energy)

- This means that both exo- and endothermic reaction can occur as long as they lead to an increase in disorder.

- To be feasible the free energy of the reaction must be equal to or less than zero, i.e. $\Delta G^{\ominus}$ must be negative or zero.

For $\Delta H > T\Delta S$ T must be low and for $T\Delta S > \Delta H$ T must be high.

$\Delta H^{\ominus}$	$\Delta S^{\ominus}$	$\Delta G^{\ominus}$	Feasible or not?
−ve	+ve	always −ve	always feasible
+ve	−ve	always +ve	never feasible
−ve	−ve	−ve when $\Delta H > T\Delta S$	feasible when $\Delta H > T\Delta S$
+ve	+ve	−ve when $T\Delta S > \Delta H$	feasible when $T\Delta S > \Delta H$

- Feasibility tells us nothing of the rate of a reaction. A feasible reaction could be so slow that you may not even detect it.

- For a system in equilibrium such as a mixture of melting ice/water $\Delta G^{\ominus} = 0$ so $\Delta S^{\ominus} = \Delta H^{\ominus}/T$

Remember a reaction can be feasible in theory, but it may not occur if the reaction has a very high activation energy.

Worked example

a) Calculate the standard Gibbs free energy change when 1 mole of calcium carbonate decomposes at 298K b) Calculate the minimum temperature required to cause a spontaneous reaction (Given $\Delta H_r^{\ominus}$ =+179 kJ mol^{-1} and $\Delta S^{\ominus}$ = +160 J K^{-1} mol^{-1})

a)
$$CaCO_3(s) \rightarrow CaO(s) + CO_2(g)$$
$$\Delta G^{\ominus} = \Delta H^{\ominus} - T\Delta S^{\ominus}$$
$$\Delta G^{\ominus} = 179 - (298 \times 160/1000) \text{ kJ mol}^{-1}$$
$$= +131.3 \text{ kJ mol}^{-1} \text{ (i.e. not feasible at 298 K)}$$

b) **For a feasible reaction: $\Delta G^{\ominus} < 0$ or $\Delta H^{\ominus} - T\Delta S^{\ominus} < 0$**

A reaction will occur when $T\Delta S^{\ominus} \geqslant \Delta H^{\ominus}$ or $T \geqslant \Delta H^{\ominus}/\Delta S^{\ominus}$

$$T \geqslant \frac{179}{160/1000}$$

$$\mathbf{T \geqslant 1119 \text{ K}}$$

The reaction must be heated to around 847°C to occur. This explains why the White Cliffs of Dover don't suddenly decompose ... good to know, I guess.

You may not be given $\Delta H_r^{\ominus}$ and $\Delta S^{\ominus}$ so you may have to use
$\Delta H_r^{\ominus} = \Sigma \Delta H^{\ominus}_{(reactants)} - \Sigma \Delta H^{\ominus}_{(products)}$
and $\Delta S^{\ominus} = \Sigma S^{\ominus}_{(products)} - \Sigma S^{\ominus}_{(reactants)}$

? Quick check questions

1 Give an equation that can be used to calculate the entropy change, ΔS, for a reaction.

2 Write an equation that shows the relationship between the total free energy change, the enthalpy change and the entropy change for a reaction.

3 What condition must be met for a reaction to be spontaneous?

4 Write an equation, including state symbols, for a reaction in which the entropy change is likely to be positive.

5 What is the sign of the enthalpy change and the entropy change when a substance melts?

6 If the enthalpy change for a reaction is +25.0 kJ mol^{-1} and the entropy change is +80.1 J K^{-1} mol^{-1} is the reaction spontaneous at standard temperature? What has led you to your decision?

Period 3 reactions and trends

In AS Chemistry you looked at the physical properties of period 3, now in A2, you look at the **chemical properties of period 3** especially the reactions of the elements and their oxides and chlorides.

> ▶ It's worth making sure you follow the **physical properties** of period 3 you studied at AS level. It's important to know about melting points, boiling points and electrical conductivity of the elements.

Properties of Period 3

Element	Atomic Number	Electronic configuration	Type of element	Bonding in the oxides	Bonding in the chlorides
Na	11	$1s^2 2s^2 2p^6 3s^1$	metal	ionic	ionic
Mg	12	$1s^2 2s^2 2p^6 3s^2$	metal	ionic	ionic
Al	13	$1s^2 2s^2 2p^6 3s^2 3p^1$	metal	ionic	covalent
Si	14	$1s^2 2s^2 2p^6 3s^2 3p^2$	non-metal	covalent	covalent
P	15	$1s^2 2s^2 2p^6 3s^2 3p^3$	non-metal	covalent	covalent
S	16	$1s^2 2s^2 2p^6 3s^2 3p^4$	non-metal	covalent	covalent
Cl	17	$1s^2 2s^2 2p^6 3s^2 3p^5$	non-metal	covalent	covalent
Ar	18	$1s^2 2s^2 2p^6 3s^2 3p^6$	non-metal	—	—

A number of conclusions can be drawn from this table of data...

	In oxides and chlorides	Ions?	Bonding	Exceptions
Metals	Lose electrons in the third principal energy level when they react	Form positively charged ions	Compounds, are predominantly **ionic**	Al forms some covalent compounds e.g. aluminium chloride. The Al^{3+} ion has a very high charge density – chloride ion is **polarised** so there is sharing of the electrons.
Non-metals	Share electrons	Can form negative ions but not in oxides or chlorides	Oxide and chlorides mainly covalent	See above

Reactions of period 3 elements

Make sure that you

- learn the reaction details
- can write the equations
- know and understand the pattern or trend across the period.

1 with oxygen – the oxides form

✓ *Quck check 1,2*

Element	Reaction details		Equations
	Colour	Reaction	
Na	Burns with a yellow flame	Vigorous	$4Na + O_2 \rightarrow 2Na_2O$
Mg	Bright white light and white smoke	Vigorous, leaves white ash	$2Mg + O_2 \rightarrow 2MgO$

Element	Reaction details		Equations
	Colour	Reaction	
Al	Bright white light and white smoke	Starts well and then stops due to Al_2O_3 formation	$4Al + 3O_2 \rightarrow 2Al_2O_3$
Si	Bright white light and white smoke	Slow forms SiO_2	$Si + O_2 \rightarrow SiO_2$
P	Bright white light and white smoke	Vigorous	$P_4 + 5O_2 \rightarrow P_4O_{10}$ (P_4O_6 can form if limited O_2)
S	Blue flame	Melts and burns forms SO_2, a choking gas	$S + O_2 \rightarrow SO_2$ (some SO_3 may also form if enough O_2)

- All elements form the oxide when heated in oxygen

2 with chlorine – the chlorides form

✓ Quick check 3

Element	Reaction details	Equations
Na	Very vigorous when heated	$2Na + Cl_2 \rightarrow 2NaCl(s)$
Mg	Vigorous when heated	$Mg + Cl_2 \rightarrow MgCl_2(s)$
Al	Vigorous (if heated under anhydrous conditions)	$2Al + 3Cl_2 \rightarrow 2AlCl_3(s)$
Si	Slow (even when heated)	$Si + 2Cl_2 \rightarrow SiCl_4(l)$
P	Slow (even when heated)	$P_4 + 10Cl_2 \rightarrow 4PCl_5(s)$

- All elements above form the chloride
- Reactions become less vigorous across the period

3 with water – hydroxides and oxides form (usually)

✓ Quick check 4,5

Element	Reaction details	Equations
Na	Vigorous	$2Na(s) + 2H_2O(l) \rightarrow 2NaOH(aq) + H_2(g)$
Mg	Slowly with cold water but more vigorous, with steam	$Mg(s) + H_2O(l) \rightarrow MgO(s) + H_2(g)$
Al	No reaction	–
Si	No reaction	–
P	No reaction	–
S	No reaction	–
Cl	Forms chlorine water	$Cl_2 + H_2O \rightleftharpoons HClO + HCl$

- Sodium reacts more vigorously than magnesium
- Chlorine reacts to make light green chlorine water

? Quick check questions

1 Sketch the change in melting points of the oxides of the Period 3 elements from sodium to phosphorus.(see page 64 for data)

2 How is the boiling point of the oxides related to the type of bonding?

3 Give the formula of the product of the reaction between chlorine and the elements of Period 3 from sodium to phosphorus and state the type of bonding present.

4 Give an equation for the reaction between sodium and water.

5 How does the reaction between magnesium and water differ from that between sodium and water?

Period 3 chlorides and oxides

It is the structure and bonding of the Period 3 oxides and chlorides which determine their properties. Make sure you know the trends in structure and bonding and can explain their acid/base behaviour.

Oxides

Oxide	Na_2O	MgO	Al_2O_3	SiO_2	P_4O_{10}	SO_3
Structure	Giant ionic lattice	Giant ionic lattice	Giant molecular	Giant molecular	Simple molecular	Simple molecular
bonding	ionic	ionic	covalent	covalent	covalent	covalent
m.p./°C	1275	2852	2072	1610	580	17
Type of oxide	basic	basic	amphoteric	acidic	acidic	acidic
pH in water	~14	~9	insoluble	insoluble	~0	~0
General trend	Alkaline oxides → acidic oxides					

1 Bonding in the period 3 oxides

- The oxides of the metals (on the left of the Period) are ionic. Therefore sodium and magnesium oxides are ionic.
- The non-metals (on the right of the period) have covalent bonding.

> Aluminium oxide is a tricky one. Some say ionic whilst others now think it is covalent.

2 Structure of the period 3 oxides

- Metal oxides have a giant ionic lattice structure and so high melting points.
- Non-metal oxides have simple molecular structures and lower melting points — except silicon dioxide.

> Al_2O_3 lattice differs from Na and Mg so its m.p. is lower than expected.

3 Acid/base nature of the period 3 oxides

✓ Quick check 1,2,6

- Metals react with oxygen to give basic oxides which dissolve in water to give alkali solutions.
- The ionic oxides are basic oxides so react with acids to form salts.

e.g.
$$Na_2O + H_2SO_4 \rightarrow Na_2SO_4 + H_2O$$
$$MgO + 2HCl \rightarrow MgCl_2 + H_2O$$
$$Al_2O_3 + 6HNO_3 \rightarrow 2Al(NO_3)_3 + 3H_2O$$

- If they dissolve in water then they form alkaline solutions.

e.g.
$$Na_2O + H_2O \rightarrow 2NaOH \text{ or } MgO + H_2O \rightarrow Mg(OH)_2$$

Al_2O_3 is amphoteric so will react with both acid and bases

Al_2O_3 as an acid: $Al_2O_3 + 3H_2O + 2OH^- \rightarrow 2[Al(OH)_4]^-$

Al_2O_3 as a base: $Al_2O_3 + 6HNO_3 \rightarrow 2Al(NO_3)_3 + 3H_2O$

> MgO is not very soluble in water and the pH of NaOH > $Mg(OH)_2$. Al_2O_3 is insoluble in water.

> SiO_2 (sand) unsurprisingly doesn't dissolve in water ...hence we have beaches.

- Non-metals react with oxygen to give acidic oxides which react with bases to form salts.

e.g.
$$SiO_2 + 2OH^- \rightarrow SiO_3^{2-} + H_2O$$
silicate ion

$$P_4O_{10} + 12OH^- \rightarrow 4PO_4^{3-} + 6H_2O$$
phosphate(V) ion

$$SO_2 + 2OH^- \rightarrow SO_3^{2-} + H_2O$$
sulphite ion

$$SO_3 + 2OH^- \rightarrow SO_4^{2-} + H_2O$$
sulphate ion

- Those which dissolve in water, give acidic solutions.

e.g.
$$P_4O_{10} + 6H_2O \rightarrow 4H_3PO_4$$
phosphoric(V) acid (pH $\approx$ 0)

$$SO_2 + H_2O \rightarrow H_2SO_3$$
sulphuric(IV) acid (pH $\approx$ 3)

$$SO_3 + H_2O \rightarrow H_2SO_4$$
sulphuric(VI) acid (pH $\approx$ 0)

Chlorides

✓ *Quick check 3,4*

Chloride	NaCl	$MgCl_2$	$AlCl_3$	$SiCl_4$	PCl_5
Structure	Giant ionic lattice	Giant ionic lattice	Simple molecular	Simple molecular	Simple molecular
Bonding	ionic	ionic	covalent	covalent	covalent
m.p./°C	801	714	177 (sub.)	–70	162 (sub.)
With water	dissolves	dissolves (bit of hydrolysis)	hydrolysed	hydrolysed	hydrolysed
Observations of the solution	colourless	colourless	dissolves giving out heat and giving off a gas	dissolves giving out a lot of heat. HCl and a white ppt forms	dissolves giving out heat with fumes of HCl
Species present	$Na^+(aq)$	$Mg^{2+}(aq)$	$[Al(H_2O)_6]^{3+}(aq)$	$Si(OH)_4(s)$	$H_3PO_4(aq)$
pH of solution	7	6	3	1	0
General trend	Neutral chlorides $\rightarrow$ Acidic chlorides				

1 Bonding in the period 3 chlorides

- Metal chlorides bond ionically.
- Non-metal chlorides bond covalently.
- $AlCl_3$ is the exception to this rule.

❶ Guess what? These two trends are similar to those in the oxides.

2 Structure of the period 3 chlorides

- Metal chlorides have giant ionic lattice structures and so have high melting points
- Non-metal chlorides have simple molecular structures so have low melting points — except $AlCl_3$

3 Acid/base nature of the period 3 chlorides

NaCl ions become **hydrated** as polar water molecules surround them. Only slight **hydrolysis** occurs with $MgCl_2$ giving a weakly acidic solution

Hydration = water molecules added

Hydrolysis = a reaction where water molecules are split up

- The metal chlorides dissolve to give neutral (or very slightly acidic) aqueous solutions. There is no or only slight **hydrolysis.**

$$NaCl(s) + water \rightarrow Na^+(aq) + Cl^-(aq)$$

$$MgCl_2(s) + water \rightarrow Mg^{2+}(aq) + 2Cl^-(aq)$$

- The non-metal chlorides dissolve to give acidic solutions, as covalent chlorides undergo a hydrolysis reaction in water.

$$AlCl_3(s) + 6H_2O(l) \rightarrow [Al(H_2O)_6]^{3+}(aq) + 3Cl^-(aq)$$

This continues: $[Al(H_2O)_6]^{3+}(aq) + H_2O(l) \rightarrow [Al(H_2O)_5(OH)]^{2+}(aq) + H_3O^+(aq)$

This explains the low pH of $AlCl_3$. $SiCl_4$ and PCl_5 however, undergo complete hydrolysis.

$$SiCl_4(l) + 4H_2O(l) \rightarrow Si(OH)_4(s) + 4HCl(aq)$$
White ppt

$$PCl_5(s) + 4H_2O(l) \rightarrow H_3PO_4(aq) + 5HCl(aq)$$

> ▶ $AlCl_3$ exists as a dimer Al_2Cl_6 so don't worry if you see it in books

> ✓ *Quick check 5*

? *Quick check questions*

1 What is the difference between hydration and hydrolysis?

2 Give equations for the reaction between water and
   (a) $AlCl_3$,
   (b) $SiCl_4$
   (c) PCl_5.

3 What would you expect the pH of the solution formed when one mole of each of the following was added to a litre of water:
   (a) sodium chloride,
   (b) aluminium chloride,
   (c) phosphorus pentachloride?

4 What is the relationship between the pH of the solution of the chlorides of Period 3 and the type of bonding present?

5 Draw a diagram to show the arrangement of the bonds in solid Al_2Cl_6.

6 Why is aluminium oxide said to be amphoteric?

Redox equilibria

This section of the specification involves some revision of concepts met at AS level. You should already know...

Oxidation is...

- addition of oxygen
- loss of hydrogen
- loss of electrons
- an increase in oxidation state

Reduction is...

- loss of oxygen
- addition of hydrogen
- addition of electrons
- a decrease in oxidation state

> ▶ Remember OIL RIG

Variable oxidation states

The oxidation state of an element is the number of electrons used by the element in bonding. Many transition metals have more than one oxidation state.

> ▶ Examples of TMs with variable oxidation states include Cu^+ and Cu^{2+} Fe^{2+} and Fe^{3+}

The rules for assigning oxidation states

1 Uncombined elements — oxidation state is **zero.**

2 Compounds — the sum of oxidation states is **zero.**

3 Simple ions — the oxidation state is the **charge on the ion.**

4 Complex ions — the **sum** of the oxidation states is the **charge on the ion.**

5 Covalent molecules are treated as ionic — where the **more electronegative** element in the bond is assigned the **more negative** oxidation state.

6 Transition elements have **variable oxidation states** — they have more than one oxidation state.

Species	Oxidation state	Examples
H^+ and Group I ions	+1	Na^+, K^+, Ag^+
Group II ions	+2	Mg^{2+}
Group III ions	+3	Al^{3+}
Group VI ions	−2	O^{2-}
Group VII ions / Hydrogen in metal hydrides	−1	F^-, Cl^-, Br^-, H^-

Worked example: Find the oxidation state of manganese in $KMnO_4$

Formula	Calculation of oxidation state	Oxidation state	Name	Comments
$\overset{+1}{K}\overset{-2}{MnO_4}$	$(1 \times Mn) + (4 \times 0) = -1$ $(1 \times Mn) + (4 \times -2) = -1$ $(1 \times Mn) + (-8) = -1$ $Mn = (-1) + 8$ $Mn = +7$	+7	MnO_4^- is the manganate(VII) ion $KMnO_4$ is potassium manganate(VII)	Potassium is in group I so has Ox = +1 which means the manganate ion is −1

> ▶ Loads more examples can be seen in the AS book. Note Ox = oxidation state, some books use ON or even OS

Half-equations

✓ *Quick check 3*

- Half-equations show the loss or gain of electrons involved when a species undergoes oxidation or reduction.
- One element in a half equation changes oxidation state.
- Half-equations can be combined to give the equation for the overall redox process.
- All redox reactions or half-reactions can be illustrated using half-equations.
- Metals are often put in their metal ions solutions (salts) to form a half-cell.

> ▶ Just can't fit in the details of how to write and combine half equations, so see the AS book, sorry.

Standard Electrode Potentials — $E^\ominus$

In redox reactions electrons are lost and gained as the reaction occurs. If the electrons from one reaction are allowed to flow via an external circuit to the other, a **potential difference** between the two half cells can be measured. This allows a list to be built up (the electrochemical series) which shows us how readily electrons are lost and allows us to predict the feasibility of reactions. The potential difference measured under standard conditions is known as the **standard electrode potential** ($E^\ominus$) — **SEP** measured in V.

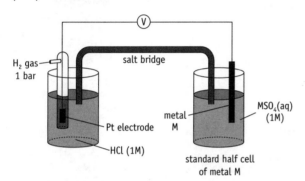

> ▶ Standard electrode potential is also known as redox potential or reduction potential. The electrochemical series is a bit like the reactivity series but just for aqueous reactions. e.m.f. = electromotive force which is the potential difference with zero current.

The standard hydrogen electrode (SHE)

> **The standard electrode potential of a half-cell is the e.m.f. of the half-cell compared to a standard hydrogen half-cell.**
> *(At 298K with 1 mol dm^{-3} solutions and pressure of 100 kPa (1 atm or 1 bar))*

The standard hydrogen electrode conditions are:

- Concentration: 1 mol dm^{-3} HCl (i.e. at pH = 0), pressure: 100 kPa H_2, temperature 298K, with a salt bridge (completes the circuit e.g. a saturated solution of KCl or KNO_3) and zero current (i.e. use a high resistance voltmeter).

Other details

- The electrode potential will change if conditions change
- $H^+(aq)/H_2(g)$ is the redox couple
- The electrode potential of the hydrogen electrode is set at 0.00 V

$$H^+(aq) + e^- \rightarrow \tfrac{1}{2}H_2(g) \quad E^\ominus = 0.00 \text{ V}$$

> ▶ **A redox couple** consists of metal atoms (or hydrogen) in equilibrium with its ions, here in aqueous solution.

- The SEP of a couple can be determined by connecting it to the SHE as the left electrode.

$$E^{\ominus}(\text{cell}) = E^{\ominus}_{(R)} - E^{\ominus}_{(L)}$$

- The SHE can be dangerous and a fiddle to use.

- You can write out a **cell diagram** to represent the cell setup. The cathode (the more +ve EP) on the right.

> Oxidation occurs with the more negative SEP. Reduction occurs with the less negative (more positive) SEP.

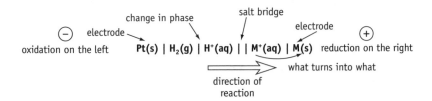

change in phase   salt bridge

⊖ electrode   electrode ⊕

oxidation on the left   Pt(s) | H₂(g) | H⁺(aq) | | M⁺(aq) | M(s)   reduction on the right

what turns into what

direction of reaction

Secondary electrodes

Secondary electrodes, like the standard calomel electrode, are easier to set up and use than the SHE.

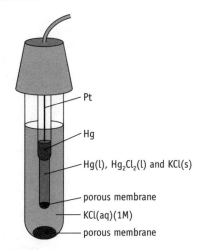

Pt
Hg
Hg(l), Hg₂Cl₂(l) and KCl(s)
porous membrane
KCl(aq)(1M)
porous membrane

- Its $E^{\ominus}$ = +0.27

- It uses mercury(I) chloride: $Hg_2Cl_2(s) + 2e^- \rightleftharpoons 2Hg(l) + 2Cl^-(aq)$

- You do not need to learn the details of secondary electrodes.

? Quick check questions

1 Give a definition of reduction in terms of electrons.

2 What is the oxidation state of the transition elements in each of the following (a) V_2O_5, (b) VO_2^+, (c) VO^{2+}, (d) $[Cr(H_2O)_5OH]^{2+}$, (e) $[Fe(CN)_6]^{3-}$ and (f) CrO_4^{2-}.

3 Write the half-equation for the oxidation of thiosulphate ions, ($S_2O_3^{2-}$) to tetrathionate ions, ($S_4O_6^{2-}$).

Using electrode potentials

The IUPAC convention for writing half-equations is for the reduction reaction.

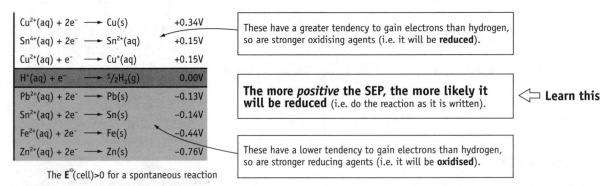

$Cu^{2+}(aq) + 2e^- \longrightarrow Cu(s)$	+0.34V
$Sn^{4+}(aq) + 2e^- \longrightarrow Sn^{2+}(aq)$	+0.15V
$Cu^{2+}(aq) + e^- \longrightarrow Cu^+(aq)$	+0.15V
$H^+(aq) + e^- \longrightarrow \frac{1}{2}H_2(g)$	0.00V
$Pb^{2+}(aq) + 2e^- \longrightarrow Pb(s)$	−0.13V
$Sn^{2+}(aq) + 2e^- \longrightarrow Sn(s)$	−0.14V
$Fe^{2+}(aq) + 2e^- \longrightarrow Fe(s)$	−0.44V
$Zn^{2+}(aq) + 2e^- \longrightarrow Zn(s)$	−0.76V

These have a greater tendency to gain electrons than hydrogen, so are stronger oxidising agents (i.e. it will be **reduced**).

The more *positive* the SEP, the more likely it will be reduced (i.e. do the reaction as it is written). ⇐ **Learn this**

These have a lower tendency to gain electrons than hydrogen, so are stronger reducing agents (i.e. it will be **oxidised**).

The $E^\ominus$(cell)>0 for a spontaneous reaction

Calculating the e.m.f. of a cell

✓ *Quick check 1,2*

In tackling redox calculations, remember 1) put the SEP equations with most negative above the other one; 2) use the anticlockwise rule *(i.e. reduction occurs at the cathode (the more +ve SEP) and oxidation occurs at the anode (the most -ve SEP)*; 3) in cell diagrams **R**eduction goes on the **R**ight, (see below).

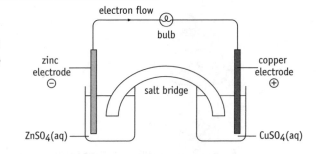

Worked example

Predict if the reaction will occur and calculate the e.m.f. of the cell when the following half cells are combined, (the Daniell cell).

$$Cu^{2+}(aq) + 2e^- \rightarrow Cu(s) \quad +0.34 \text{ V}$$
$$Zn^{2+}(aq) + 2e^- \rightarrow Zn(s) \quad -0.76 \text{ V}$$

Step 1 Write out the reduction potentials putting the more negative one above the other.

$$Zn^{2+}(aq) + 2e^- \rightarrow Zn(s) \quad \boxed{-0.76 \text{ V}}$$

More negative on top

$$Cu^{2+}(aq) + 2e^- \rightarrow Cu(s) \quad +0.34 \text{ V}$$

Step 2 Draw in anti-clockwise arrows (to indicate the direction of the reaction)

$$Zn^{2+}(aq) + 2e^- \rightarrow Zn(s)$$

$$Cu^{2+}(aq) + Zn(s) \rightarrow Cu(s) + Zn^{2+}(aq)$$

$$Cu^{2+}(aq) + 2e^- \rightarrow Cu(s)$$

Step 3 Write out the cell diagram

$$Zn(s) \mid Zn^{2+}(aq) \mid\mid Cu^{2+}(aq) \mid Cu(s)$$

Step 4 Use $E^\ominus$ (cell) = $E^\ominus_{(R)} - E^\ominus_{(L)}$ to work out the e.m.f. of the cell.
$E^\ominus$ (cell) = +0.34 − (−0.76) = **+1.10 V** so reaction proceeds as written in step 2.
Step 5 Smile ☺ (Optional)

This pretty much says oxidation occurs at the one with the more negative SEP and so is the negative electrode. Reduction occurs at the more positive SEP and is the positive electrode.

If the e.m.f. was −1.10 V then the reaction wouldn't proceed in the direction written.

Predicting the results of redox reactions

There are so many questions AQA could ask you on this topic, it is best to see how a general method is adapted using as many worked examples as you can. Practice, as they say, makes perfect...whilst not always true, experience goes a long way.

Worked example

Explain why copper will not react with HCl (aq) but will react with conc. HNO_3?

Given: $Cu^{2+} + 4HNO_3(aq) \rightarrow Cu(NO_3)_2(aq) + 2NO_2(g) + 2H_2O(l)$

Step 1 Write out the reduction potentials putting the more negative one above the other.

$$2H^+(aq) + 2e^- \rightarrow H_2(g) \qquad\qquad E^\ominus = 0.00 \text{ V}$$
$$Cu^{2+}(aq) + 2e^- \rightarrow Cu(s) \qquad\qquad E^\ominus = +0.34 \text{ V}$$
$$NO_3^-(aq) + 2H^+(aq) + e^- \rightarrow NO_2(g) + H_2O(l) \quad E^\ominus = +0.80 \text{ V}$$

Step 2 Draw in anti-clockwise arrows (to indicate the direction of the reaction)

$$2H^+(aq) + 2e^- \rightarrow H_2(g) \qquad Cu^{2+}(aq) + 2e^- \rightarrow Cu(s)$$
$$Cu^{2+}(aq) + 2e^- \rightarrow Cu(s) \qquad NO_3^-(aq) + 2H^+(aq) + e^- \rightarrow NO_2(g) + H_2O(l)$$

Step 3 Write out the cell diagrams

$$H_2(g) \mid H^+ \mid\mid Cu^{2+}(aq) \mid Cu(s) \qquad\qquad Cu^{2+}(aq) \mid Cu(s) \mid\mid NO_3^-(aq) \mid NO_2(g)$$

Step 4 Use $E^\ominus(\text{cell}) = E^\ominus_{(R)} - E^\ominus_{(L)}$ to work out the e.m.f. of each cell.

$$E^\ominus(\text{cell}) = 0.00 - (+0.34) \qquad\qquad E^\ominus(\text{cell}) = +0.80 - (+0.34)$$
$$= -0.34 \text{ V} \qquad\qquad\qquad\qquad = \underline{+0.46 \text{ V}}$$

Since the SEP of the cell needs to be greater than zero, the copper hydrochloric acid reaction would not proceed as written, but the copper would react with nitric acid. Electrons would flow from the less positive copper half cell to the more positive nitrate half cell.

Limitation of predictions using $E^\ominus$

- $E^\ominus$ values only predict whether a reaction is feasible or not, they do not predict the rate of reaction — they are not derived from experimentation or kinetics.

- Other reactions may proceed in preference to the one in question.

- $E^\ominus$ values are true only for standard conditions, this may affect the feasibility of a reaction (however, not usually very much).

Side notes:

▶ These may or may not be given.

✓ *Quick check 5*

▶ Reduction on the **R**ight
Reduction is gain of e^-

▶ Here copper is the reductant and the nitrate is the oxidant.

▶ If you forget everything, just remember the answer to the Daniell cells and work back from it.

? Quick check questions

1. Which electrode is used as the standard and what is its standard electrode potential?

2. Why are secondary standards more commonly used than the standard electrode?

3. Why is it important to quote the conditions when giving an electrode potential and what are the standard conditions?

4. What is the electrochemical series?

5. Use the information given to decide what will happen when a solution containing iron(II) and iron(III) ions is added to a solution of iodine in potassium iodide and give an equation for any reaction, if any, that may occur.

$$Fe^{3+}(aq) + e^- \rightarrow Fe^{2+}(aq) \qquad E^\ominus = +0.77 \text{ V}$$
$$\tfrac{1}{2} I_2(aq) + e^- \rightarrow I^-(aq) \qquad E^\ominus = +0.54 \text{ V}$$

Transition metals (I)

The transition metals (TM) you study are found in the first row of the d block of the Periodic Table (see inside front cover). You will already know that their atoms have partially filled 3d energy sub-level and it is this which governs most of their properties that are listed opposite.

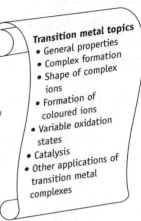

Transition metal topics
- General properties
- Complex formation
- Shape of complex ions
- Formation of coloured ions
- Variable oxidation states
- Catalysis
- Other applications of transition metal complexes

General properties

Element	Symbol	Electronic configuration
Scandium	Sc	$1s^2 2s^2 2p^6 3s^2 3p^6 3d^1 4s^2$
Titanium	Ti	$1s^2 2s^2 2p^6 3s^2 3p^6 3d^2 4s^2$
Vanadium	V	$1s^2 2s^2 2p^6 3s^2 3p^6 3d^3 4s^2$
Chromium	Cr	$1s^2 2s^2 2p^6 3s^2 3p^6 3d^5 4s^1$
Manganese	Mn	$1s^2 2s^2 2p^6 3s^2 3p^6 3d^5 4s^2$
Iron	Fe	$1s^2 2s^2 2p^6 3s^2 3p^6 3d^6 4s^2$
Cobalt	Co	$1s^2 2s^2 2p^6 3s^2 3p^6 3d^7 4s^2$
Nickel	Ni	$1s^2 2s^2 2p^6 3s^2 3p^6 3d^8 4s^2$
Copper	Cu	$1s^2 2s^2 2p^6 3s^2 3p^6 3d^{10} 4s^1$

> ▶ Remember that the 4s sub-level fills *before* the 3d sub-level and the electronic structures of **Cr** and **Cu** are weird as they have a half-filled 3d/4s and a full 3d respectively.

When ions form, the 4s electrons are lost *before* they lose the 3d electrons.

$$Sc \rightarrow Sc^{3+} \quad [Ar]3d^1 4s^2 \quad \rightarrow \quad 1s^2 2s^2 2p^6 3s^2 3p^6$$
$$Fe \rightarrow Fe^{2+} \quad [Ar]3d^6 4s^2 \quad \rightarrow \quad [Ar]3d^6$$
$$Fe \rightarrow Fe^{3+} \quad [Ar]3d^6 4s^2 \quad \rightarrow \quad [Ar]3d^5$$
$$Co \rightarrow Co^{3+} \quad [Ar]3d^7 4s^2 \quad \rightarrow \quad [Ar]3d^6$$
$$Cu \rightarrow Cu^+ \quad [Ar]3d^{10} 4s^1 \quad \rightarrow \quad [Ar]3d^{10}$$

Sc^{3+} and Cu^+ and Zn are *not* transitional since they do not have a partially filled 3d energy sub-level.

✓ *Quick check 1,2*

> ▶ The ligand can also be thought of as a nucleophile, but nucleophiles donate e^- to C^+ or δ^+C.

Complex formation

- A complex is where a transition metal, or its ion forms co-ordinate bonds with **ligands**.

- **Ligands** donate pairs of electrons to vacant orbitals on the metal atom or ion.

- The TM ion is a **Lewis acid** (as it accepts a pair of electrons) and the ligand is a **Lewis base** (as it donates a pair of electrons).

- The number of co-ordinate bonds formed in the complex is the co-ordination number.

- Ligands are classified by the number of co-ordinate bonds that they are able to form in complexes.

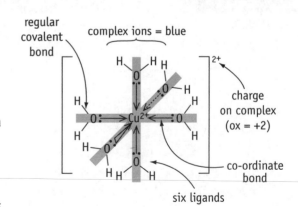

regular covalent bond

complex ions = blue

charge on complex (ox = +2)

co-ordinate bond

six ligands

Ligands	No. of co-ordinate bonds per ligand	Examples	Co-ordination number
Unidentate	one	Cl^-, NH_3, CN^-	number of ligands attached
Bidentate	two	1,2-diaminoethane, $H_2NCH_2CH_2NH_2$, (or 'en')*, and the ethanedioate ion, $C_2O_4^{2-}$	twice the number of ligands
Multidentate	three or more	Bis[di(carboxymethyl)amino]ethane (the EDTA^{4-} ion)	6

* en is drawn as N ⌣ N

The charge on the ligand and the oxidation state of the metal ion determines the overall charge of the complex, e.g. $[Cu(H_2O)_6]^{2+}$ $[Cu(NH_3)_4(H_2O)_2]^{2+}$ and $CuCl_4^{2-}$. Here copper(II) is joined to neutral water or ammonia and -1 chloride so the charge on the ions changes.

Multidentate ligands are chelating ligands. **Chelating** means that they form very stable complexes (more so than unidentate ligands) since their formation has a favourable entropy change (i.e. it increases entropy). See page 83 for more on EDTA.

$$\text{e.g. } [Cu(H_2O)_6]^{2+} + EDTA^{4-} \rightarrow [Cu(EDTA)]^{2-} + 6H_2O$$

Shapes of complex ions

Shapes are worked out using the rules you met in AS chemistry (see AS book).

No. ligands	Shape	Example
2	linear	$[Ag(CN)_2]^-$
4	tetrahedral	$FeCl_4^{2-}$
4	square planar (less common)	$[Pt(NH_3)_2Cl_2]$
6	octahedral	$[Co(NH_3)_6]^{3+}$

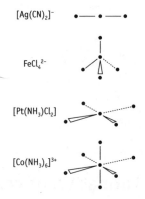

$[Ag(CN)_2]^-$

$FeCl_4^{2-}$

$[Pt(NH_3)Cl_2]$

$[Co(NH_3)_6]^{3+}$

- If the ligand is small and uncharged like water or ammonia, the shape is usually octahedral (co-ordination number 6) and if it is larger and charged like chloride then it is tetrahedral.

- Chloride ions are larger than water and ammonia molecules, and repel each other so only four can fit around the TM ion. But F^- is small enough to fit six around an ion.

- Silver (Ag^+) and copper(I) (Cu^+) are often linear (co-ordination number 2)

? *Quick check questions*

1 What is a complex?

2 Define the term ligand.

3 What is the co-ordination number of a complex?

4 Name the type of bond formed between a ligand and a metal ion and explain how it forms.

Transition metals (II)

There are four parts to the name of a complex.

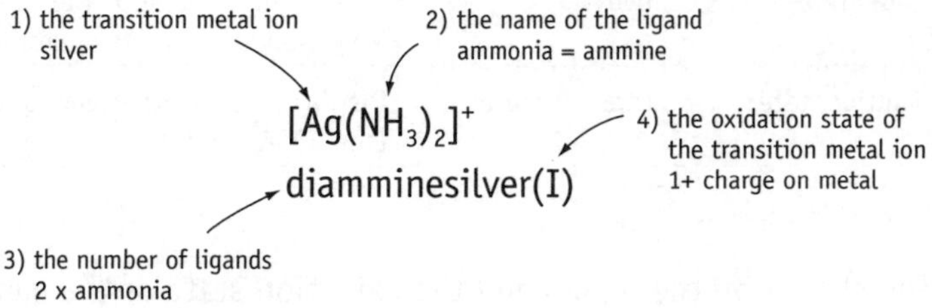

Rules for naming transition metal complexes

1 Name the transition metal: The easy bit.

2 Name the ligand: The common ones are shown opposite. Ligands are named in alphabetical order ignoring prefixes below.

3 Note how many ligands: e.g. hexa = 6 ligands, tetra = 4, di = 2

4 Negative complexes end with –ate: They use the Latin name and the oxidation state of the metal at the end in brackets.

1) the transition metal ion
silver

2) the name of the ligand
ammonia = ammine

$$[Ag(NH_3)_2]^+$$

diamminesilver(I)

3) the number of ligands
2 x ammonia

4) the oxidation state of the transition metal ion
1+ charge on metal

Ligand	Formula	Name
water	$:OH_2$	aqua-
chloride	$:Cl^-$	chloro-
ammonia	$:NH_3$	ammine-
cyanide	$:CN^-$	cyano-
ethane-1,2-diamine	$H_2NCH_2CH_2NH_2$	ethane-1,2-diamine-

Positively charged complex	Negatively charged complex
Chromium	chromate
Cobalt	cobaltate
Copper	cuprate
Iron	ferrate
Manganese	manganate
Silver	argentate
Vanadium	vanadate

Examples	
$[Ag(S_2O_3)_2]^{3-}$	dithiosulphateargentate(I)
$[Cr(NH_3)_6]^{3+}$	hexamminechromium(III)
$[CoCl_2(NH_3)_4]^+$	tetraamminedichlorocobalt(III)
$[Cu(NH_3)_4(H_2O)_2]^{2+}$	tetraamminediaquacopper(II)
$[Fe(CN)_6]^{4-}$	hexacyanoferrate(II)

Formation of coloured ions

✓ **Quick check 1,2**

Complex ions can be identified by their colour. The colour depends on the:

- Oxidation state of the TM ion
- The co-ordination number
- The ligand present

Changing any of these factors will change the colour.

Colour occurs when light energy is reflected. When visible light falls on a TM some light is absorbed and some is reflected. The light which is not absorbed is known as the complementary colour and, if visible, is seen. The light absorbed causes unpaired electrons to become excited (they are promoted to a higher energy level).

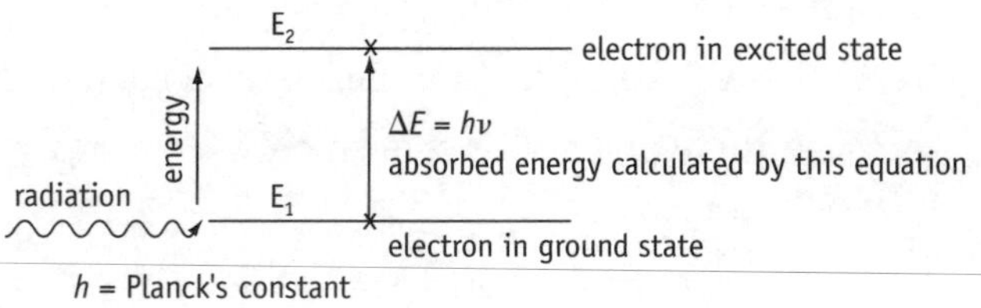

E_2 ———×———— electron in excited state

energy

radiation

$\Delta E = h\nu$
absorbed energy calculated by this equation

E_1 ———×———— electron in ground state

h = Planck's constant

- The wavelength of the light absorbed depends on the energy difference between the ground state and the excited state.

- This energy difference is determined by the ligands present and the shape of the complex.

Changing the oxidation state

Change the oxidation state and you change the colour.

$$[Fe(H_2O)_6]^{2+} \rightarrow [Fe(H_2O)_6]^{3+}$$

Green $\quad\quad\quad$ Violet

Changing the co-ordination number

Changing the co-ordination number usually involves changing the ligand

e.g. $\quad\quad\quad\quad\quad\quad\quad\quad\quad$ $[Cu(H_2O)_6]^{2+}$ $\quad\rightarrow\quad$ $[CuCl_4]^{2-}$

Ox (Cu) =	+2	+2
Shape =	Octahedral	tetrahedral
Colour =	blue	yellow
		(solution looks green)

> ► You may not be asked this but the trend in ligand strength is:
> EDTA > NH_3 > Cl^- > H_2O

Changing the ligand

Changing the ligand but not the oxidation number or co-ordination number e.g.

		$Cr(H_2O)_6]^{3+}$	$\rightarrow$	$[Cr(NH_3)_6^{3+}$	or	$[Co(H_2O)_6]^{2+}$	$\rightarrow$	$[Co(NH_3)_6]^{2+}$
Ox	=	+3		+3		+2		+2
Shape	=	Octahedral		Octahedral		Octahedral		Octahedral
Colour	=	red/violet		violet/purple		pink		pale brown

Examples

✓ Quick check 3

Ion	H_2O	NH_3	OH^-	Cl^-	CN^-
Cr^{3+}	$[Cr(H_2O)_6]^{3+}$ dull red blue (green solution)	$[Cr(NH_3)_6]^{3+}$ violet/purple	$[Cr(H_2O)_3(OH)_3]$ grey/green	$[Cr(H_2O)_x(Cl)_y]$ x + y = 6 green	–
Mn^{2+}	$[Mn(H_2O)_6]^{2+}$ pink	–	$[Mn(H_2O)_4(OH)_2]$ off white	–	–
Fe^{2+}	$[Fe(H_2O)_6]^{2+}$ green	–	$Fe(OH)_2$ green ppt	–	–
Fe^{3+}	$[Fe(H_2O)_6]^{3+}$ yellow	–	$Fe(OH)_3$ red/brown ppt	–	–
Co^{2+}	$[Co(H_2O)_6]^{2+}$ pink	$[Co(NH_3)_6]^{2+}$ pale yellow/brown	$Co(OH)_2$ blue ppt then pink	$[CoCl_4]^{2-}$ blue	–
Co^{3+}	–	$[Co(NH_3)_6]^{3+}$ dark brown	$Co(OH)_3$ black ppt	–	–
Cu^+	–	$[Cu(NH_3)_4]^+$ colourless	–	$[CuCl_2]^-$ colourless	–
Cu^{2+}	$[Cu(H_2O)_6]^{2+}$ blue	$[Cu(NH_3)_4(H_2O)_2]^{2+}$ deep blue	$Cu(OH)_2$ blue ppt	$[CuCl_4]^{2-}$ yellow/green	–
Ag^+	–	$[Ag(NH_3)_2]^+$ colourless	Ag_2O brown ppt	$[AgCl_2]^-$ colourless	$[Ag(CN)_2]^-$ colourless

? Quick check questions

1 Give the structure of the tetraamminediaquacopper(II) complex showing the arrangement of the ligands.

2 What can cause a change of colour in a complex?

3 Give an equation for a reaction in which there is a change of ligand and shape of a cobalt(II) complex.

Transition metals (III)

Variable oxidation states

The electrons in the d-orbitals are close together in energy, so it is easy to remove more than one electron. This is in contrast to other groups where orbitals exist in very different energy levels, so just one ion forms.

TMs can lose electrons from...

1 The 4s orbital only

2 The 4s and half the 3d

3 The 3d and the 4s orbitals

Facts you should know about oxidation states

- All TMs have Ox = +2 (but not Sc) where the 4s electrons are removed.

- The stability of the +2 state increases across the series (as the nuclear charge increases it holds the inner 3d electrons more tightly).

- From Sc to Mn the oxidation states increase to a maximum of +7. From Fe to Zn the +3 and +2 oxidation states are more common.

- Simple ions give the lower oxidation states like M^{2+} or M^{3+}.

- When the TM is covalently bonded to a very electronegative element like oxygen — the higher oxidation states (+5 or +7) can be seen e.g. VO_3^-, VO_2^+, CrO_4^{2-}, and $Cr_2O_7^{2-}$, MnO_4^-.

- Generally:
  High oxidation states = **oxidising agents** e.g. MnO_4^- ions and $Cr_2O_7^{2-}$
  Lower oxidation states = **reducing agents** e.g. Cr^{2+} and Fe^{2+}

Reaction summaries for the main transition metals

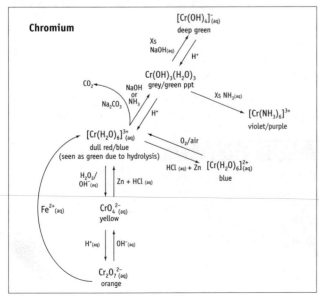

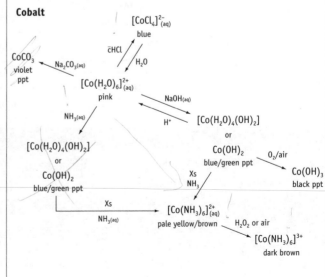

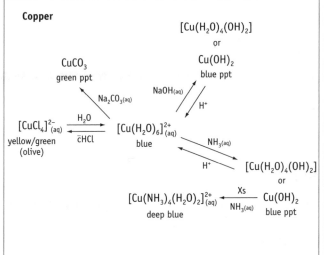

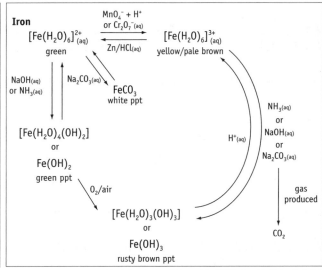

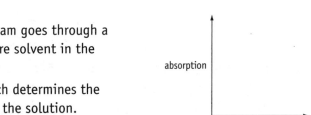

UV and visible spectrophotometry

Solutions of TM ions are coloured, and the more concentrated a solution, the more light is absorbed. Colour can therefore be used to determine concentration. A **colorimeter** measures the absorption of visible radiation and is a simple type of visible spectrophotometer.

- A visible light source passes through a coloured filter chosen so that the light it transmits is absorbed by the sample.

- The radiation is then split into two beams. 1) One beam goes through a solution of the sample; 2) the other goes through pure solvent in the reference cell.

- The two beams are compared by a light detector which determines the percentage of radiation absorbed (or transmitted) by the solution.

- The more concentrated the solution, the more light it will absorb.

- If the absorption values are measured for solutions of known concentrations, a calibration graph of concentration against absorption can be plotted.

- If a solution of unknown concentration is then placed in the colorimeter and its absorption recorded, the concentration of this solution can be found from the calibration graph.

❓ Quick check questions

1 Which oxidation state is shown by most transition metals?

2 What happens to the stability of this oxidation state as we go across the series?

3 How can chromium(VI) be reduced to chromium(II)?

4 Name the two transition metal ions that are commonly used as oxidising agents in acidic solution in titration work.

5 Give an equation in which an iron ion can act as a reducing agent with one of the ions you have named in 4.

6 Why are copper(II) complexes coloured but copper(I) complexes white/colourless?

Redox titrations

Potassium manganate(VII) and potassium dichromate(VI) are used as quantitative oxidising agents in redox titrations.

$$MnO_4^- + 8H^+ + 5e^- \rightarrow Mn^{2+} + 4H_2O$$
$$Cr_2O_7^{2-} + 14H^+ + 6e^- \rightarrow 2Cr^{3+} + 7H_2O$$

Often questions arise which ask you to find out the amount of iron(II) in a sample.

✓ *Quick check 1,2*

Ye olde favourite — Analysis of iron tablets

One method

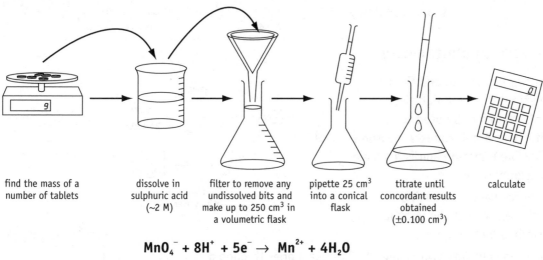

| find the mass of a number of tablets | dissolve in sulphuric acid (~2 M) | filter to remove any undissolved bits and make up to 250 cm³ in a volumetric flask | pipette 25 cm³ into a conical flask | titrate until concordant results obtained (±0.100 cm³) | calculate |

$$MnO_4^- + 8H^+ + 5e^- \rightarrow Mn^{2+} + 4H_2O$$
$$5Fe^{2+}(aq) \rightarrow 5Fe^{3+}(aq) + 5e^-$$
$$5Fe^{2+}(aq) + MnO_4^-(aq) + 8H^+(aq) + 5e^- \rightarrow 5Fe^{3+}(aq) + Mn^{2+}(aq) + 4H_2O(l) + 5e^-$$
$$5Fe^{2+}(aq) + MnO_4^-(aq) + 8H^+(aq) \rightarrow 5Fe^{3+}(aq) + Mn^{2+}(aq) + 4H_2O(l)$$
🖝 The important equation
$$\therefore 5Fe^{2+}(aq) \equiv MnO_4^-(aq)$$

Worked example

People who suffer from anaemia often take iron tablets which they can buy in supermarkets and pharmacists. Analysis of a 0.800 g tablet found that it contained iron(II) sulphate and one tablet dissolved in sulphuric acid decolorised 27.5 cm³ of 0.0200 M KMnO₄. Calculate the % mass of Fe^{2+} in the tablet. (A_r Fe = 55.9)

> Dilute sulphuric acid is used to acidify. If HCl is used, Cl⁻ is oxidised by MnO_4^- and chlorine is liberated. This doesn't happen with $Cr_2O_7^{2-}$. HNO_3 is an oxidising agent so can't be used.

Step 1 Calculate the number of moles of MnO_4^- used

$$\text{No. moles} = \frac{\text{concn} \times \text{vol}}{1000} = \frac{0.0200 \times 27.5}{1000} = 0.000550 \text{ moles}$$

Step 2 Calculate the number of moles of iron(II)
Looking at the above equation you see that $5Fe^{2+}(aq) = MnO_4^-(aq)$

$$\therefore \text{ moles of iron(II)} = 5 \times 0.000550 = 0.00275 \text{ moles}$$

Step 3 Calculate the mass of iron(II) present in the tablet

$$n = m/M_r = 0.00275 = m/55.9$$
$$m = 0.00275 \times 55.9 = 0.154 \text{ g}$$

Step 4 Calculate the % mass of iron(II) present in the tablet.

$$0.154/0.800 \times 100 = 19.2\%$$

Analysis of moss killer (lawn sand)

✓ *Quick check 3*

Manganate(VII) or dichromate(VI) are commonly used in these types of questions. The difficult bit is unpicking the chemistry from the rest of the blurb in the question.

▶ Same sort of questions are asked about composition of moss killer or lawn sand. Again it is familiar chemistry just devilishly put in an unfamiliar way.

For dichromate(VI) and iron(II)

$$Cr_2O_7^{2-}(aq) + 14H^+(aq) + 6e^- \rightarrow 2Cr^{3+}(aq) + 7H_2O(l)$$
$$6Fe^{2+}(aq) \rightarrow 6Fe^{3+}(aq) + 6e^-$$

$$6Fe^{2+}(aq) + Cr_2O_7^{2-}(aq) + 14H^+(aq) \rightarrow 6Fe^{3+}(aq) + 2Cr^{3+}(aq) + 7H_2O(l)$$
$$\therefore 6Fe^{2+}(aq) \equiv Cr_2O_7^{2-}(aq)$$

Worked example

Anyone with a lawn knows that moss growing in the grass is a real pain. The active ingredient of moss killer is often iron sulphate, but it contains other chemicals such a lawn feed, sand and weed killer.

Analysis: 4.00 g of moss killer bought from a local garden centre was dissolved in sulphuric acid. The solution required 28.0 cm^3 of 0.0200 mol dm^{-3} $K_2Cr_2O_7$ using sodium diphenylaminesulphonate as the indicator. Calculate the % mass of Fe^{2+} in the moss killer.

Step 1 Calculate the number of moles of $Cr_2O_7^{2-}$ used.

$$\text{No. moles} = \frac{\text{concn} \times \text{vol}}{1000} = \frac{0.02 \times 28.0}{1000} = 0.000560 \text{ moles}$$

▶ Sodium diphenylamine sulphonate turns colourless to purple at the end point.

Step 2 Calculate the number of moles of iron(II)
Looking at the above equation you see that $6Fe^{2+}(aq) \equiv Cr_2O_7^{2-}$ (aq)
moles of iron(II) = 6 x 0.000560 = 0.00336 moles
Step 3 Calculate the mass of iron(II) present in the moss killer.

$$n = m/M_r = 0.00336 = m/55.9$$
$$m = 0.00336 \times 55.9 = 0.188 \text{ g}$$

Step 4 Calculate the % mass of iron(II) present in the moss killer.

$$0.188/4.00 \times 100 = 4.70\%$$

? **Quick check questions**

1 What is the colour change in a titration when potassium manganate(VII) is added to an acidified reducing agent from a burette?

2 Which acid is used to supply the H$^+$ ions needed?

3 Explain why the colour change, from orange to green, when $Cr_2O_7^{2-}$ is reduced to Cr^{3+}, cannot be used to detect the end-point of the titration.

Catalytic behaviour

Transition metals (and their compounds) are good catalysts because...

- They have variable oxidation states: so they can gain and lose electrons easily and can transfer electrons to speed up a reaction.
- They provide a surface for a reaction or sites where a reaction can occur.
- They bond readily to ions and molecules when in solution and as solids.

> ▶ Surface adsorption theory involves 3 steps
> 1 adsorption of a reactant
> 2 reaction
> 3 desorption of a product

Heterogeneous catalysts — different phase to reactants

1 Increase the chance of collision: The availability of d and p electrons helps gaseous reactants adsorb to them and so become more likely to collide with another reactant.

2 Decrease the activation energy: The adsorbed reactants may be in a good orientation for a successful collision, so require less energy to kick-start the reaction. The adsorbed reactants may also rearrange their structure, or bonds may break whilst adsorbed, and so the reaction may require less energy to start.

Problems with adsorption

> ✓ Quick check 4

A balanced adsorption is needed for an efficient catalysis.

- If the metal bonds too strongly, the reactants cannot break free and this blocks active sites — the catalyst is **poisoned**.
- If the metal adsorbs too weakly, then the chance of a reaction is not much higher than without a catalyst.

> ▶ For information on catalytic converters see Module 3 in the AS book.

Strong adsorption
e.g. W

Good adsorption
Ni, Pd, Pt

Weak adsorption
Ag

Catalytic conditions in a catalytic converter

- Usually the catalyst is in the form of a mesh or fine powder to provide a large surface area.
- Spread over some kind of inert support medium. This may increase cost but maximizes the surface area.

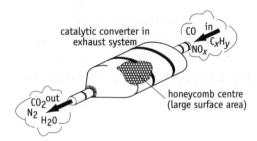
catalytic converter in exhaust system
CO in
NO_x C_xH_y
honeycomb centre (large surface area)
CO_2 out
N_2 H_2O

Other exciting examples of heterogeneous catalysts

1 Haber Process:
   Finely divided Fe or Fe_2O_3 in the production of ammonia

   $$N_2(g) + 3H_2(g) \rightleftharpoons 2NH_3(g)$$

2 Contact Process:
   Solid V_2O_5 in the production of sulphuric acid

   $$2SO_2(g) + O_2(g) \rightleftharpoons 2SO_3(g)$$

Homogenous catalysts — same phase as reactants

In solution TM ions catalyse reactions proceeding via **intermediates**

> ✓ Quick check 1,2,3

$$\textbf{e.g. } S_2O_8{}^{2-}(aq) \quad + \quad 2I^-(aq) \xrightarrow{Fe^{2+} \text{ or } Fe^{3+}} 2SO_4{}^{2-}(aq) \quad + \quad I_2(aq)$$

peroxodisulphate ions +   iodide ions $\rightarrow$ sulphate   +   iodine

Both ions are negative so uncatalysed reaction is very slow as the particles repel each other, making the chance of a collision slim.

1   Iron(II) or iron(III) solution catalyse the reaction. The positively charged $Fe^{2+}(aq)$ reacts with the negatively charged $S_2O_8^{2-}(aq)$.

$$S_2O_8^{2-}(aq) + 2Fe^{2+}(aq) \rightarrow 2SO_4^{2-}(aq) + 2Fe^{3+}(aq)$$

2   Then the $Fe^{3+}(aq)$ formed reacts with the negatively charged $I^-(aq)$:

$$2Fe^{3+}(aq) + 2I^-(aq) \rightarrow 2Fe^{2+}(aq) + I_2(aq)$$

Autocatalysis by Mn^{2+} for $C_2O_4^{2-}$ with MnO_4^-

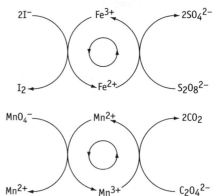

- The initial rate of the reaction is slow, as reactants repel due to their negative charge.

- As $Mn^{2+}(aq)$ concentration builds up the rate of reaction increases. Why?

- The $Mn^{2+}(aq)$ acts as an **autocatalyst** — it reacts with MnO_4^- ions to form Mn^{3+} ions, which react with the ethanedioate ions regenerating Mn^{2+} ions.

$$4Mn^{2+} + MnO_4^- + 8H^+ \rightarrow 5Mn^{3+} + 4H_2O$$
$$2Mn^{3+} + C_2O_4^{2-} \rightarrow 2CO_2 + 2Mn^{2+}$$

$$2MnO_4^-(aq) + 5C_2O_4^{2-}(aq) + 16H^+(aq) \rightarrow 2Mn^{2+}(aq) + 5CO_2(g) + 8H_2O(l)$$

Other applications of transition metals

- **Haem** is an iron(II) complex with a multidentate ligand, found naturally in blood. The complex has four nitrogen atoms co-ordinately bonded to the iron but the co-ordination number is six. The two other bonds combine with a protein called globin making **haemoglobin** and either a water molecule (making deoxyhaemoglobin) or an oxygen molecule (making oxyhaemoglobin).

- Haemoglobin plays a vital role in transporting oxygen around the body. This complex easily co-ordinates with oxygen or water, but also more strongly with carbon monoxide. This carboxyhaemoglobin is very stable and reduces the oxygen-carrying capacity of the blood, which is why carbon monoxide is so dangerous.

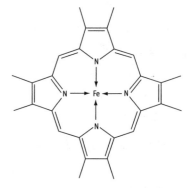

Haem

Cisplatin

- Pt is used in chemotherapy treatment for cancer. **Cisplatin** bonds with guanine, a base in DNA and stops cancer cells replicating.

- $[Ag(NH_3)_2]^+$ Tollen's reagent: used to distinguish between aldehydes and ketones. (see AS book Module 3)

- $[Ag(S_2O_3)_2]^{3-}$ Used in photography as a fixer: Silver bromide is dissolved away from film by sodium thiosulphate leaving silver on the negative.

✓ *Quick check 3*

- $[Ag(CN_2)_2]^-$ a solution used in electroplating.

? Quick check questions

1   Define homogeneous catalysis, give an example of a transition metal compound acting as a homogeneous catalyst.

2   What are the three stages in heterogeneous catalysis?

3   Why is tungsten generally an inefficient catalyst?

4   Give an example of a reaction in which silver is used as a heterogeneous catalyst because it is inefficient.

Inorganic compounds in solution

A Lewis acid — accepts a pair of electrons e.g. the transition metal

A Lewis base — donates a lone pair of electrons e.g. a ligand

In aqueous solution transition metal ions act as a Lewis acid accepting a pair of electrons to form metal-aqua ions with a co-ordination number of six, e.g.

$$CuSO_4(s) + 6H_2O(l) \rightarrow [Cu(H_2O)_6]^{2+}(aq) + SO_4^{2-}(aq)$$
$$\text{white} \qquad\qquad\qquad \text{blue}$$

White anhydrous copper (II) sulphate dissolves in water exothermically as bonds form when the blue hydrated ion forms. In this Cu^{2+} aqueous solution the $[Cu(H_2O)_6]^{2+}(aq)$ is the species present.

✓ *Quick check 1*

Metal-aqua ions you need to know about

You need to know about seven metal aqua ions: three $[M(H_2O)_6]^{2+}$ and four $[M(H_2O)_6]^{3+}$ aqueous ions. See the table opposite.

TM	Hexaqua ions	Colour
Iron	$[Fe(H_2O)_6]^{2+}$	pale green
Copper	$[Cu(H_2O)_6]^{2+}$	blue
Cobalt	$[Co(H_2O)_6]^{2+}$	pink
Iron	$[Fe(H_2O)_6]^{3+}$	yellow/orange
Chromium	$[Cr(H_2O)_6]^{3+}$	dull red/blue, often seen as green (aq)
Aluminium	$[Al(H_2O)_6]^{3+}$	colourless
Vanadium	$[V(H_2O)_6]^{3+}$	green

Some aqua ions exist in the solid crystalline transition metal compounds, e.g.

- hydrated copper(II) sulphate, $CuSO_4.5H_2O$

- hydrated iron(II) sulphate, $FeSO_4.7H_2O$

- hydrated cobalt(II) chloride, $CoCl_2.6H_2O$

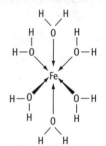

Iron hexaqua ion showing the co-ordinate bonds

The colour of the TM crystalline compounds is due to the hexaqua ion, e.g. $[Co(H_2O)_6]^{2+}$

✓ *Quick check 2*

▶ Remember
HO-MO-REDOX

When M(II) ions are hydrated energy is released which overcomes the lattice energy. For M(III) ions the energy released is larger making their hydration more exothermic, e.g. $AlCl_3$.

Reactions of metal-aqua ions

- The H—O bond of a co-ordinated water molecule is broken — an acidity or hydrolysis reaction.

- The M—O bond is broken; this is ligand replacement = a substitution reaction.

- Redox — no bonds around the hexaqua ion are broken; electrons are gained or lost by the metal ion.

✓ *Quick check 3,4*

broken in ligand substitution

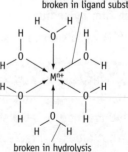

broken in hydrolysis

Metal-aqua ion showing which bonds break in hydrolysis and substitution reactions

Acidity or hydrolysis reactions (H—O)

You need to understand the equilibria for M³⁺

1 $\qquad [M(H_2O)_6]^{3+} + H_2O \rightleftharpoons [M(H_2O)_5(OH)]^{2+} + H_3O^+$

- The position of equilibrium lies to the LHS so the pH of the resulting solutions is about 3.

- If pH = 3 → $[H^+] = 0.001$ mol dm⁻³. In a 1 M solution of $[M(H_2O)_6]^{3+}$ it's concentration is $1000 \times$ that of $[M(H_2O)_5(OH)]^{2+}$, so equilibrium lies to the LHS.

- Tri-positive metal ions are small and have large charge/size ratios, so are very polarising. They polarise the water ligands so much that the O–H bond is weakened and donates a hydrogen to the water solvent.

- The metal-aqua ion behaves as a Brønsted–Lowry acid, pH ~ 3.

- Further hydrolysis steps are possible in the presence of a base:

2 $\qquad [M(H_2O)_5(OH)]^{2+} + H_2O \rightleftharpoons [M(H_2O)_4(OH)_2]^+ + H_3O^+$

3 $\qquad [M(H_2O)_4(OH)_2]^+ + H_2O \rightleftharpoons [M(H_2O)_3(OH)_3] + H_3O^+$

- These further steps require equilibria to be over to the RHS. This is achieved by adding something which removes H_3O^+ ions, e.g. bases, like NaOH, NH_3 or Na_2CO_3, or a reactive metal like magnesium or zinc.

- This leads to the formation of the metal hydroxide precipitate, $[M(H_2O)_3(OH)_3]$.

You also need to understand this equilibria for M²⁺

1 $\qquad [M(H_2O)_6]^{2+} + H_2O \rightleftharpoons [M(H_2O)_5(OH)]^+ + H_3O^+$

- The position of the equilibrium here is very much more to the LHS so di-positive transition metal-aqua ions are **weaker** acids than the tri-positive metal aqua-ions

- The **charge /size ratio** is much smaller and therefore less polarising, so di-positive metal ions in solution are only very weakly acidic. pH ~ 6

- If pH was 6 it would mean that 1 in 1,000,000 molecules would be hydrolysed. This is why the equilibrium lies way to the LHS.

- Further hydrolysis occurs when a base is added.

2 $\qquad [M(H_2O)_5(OH)]^+ + H_2O \rightleftharpoons [M(H_2O)_4(OH)_2] + H_3O^+$

The overall equation (simplified) $M^{2+}(aq) + 2OH^-(aq) \rightarrow M(OH)_2(s)$

▶ $[M(H_2O)_6]^{3+}$ is a polarising ion and acidic. $[M(H_2O)_6]^{2+}$ is less polarising and is less acidic. Acidity increases the higher the charge and smaller the ion.

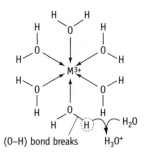

(O–H) bond breaks H_3O^+

▶ When $NH_3(aq)$ or NaOH are added to M(II) or M(III) (aq) ions the insoluble metal hydroxide forms. $[M(H_2O)_6]^{4+}$ does not usually exist and totally hydrolyses to $M(OH)_4$.

? Quick check questions

1 What is a Lewis acid and a Lewis base?

2 Give the formula of seven metal-aqua ions you need to know about.

3 Name the three types of reactions that metal-aqua ions undergo.

4 Why do M³⁺ ions have different acidic properties to M²⁺ ions?

Reactions of M^{2+} and M^{3+} (aq)

There are a number of simple test-tube reactions you need to know about. The metal ion reactions where: M^{3+} = Al, Cr or Fe and M^{2+} = Fe, Co or Cu, with OH^-, NH_3 and CO_3^{2-}.

With NaOH — both M^{3+} and M^{2+} form the metal hydroxide ppt.

Metal-aqua ion	Addition of NaOH solution	Addition of excess NaOH solution	On standing in air
$[Al(H_2O)_6]^{3+}$ colourless solution	$[Al(H_2O)_3(OH)_3]$ white ppt	$[Al(H_2O)_2(OH)_4]^-$ colourless solution	no change
$[Cr(H_2O)_6]^{3+}$ red/blue solution (appears green on hydrolysis)	$[Cr(H_2O)_3(OH)_3]$ grey/green ppt	$[Cr(OH)_6]^{3-}$ green solution	no change
$[Fe(H_2O)_6]^{3+}$ pale violet (appears yellow on hydrolysis)	$[Fe(H_2O)_3(OH)_3]$ red/brown ppt	no change	no change
$[Fe(H_2O)_6]^{2+}$ pale green solution	$[Fe(H_2O)_4(OH)_2]$ green ppt	no change	$[Fe(H_2O)_3(OH)_3$ red/brown ppt
$[Co(H_2O)_6]^{2+}$ pink solution	$[Co(H_2O)_4(OH)_2]$ blue/green ppt	no change	$[Co(H_2O)_3(OH)_3]$ pink then beige ppt
$[Cu(H_2O)_6]^{2+}$ blue solution	$[Cu(H_2O)_4(OH)_2]$ pale blue ppt	no change	no change

- Three successive reactions occur with M^{3+} (the OH^- removes H_3O^+ and shifts the equilibrium, e.g.

1 $[Fe(H_2O)_6]^{3+} + H_2O \rightleftharpoons [Fe(H_2O)_5(OH)]^{2+} + H_3O^+$

2 $[Fe(H_2O)_5(OH)]^{2+} + H_2O \rightleftharpoons [Fe(H_2O)_4(OH)_2]^+ + H_3O^+$

3 $[Fe(H_2O)_4(OH)_2]^+ + H_2O \rightleftharpoons [Fe(H_2O)_3(OH)_3] + H_3O^+$
       red/brown ppt

Simplified to: $Fe^{3+} + 3OH^- \rightleftharpoons Fe(OH)_3$

- If strong acid is added, the reactions are reversed, e.g.

$$[Fe(H_2O)_3(OH)_3] + 3H_3O^+ \rightleftharpoons [Fe(H_2O)_6]^{3+} + 3H_2O$$

- Two successive reactions can occur with M^{2+} (again the OH^- removes H_3O^+) e.g.

1 $[Cu(H_2O)_6]^{2+} + H_2O \rightleftharpoons [Cu(H_2O)_5(OH)]^+ + H_3O^+$
       blue solution

2 $[Cu(H_2O)_5(OH)]^+ + H_2O \rightleftharpoons [Cu(H_2O)_4(OH)_2] + H_3O^+$
       pale blue ppt

Simplified to: $Cu^{2+} + 2OH^- \rightleftharpoons Cu(OH)_2$

- Further hydrolysis reaction (above) only occur if a large excess of *concentrated* NaOH solution is added.

▶ This does not happen very much.

With $NH_3(aq)$ — both M^{3+} and M^{2+} form the metal hydroxide ppt (initially)

The hydroxide precipitates form (initially, as with addition of NaOH) because aqueous ammonia is a weak base, but some form an **ammine complex.**

Ammonia produces OH^-

$$NH_3 + H_2O \rightleftharpoons NH_4^+ + OH^-$$

The OH^- reacts with the H_3O^+

✓ *Quick check 1*

Metal-aqua ion	Addition of NH_3 solution (initial reaction)	Addition of excess NH_3 solution	On standing in air
$[Al(H_2O)_6]^{3+}$	$[Al(H_2O)_3(OH)_3]$ white ppt	no change	no change
$[Cr(H_2O)_6]^{3+}$	$[Cr(H_2O)_3(OH)_3]$ grey/green ppt	$[Cr(NH_3)_6]^{3+}$ purple solution	no change
$[Fe(H_2O)_6]^{3+}$	$[Fe(H_2O)_3(OH)_3]$ red/brown ppt	no change	no change
$[Fe(H_2O)_6]^{2+}$	$[Fe(H_2O)_4(OH)_2]$ green ppt	no change	$[Fe(H_2O)_3(OH)_3]$ red/brown ppt
$[Co(H_2O)_6]^{2+}$	$[Co(H_2O)_4(OH)_2]$ blue ppt	$[Co(NH_3)_6]^{2+}$ pale yellow solution	$[Co(NH_3)_6]^{3+}$ brown solution
$[Cu(H_2O)_6]^{2+}$	$[Cu(H_2O)_4(OH)_2]$ blue ppt	$[Cu(NH_3)_4(H_2O)_2]^{2+}$ deep blue solution	no change

- The equations for the reactions are similar to those of NaOH. Simplified:

$$[M(H_2O)_6]^{3+} + 3OH^- \rightleftharpoons [M(H_2O)_3(OH)_3] + 3H_2O$$
$$[M(H_2O)_6]^{2+} + 2OH^- \rightleftharpoons [M(H_2O)_4(OH)_2] + 2H_2O$$

- If excess ammonia solution is added a ligand substitution reaction can occur forming an **ammine** complex.

With Na_2CO_3 — M^{3+} forms hydroxide ppts and M^{2+} form carbonate ppts

- The M^{3+} forms hydroxide ppt and M^{2+} forms carbonate ppt.
- The acidic, M^{3+} aqua ions, like Al^{3+} Cr^{3+} and Fe^{3+}, react with sodium carbonate to form the metal hydroxide precipitate and carbon dioxide.
- The H_3O^+ are removed by reaction with the carbonate ions moving the equilibrium to the RHS forming the metal hydroxide precipitate.

$$[M(H_2O)_6]^{3+} + H_2O \rightleftharpoons [M(H_2O)_5(OH)]^{2+} + H_3O^+$$

$$H_3O^+ + CO_3^{2-} \rightarrow H_2O + HCO_3^- \text{ then } H_3O^+ + HCO_3^- \rightarrow H_2O + CO_2$$

- The M^{2+} aqua ions do **not** release carbon dioxide as they are weaker acids than carbonic acid. They form the **metal carbonate** precipitates.

Generally: $$M^{2+}(aq) + CO_3^{2-}(aq) \rightarrow MCO_3(s)$$

Metal-aqua ion	Addition of Na_2CO_3 solution
$[Al(H_2O)_6]^{3+}$	$[Al(H_2O)_3(OH)_3] + CO_2 + H_2O$ white ppt
$[Cr(H_2O)_6]^{3+}$	$[Cr(H_2O)_3(OH)_3] + CO_2 + H_2O$ grey/green ppt
$[Fe(H_2O)_6]^{3+}$	$[Fe(H_2O)_3(OH)_3] + CO_2 + H_2O$ red/brown ppt
$[Fe(H_2O)_6]^{2+}$	$FeCO_3 + H_2O$ green ppt
$[Co(H_2O)_6]^{2+}$	$CoCO_3 + H_2O$ pink/lilac ppt
$[Cu(H_2O)_6]^{2+}$	$CuCO_3 + H_2O$ green/blue ppt

▶ Here M^{3+} release CO_2 and some M^{3+} ions like Al^{3+} are acidic enough to react with metals like Zn and Mg to release hydrogen gas.

▶ $[M(H_2O)_5(OH)]^{2+}$ then goes on to form $[M(H_2O)_3(OH)_3]$

▶ $M_2(CO_3)_3$ where M = Al or Fe or Cr does NOT form in soln and are unknown.

Amphoteric nature of M^{3+} hydroxides

Some metal (III) hydroxides like Al^{3+} and Cr^{3+} are amphoteric and dissolve in both acids and bases.

Generally: $$[M(H_2O)_6]^{3+} \underset{H^+}{\overset{OH^-}{\rightleftharpoons}} [M(H_2O)_3(OH)_3] \underset{H^+}{\overset{OH^-}{\rightleftharpoons}} [M(OH)_4]^-$$

$$\text{Acid} \qquad\qquad \text{Neutral} \qquad\qquad \text{Alkaline}$$

In high oxidation sates TMs can often been seen as anions. As an example you must know the equilibrium below:

e.g. $$2CrO_4^{2-} \underset{2OH^-}{\overset{2H^+}{\rightleftharpoons}} 2Cr_2O_7^{2-} + H_2O$$

$$\text{Yellow} \qquad\qquad \text{orange}$$
$$\text{Chromate (VI)} \qquad \text{dichromate (VI)}$$

❓ Quick check questions

1 Which three complexes give a precipitate with ammonia solution that dissolve when excess ammonia(aq) is added?

2 Write an equation in which the complex $[M(H_2O)_6]^{3+}$ acts as an acid with carbonate ions.

Ligand substitution reactions

Transition metal hexaqua ions undergo ligand substitution reactions where the water ligands are replaced by chloride or ammonia ligands by breaking one or more of the M—O bonds.

- H_2O and NH_3 ligands are of a similar size and are uncharged so ligand exchange may take place with no change in shape.

$$[Co(H_2O)_6]^{2+} + 6NH_3 \rightleftharpoons [Co(NH_3)_6]^{2+} + 6H_2O$$
$$\text{Pink} \qquad\qquad \text{pale yellow}$$

- Substitution reactions are sometimes incomplete, e.g. only four of the six water ligands are replaced with copper(II) ions.

$$[Cu(H_2O)_6]^{2+} + 4NH_3 \rightleftharpoons [Cu(NH_3)_4(H_2O)_2]^{2+} + 4H_2O$$

- These ligand substitution reactions are reversible.

- The size and charge of the ligand determines the co-ordination number and the shape of the complex (see page 68). Chloride ligands are larger than water and negatively charged, so six cannot fit around the ion; that and repulsion causes a tetrahedral shape to be favoured.

$$[Co(H_2O)_6]^{2+} + 4Cl^- \underset{\textit{dilute in } H_2O}{\overset{\textit{Add } Cl^- \textit{ (e.g. conc HCl)}}{\rightleftharpoons}} [CoCl_4]^{2-} + 6H_2O$$

$$\text{Pink} \qquad\qquad\qquad \text{blue}$$
$$\text{octahedral} \qquad\qquad \text{tetrahedral}$$

> This complete substitution equation is a summary of six individual reactions where an ammonia ligand replaces a water ligand.

Examples

✓ Quick check 1

Transition metal	Hexaqua ion	Ammine ion	Tetrachloro ion
Copper (II)	$[Cu(H_2O)_6]^{2+}$ octahedral blue	$[Cu(NH_3)_4(H_2O)_2]^{2+}$ octahedral deep blue	$[CuCl_4]^{2-}$ tetrahedral yellow/green soln.
Cobalt (II)	$[Co(H_2O)_6]^{2+}$ octahedral pink	$[Co(NH_3)_6]^{2+}$ octahedral pale yellow/brown	$[CoCl_4]^{2-}$ tetrahedral blue
Cobalt(III)	–	$[Co(NH_3)_6]^{3+}$ octahedral brown	–
Chromium (III)	$[Cr(H_2O)_6]^{3+}$ octahedral violet	$[Cr(NH_3)_6]^{3+}$ octahedral purple	–

The ligand and the oxidation state of the central metal ion determines the colour of the complex (see the Co(II) and Co(III) ions with ammonia below).

Stability of complex ions

The ligand determines the stability of the complex. A complex is more stable if a bidentate or multidentate ligand replaces a monodentate due to the increase in entropy.

e.g. $\qquad\qquad [M(H_2O)_6]^{2+} + EDTA^{4-} \rightleftharpoons [M(EDTA)]^{2-} + 6H_2O$

Six water ligands are released into the solution when they are replaced by one EDTA ligand. This increases the entropy, so the forward reaction is favoured.

The structure of $EDTA^{4-}$ showing the six lone pairs.

Use of EDTA

- EDTA can be used in the treatment of poisoning to remove metal from the body. The EDTA ligand is so favoured that metal-aqua ions are kept in solution even if the anion would normally cause a ppt. This is called **sequestering.**

- The concentration of a metal ion in solution can be determined by titration with a standard solution of EDTA (reacting ratio = 1: 1).

> EDTA-metal complexes are less stable at low pH so EDTA titrations are buffered to pH ~ 10

Worked example

25.00 cm^3 of a cobalt (II) solution were titrated against 0.100 M EDTA and required 23.50 cm^3 of EDTA. Calculate the concentration of the cobalt (II) solution.

Step 1 Write down what you know

$$[Co(H_2O)_6]^{2+} + EDTA^{4-} \rightleftharpoons [Co(EDTA)]^{2-} + 6H_2O$$

No. moles of EDTA
$= cV/1000$
$= (0.100 \times 23.50)/1000$
$= 0.00235$ mol

Step 2 Work out the moles of the metal ion
1 mole $EDTA^{4-}$ $\equiv$ 1 mole Co^{2+}
$\therefore$ No. moles of Co^{2+}aq) $= 0.00235$

Step 3 Calculate the concentraion of the metal in solution
Concentration of Co^{2+}(aq) $= $ (moles $\times$ 1000)/V
$= (0.00235 \times 1000)/25$
$= 0.094$ M

? Quick check questions

1 Give an equation for a reaction in which the shape of a cobalt complex changes from tetrahedral to octahedral in a ligand substitution reaction.

2 Why is the replacement of a unidentate ligand by a bidentate ligand likely to have a positive entropy change?

Module 5 Exam style questions

These questions are largely concerned with the material in module 5. But the trouble is that a lot of module 5 builds on top of material that you should already know. This means you might need to revise the concepts of AS material to help you really understand Module 5. Try the exam style questions below and then mark them using the answers on page 102. ☺

1 a Use the data in the table to calculate a value for $\Delta G^{\ominus}$ for the reaction, at 298K: (6)

$$Cu_2S(s) \rightarrow CuS(s) + Cu(s)$$

Substance	$Cu_2S(s)$	$CuS(s)$		$Cu_2S(s)$	$CuS(s)$	$Cu(s)$
$\Delta H_f^{\ominus}/kJmol^{-1}$	-79.5	-48.5	$S/JK^{-1}mol^{-1}$	121	66.5	33.3

b What does your value for $\Delta G^{\ominus}$ tell you about the feasibility of the reaction? (2)

c Explain why it is not possible to change the temperature in order to make the reaction feasible. (2)

d Produce a Born–Haber cycle for the formation of Cu_2S and state clearly the sign of the enthalpy change for each stage in the cycle. (7) Total (17)

2 When a transition metal salt was dissolved in water the resulting solution was blue due to the **complex A**. When ammonia solution was added, drop by drop, a blue precipitate, containing **complex B**, was formed. On adding an excess of ammonia solution the precipitate dissolved to give a dark blue solution, containing **complex C**. When the original solution was acidified with dilute nitric acid and barium nitrate was added a white solid formed.

a Give equations for all of the reactions described above and identify the three complexes. (6)

b Deduce the identity of the transition metal salt. (1) Total (7)

3 a Write a half-equation showing the reduction of the manganate(VII) ion in an acidic solution. (1)

b Write half-equations showing the oxidation of (i) the iron(II) ion and (ii) the ethanedioate ion. (2)

c Use your three half-equations to produce a full balanced equation for the reaction between an solution of acidified potassium manganate(VII) and a solution of iron(II) ethanedioate. (2)

d In a titration 25.0 cm³ of an iron(II) ethanedioate solution required 29.6 cm³ of a 0.0200 molar solution of potassium manganate(VII) solution for complete reaction. Calculate the molarity of the salt solution. (4) Total (9)

If you have not been able to produce a balanced equation in c) you can assume that 2.5 mol of the iron(II) ethanedioate react with 1 mol potassium manganate(VII). This is not the correct ratio.

4 a Explain why zinc, despite being placed in the d-block of elements, cannot be considered to be a transition metal. (1)

b Define the term *homogeneous catalysis*. (1)

c   Give an equation for a reaction in which a compound of a transition metal acts as a homogeneous catalyst. (2)

d   Which typical transition metal property enables the ions of these metals to act as good homogeneous catalyst? (1)

e   Give the three stages involved in heterogeneous catalysis. (3)

f   Explain why silver and tungsten are both inefficient catalysts. (2)

g   In the catalytic converter of a car the catalyst is adsorbed onto a ceramic honeycomb. Explain the advantages of this. (2)   Total (12)

5   Direct hydration of ethane to produce ethanol is an important industrial reaction. It can be represented by the equation:

$$CH_2=CH_2(g) + H_2O(g) \rightarrow CH_3CH_2OH(g).$$

a   Use the mean bond enthalpies in the table to calculate a value for ΔH for the reaction.

Bond	C-C	C=C	C-O	C-H	O-H	
Mean bond enthalpy/kJmol^{-1}	348	612	360	412	463	(5)

b   If the enthalpies of formation of gaseous ethene, steam and ethanol are +52.3 kJmol^{-1}, −242 kJmol^{-1} and −235 kJmol^{-1} respectively calculate a second value for ΔH for the reaction. (2)

c   Compare your two ΔH values and explain what they indicate about the bond enthalpies of the bonds in the three compounds. (2)

d   Use the entropy values below to calculate the entropy change for the reaction. (2)

	Entropy/JK^{-1}mol^{-1}
$CH_2 = CH_2$ (g)	219
H_2O	189
CH_3CH_2OH	282

e   Use your answers in parts b) and d) to calculate a value for $\Delta G^\ominus$, at 298K (2)

f   Calculate the maximum temperature at which the reaction is spontaneous. (3)

g   The temperature used in the industrial process is close to 600 K. Comment on this value in the light of the value you have calculated in **f**. (1)   Total (17)

6 a   Write an equation for the reaction between iron(II) ions and dichromate(VI) ions in an acidic solution. (1)

b   Explain why it is possible to use hydrochloric acid to supply the hydrogen ions for this reaction but not when potassium manganate(VII) is used as the oxidising agent. (2)

c   3.00 g of an iron(II) salt were dissolved in 50 cm^3 of dilute sulphuric acid and the solution was then made up to exactly 250 cm^3 in a volumetric flask. 25.0 cm^3 of this solution required 23.6 cm^3 of 0.0167 molar sodium dichromate(VI) solution for complete reaction.

   Calculate the formula mass of the iron(II) salt. (6)

d   Use the value you have calculated for the formula mass to suggest which anion was present in the salt. (2)   Total (11)

Synoptic questions

These are like the type of questions you'll get in exams. They can require short answers or what they call **extended prose**, which means longer answers. The questions are not as structured as you might expect, but your answers must be.

You can use diagrams, graphs, bulleted points as well as continuous written answers, but it is advisable to plan what you are going to say, before putting pen to paper. If you're not **clear**, **coherent** and use the **language of chemistry** you're going to lose valuable marks and if the examiner can't read or understand your answers, they'll probably mark it wrong. Tough but that's the way it is.

Try these, then see what the answers say.

1 Use the redox half equations and your knowledge of reaction kinetics to explain the following.

    a   When a colourless solution containing iodide ions is added to a colourless solution containing peroxydisulphate ions the mixture slowly turns brown.(5)

    b   If a solution of iron(II) sulphate is added to the reaction mixture it turns brown more quickly. (5)

    c   When a solution of magnesium sulphate is added to the reaction is turns brown a little more slowly. (4)

    d   What would happen if iron(III) chloride was added the reaction mixture? (2)

$$Mg^{2+}(aq) + 2e^- \rightarrow Mg(s) \qquad E^{\ominus} = -2.38 \text{ V}$$
$$Fe^{3+}(aq) + e^- \rightarrow Fe^{2+}(aq) \qquad E^{\ominus} = +0.77 \text{ V}$$
$$\tfrac{1}{2} I_2(aq) + e^- \rightarrow I^-(aq) \qquad E^{\ominus} = +0.54 \text{ V}$$
$$\tfrac{1}{2} Cl_2(aq) + e^- \rightarrow Cl^-(aq) \qquad E^{\ominus} = +1.36 \text{ V}$$
$$S_2O_8^{2-} + 2e^- \rightarrow 2SO_4^{2-} \qquad E^{\ominus} = +2.01 \text{ V} \qquad\qquad \text{Total (16)}$$

2 The ability to donate a lone pair allows a species to act in three different ways. Using examples of reactions from both inorganic and organic chemistry show how ammonia is able to act in each of these three ways. In your answer state clearly the way in which ammonia is acting and, drawing on your knowledge from organic chemistry, give curly arrow mechanisms for each of the reactions you choose. (10)

3 **a**   Assume that the equations below represent two industrially important reactions.

    Reaction 1 $A(g) + 2B(g) \rightleftharpoons C(g)$ $\Delta H = -100 \text{ kJmol}^{-1}$
    Reaction 2 $R(g) + S(g) \rightleftharpoons 3T(g)$ $\Delta H = +50 \text{ kJmol}^{-1}$

    Use your chemical knowledge to predict the conditions that you would expect to be used in each process explaining clearly the reasoning that leads you to your predictions. Set your answer out clearly dealing with each process separately. (20)

    b   Discuss the factors that determine whether a chemical reaction will occur or not. In your answer explain why some reactions can only occur below a certain temperature whilst others have to be heated before the reaction starts. You may use actual or hypothetical examples to illustrate your answer, if you wish. (7)

c Sulphuric acid is an important chemical as it is able to act in a number of different ways including (i) a strong acid, (ii) a catalyst and (iii) an oxidising agent. Illustrate this by using different reactions as examples. You must use at least one organic and one inorganic reaction. Give equations wherever you think they are useful. Set your answer out in a clear and logical way assuming that the reader is a pretty intelligent A-level student. (12)   Total 39

> 39 marks is unlikely in an exam question.

4 Assume that you are an A-level student (difficult I know) and that you have been set this task as part of an A-level practical assessment:

You have been provided with five solutions in unlabelled bottles plus a labelled bottle containing sodium hydroxide solution. The solutions are known to be sodium carbonate, aluminium sulphate, potassium sulphate, calcium chloride and water. You can assume that any sodium or potassium compounds that may be produced will be freely soluble. Produce a plan that could be used to identify the five solutions using only the six bottles of reagents provided and a supply of clean test tubes. It should be possible for an A-level student to follow your plan and use the reagents and test tubes to identify each of the five unlabelled solutions. Present your plan in way that is clear and easy to follow. (14)

5 Mass spectrometry can be used to provide evidence for a) the existence of isotopes, b) determining relative atomic and molecular masses and c) the identification of compounds. Explain how a mass spectrometer works and how the data it produces can be used in each of these areas. (24)

6 When we study organic chemistry we tend to look at each homologous series separately. Spider diagrams or flow charts are good ways of showing the relationships between different homologous series. A clear example is useful for revision. The production of such a diagram, whilst not exactly the most interesting exercise, does pull together the knowledge you have gained on a number of modules you have studied.
Produce a flow chart or diagram to show the relationships between alcohols, alkenes, amines and carboxylic acids. You may need to include some other homologous series to fill gaps. Your final version should include the various homologous series and the type of reaction involved in each conversion. (15)

7 When we carry out a reaction we usually have a product in mind. Some reagents react together to give more than one product. Sometimes it is not possible to influence the outcome but there are times when it is. Explain this using three examples from organic chemistry. In two of your examples it should be possible to influence the outcome and in the other it should not. (9)

8 Volumetric analysis is a relatively straight-forward procedure that uses simple pieces of apparatus but, if done carefully, can give very accurate results. Produce a worksheet that could be used to determine the molecular mass of an iron(II) salt. You do not need to include any reference to safety but, of course, a worksheet for use in a practical session should. You can decide the best method of presentation and you should assume that the method is to be followed by another member of your group who unfortunately is not a mind reader...so give detail, be clear and be explicit. (30)

Objective questions

AQA assess your knowledge of **all** modules by using **sneaky objective questions**. Just to clear things up...these are not 'multiple guess' questions, they tend to ask **multiple choice**, **matching pairs** and **multiple completion** questions. Having said this, never leave one blank.

Now if there are 40 questions and you've 1 hour (= 1½ minute per question). Each question is ¼% of your A level as the total could account for 10% of the total A level assessment. This figure could be the difference between a good grade and a not-so-good grade. Worried ? — Try these questions below and if you're unsure read the relevant bit of the module in this book or the AS book, then mark them yourself. If you're still unsure ask Teach. Have fun ☺

Select the appropriate answer to each question

1 The first ionisation enthalpies increase from left to right across Period 3 because:

    A   There is an increase in the number of electrons in the third energy level

    B   There is a decrease in the shielding effect caused by electrons in inner energy levels

    C   There in an increase in the number of protons

    D   The atoms get larger.

2 If 1 kg of sodium hydrogencarbonate is heated to constant mass what will be the mass of anhydrous sodium carbonate formed?

    A   988 g

    B   631 g

    C   1262 g

    D   1584 g

3 For the change $CH_3CH_2OH(l) \rightarrow CH_3CH_2OH(g)$:

    A ΔH is positive and ΔS is negative

    B ΔH is negative and ΔS is negative

    C ΔH is negative and ΔS is positive

    D ΔH is positive and ΔS is positive.

4 Which change brings about the biggest increase in reaction rate for a first order reaction?

    A   A temperature increase of 50°C and a halving of reactant concentration.

    B   A temperature decrease of 50°C and a halving of reactant concentration.

    C   A temperature increase of 50°C and a doubling of reactant concentration.

    D   A temperature decrease of 50°C and a doubling of reactant concentration.

5 Which ion is the strongest reducing agent?

    A At^- C Cl^-

    B F^- D I^-

6 The typical reaction of alkenes are:

   A   Free radical substitution

   B   Elimination

   C   Nucleophilic substitution

   D   Electrophilic addition

7 If the rate equation for a reaction is; rate = $k[X][Y]^2$

   A   The rate increases by a factor of 2 if the concentration of X is halved and that of Y is doubled

   B   There is no change in rate if the concentration of Y is halved and the concentration of X is doubled

   C   The reaction equation must be X + 2Y → Products

   D   The rate will increase more if [X] is trebled whilst keeping [Y] constant than if [Y] is doubled whilst keeping [X] constant.

8 The compound $CH_3CHFCH_2CH=CHCH_3$ is

   A   2-fluorohex-4-ene

   B   5-fluorohex-2-ene

   C   2-fluorohex-2-ene

   D   2-fluorohexene

9 Which compound can exhibit geometrical isomerism?

   A $CH_3CH=CH_2$

   B $CH_3CH=C(CH_3)_2$

   C $CH_3CH(OH)CH_2CH_3$

   D $CH_3CH=CHCH_3$

10 Which of the combinations of solutions could be used to make an acidic buffer?

   A   Ammonium chloride and hydrochloric acid.

   B   Sodium hydroxide and propanoic acid.

   C   Potassium hydroxide and sodium chloride

   D   Hydrochloric acid and potassium chloride.

11 In which of the following complex ions does the metal have an oxidation state of +2?

   A $[Ag(NH_3)_2]^+$

   B $[Co(NH_3)_5Cl]^{2+}$

   C $[Fe(H_2O)_4(OH)_2]$

   D $[Fe(H_2O)_5(OH)]^{2+}$

12 For which combination of acid and base can the equivalence point in the titration be detected by methyl orange?

   A   Ethanoic acid plus ammonia

   B   Methanoic acid plus potassium hydroxide

   C   Hydrochloric acid ammonia

   D   Propanoic acid plus potassium hydroxide

13 What are the units for the equilibrium constant for the reaction

$$A(g) + 2B(g) + C(g) \rightleftharpoons 3D(g)$$

A $mol\ dm^{-3}$

B $mol^{-2}\ dm^{6}$

C $mol^{-4}\ dm^{12}$

D $mol^{-1}\ dm^{3}$

14 Which change in conditions will cause the equilibrium shown below to move furthest to the left?

$$2CO(g) + NO_2(g) \rightleftharpoons CO_2(g) + N_2(g)\ \Delta H = -601.9\ kJ\ mol^{-1}$$

A   An increase in temperature and a decrease in pressure

B   A decrease in temperature and a decrease in pressure

C   An increase in temperature and a increase in pressure

D   A decrease in temperature and a increase in pressure

15 If the formula $q = m \times c \times \Delta T$

A q will be in joules if m is in kg and T in kelvin

B q will be in kilojoules if m is in g and T in kelvin

C q will be in kilojoules if m is in kg and T in kelvin

D q will be in joules if m is in kg and T in degrees Celsius

16 If the ester $CH_3CH_2CH_2COOCH_3$ is hydrolysed the products will be

A   methanoic acid and butan-1-ol

B   methanol and butanoic acid

C   methanoic acid and butan-2-ol

D   methanol and propanoic acid

Multiple completion questions

17 Ammonia is able to act as

  (i)   an electrophile

  (ii) a nucleophile

  (iii) an electron pair acceptor

  (iv) a ligand

Multiple completion questions
Answer A if (i), (ii) and (iii) are correct
Answer B if (i) and (iii) are correct
Answer C if (ii) and (iv) are correct
Answer D if (iv) only is correct

18 Which of the following is/are correct?

  (i)   Aluminium chloride solution is acidic because the chloride ions react with the water.

  (ii) Calcium hydroxide solution has a pH of 14 because it is a strong base

  (iii) Silicon dioxide forms an acidic solution due to hydrolysis of the silicon ion.

  (iv) Iron(III) sulphate solutions are acidic due to hydrolysis of the hydrated iron(III) ions.

19 Which of the following combinations of solutions give a coloured precipitate when mixed?

  (i)   Sodium iodide and silver nitrate.

  (ii) Cobalt chloride with an excess of ammonia solution

  (iii) Iron(II) chloride and sodium hydroxide

  (iv) Copper(II) sulphate and hydrochloric acid

20 Which of the following combinations of solutions can be used to estimate the concentration of Fe^{2+} ions in solution?

  (i)   Standard sodium dichromate(VI) plus sulphuric acid.

  (ii) Standard sodium dichromate(VI) plus hydrochloric acid.

  (iii) Standard potassium manganate(VII) plus sulphuric acid.

  (iv) Standard potassium manganate(VII) plus hydrochloric acid.

> Multiple completion questions
>
> Answer A if (i), (ii) and (iii) are correct
>
> Answer B if (i) and (iii) are correct
>
> Answer C if (ii) and (iv) are correct
>
> Answer D if (iv) only is correct

21 A monohydric alcohol containing 18.2% oxygen

  (i)   Is an isomer of ethyl ethanoate

  (ii) Contains five carbon atoms

  (iii) Can be made by reacting aqueous sodium hydroxide with 2-bromobutane

  (iv) Could be a product of the reaction between pentanal and sodium tetrahydridoborate(III)

22 Members of which of the following pairs of homologous series could be isomers of each other?

  (i)   Aldehydes and ketones

  (ii) Carboxylic acids and alcohols

  (iii) Carboxylic acids and esters

  (iv) Alcohols and esters

23 A solid that dissolves in water to give a coloured solution that gives a blue precipitate when sodium hydroxide is added could be;

  (i)   Chromium(III) sulphate

  (ii) Copper(II) carbonate

  (iii) Iron(II) sulphate

  (iv) Cobalt(II) chloride

24 The electrode potential of the $Ni^{2+}(aq) + 2e^- \rightarrow Ni(s)$; $E^\ominus = -0.25$ V half-cell will become less negative if:

  (i)   the concentration of Ni^{2+} in decreased

  (ii) the pressure is decreased

  (iii) more nickel metal is added

  (iv) the concentration of Ni^{2+} in increased

25 Which of the organic compounds below causes an acidified solution of potassium dichromate(VI) to change from orange to green

  (i)   ethanal

(ii) propan-1-ol

(iii) propan-2-ol

(iv) propanone

26 Which of the compounds below have a singlet and a quartet in their n.m.r. spectra?

(i) methanol

(ii) butanone

(iii) pentan-3-one

(iv) ethanol

27 Which of the alcohols below can be oxidised but does not undergo dehydration?

(i) 2-methylbutan-2-ol

(ii) 3-methylbutan-2-ol

(iii) 3-methyl,3-bromobutan-2-ol

(iv) 2-chloromethylpropan-1-ol

28 Which property increases from left to right across Period 3

(i) Electronegativity

(ii) First ionisation enthalpy

(iii) Relative atomic mass

(iv) Atomic radius

Multiple completion questions
Answer A if (i), (ii) and (iii) are correct
Answer B if (i) and (iii) are correct
Answer C if (ii) and (iv) are correct
Answer D if (iv) only is correct

Matching pairs

For questions 29 to 32 select the appropriate answer from this list

A 2 **B** 3 **C** 5 **D** 7

29 What is the oxidation state of chlorine in ClO_3^-?

30 How many molecular ion peaks are there in the mass spectrum of 1,2-dibromoethane?

31 How many branch chained isomers are there of C_4H_9Br?

32 What is the oxidation state of cobalt in $[Co(NH_3)_4Cl_2]^+$?

For questions 33 to 36 select the appropriate answer from the list below

A *Acid **B** *Base **C** Electrophile **D** Nucleophile

*Brønsted-Lowry acid and base.

33 What is the role of the hydroxide ion when potassium hydroxide reacts with 2-bromopropane to produce propene?

34 In the reaction $R-NH_2 + H_2O \rightleftharpoons R-NH_3^+ + OH^-$ what is the role of the water?

35 What name is given to a species that donates a lone pair of electrons to an electron deficient carbon atom?

36 When an acyl chloride reacts with an alcohol what is the role of the acyl chloride.

For questions 37 to 40 select the appropriate answer from the list below

A Benzene **B** Calcium **C** Graphite **D** Sodium chloride

37 Which does not have delocalised electrons in the solid state?

38 Which reacts with cold water to give a solution with a pH greater than seven?

39 Which is used in the extraction of aluminium?

40 Which is an electrical insulator when solid but a conductor when molten?

Answers

Answers to quick check questions

Module 4: Further physical and organic chemistry

These answers are here to give you an idea of the level that is required at A2. They have not been through the rigorous scrutiny that exam mark schemes go through, as they are not produced by AQA, so may be different from what the board produces. Also, AQA will assess the quality of **written communication**, where extended written material is required. This is not covered here in these answers. But these answers should give you some idea that you are heading in the right direction.

Page 3

1 mol dm^{-3}

2 rate constant or velocity constant

3 x for A and y for B

4 x + y

5 increases × 8

Page 5

1 **a** rate doubles; **b** rate increases × 27; **c** rate increases × 81; **d** mol^{-3}dm^{9}s^{-1}

2 doubling [W]

3 rate increases

4 k

Page 7

1 $[I_3^-]/[I_2][I^-]$ mol^{-1}dm^3

2 L to R

3 Move to R to remove OH$^-$

4 No change in K_c when NaOH added. K_c increases when temp. decreases.

Page 9

1 $[HI]^2/[H_2][I_2]$

2 0.0228 mol H_2 and I_2, 0 mol HI

3 0.00515 mol H_2 and I_2, 0.0353 mol HI

4 0.002575 mol dm^{-3} H_2 and I_2, 0.01765 mol dm^{-3} HI

5 46.98 i.e. 47.0 no units

Page 11

1 **a** 0.167 **b** 0.5

2 25 kPa

3 $K_p = \dfrac{(p\text{PCl}_3)(p\text{Cl}_2)}{(p\text{PCl}_5)}$ kPa

4 Temperature change

Page 13

1 Proton acceptor; pH = $-\log_{10}[H^+]$

2 Strong = fully ionised; weak = only partly ionised

Page 15

1 **a** $- 3.0 \times 10^{-1}$; **b** 0.0

2 **a** 1.0×10^{-2}; **b** 1×10^{-9}; **c** 1×10^{-7} (all mol dm^{-3})

3 Ionic product of water; 1.0×10^{-14} mol^{2}dm^{-6}

4 7.11 at 18°C, 6.77 at 40°C

Page 17

1 2.00

2 12.3

3 11.7

4 5.37×10^{-9} mol dm^{-3}

Page 19

1 Similarities: both rise sharply through the equivalence point to about pH 11 and then rise steadily to about 13.

 Differences: HCl curve starts at pH 1 and stays low until near equivalence point. CH_3COOH starts at pH 2.88, rises, then levels off before rising again close to equivalence point.

2 Weak acid + weak base.

3 Ethanoic acid has two protons so two equivalence points.

Page 22

1 $pK_a = -\log_{10}K_a$

2 Indicator = usually a weak organic acid which has different colours for dissociated and undissociated forms. Changes colour over narrow pH range. Buffer = maintains an almost constant pH when small amounts of acid or base are added or when diluted.

3 Weak base + salt of that base.

4 When an acid is added the H$^+$ reacts with A$^-$ to form HA – when a base is added OH$^-$ reacts with H$^+$

causing more HA to dissociate. In either case [H$^+$] changes little. When diluted the ratio [HA]/[A$^-$] is not changed and neither is K_a so pH does not change.

5 1.42

6 a 5.42 **b** 5.40

Page 25

1 a (CH$_3$)$_2$C=CH$_2$ **b** (CH$_3$)$_2$C(OH)CH$_2$CH$_3$ **c** CH$_3$COCH(CH$_2$CH$_3$)CH$_2$CH$_3$ **d** HCOOCH$_2$CH$_2$CH$_2$CH$_3$

2 a pent–2-ene **b** 3-methylbutan-2-ol **c** 2-bromo, 3-chlorohexane **d** propylpropanoate

3 Compounds with the same molecular and structural formulae but differing arrangement of bonds.

4 A carbon atom bonded to four different groups.

5 One with equal amounts of each of a pair of optical isomers.

Page 27

1 Warm with (i) ammoniacal AgNO$_3$(aq) (Tollen's) or (ii) Fehling's soln. Look for silver mirror or red, solid/precipitate.

2 Primary

3 Add alcohol, drop-wise to oxidising agent at a temperature between the boiling points of the alcohol and the aldehyde.

4 H$_2$ + Ni-catalyst or NaBH$_4$

5 Nucleophilic addition

6 HCN is highly toxic. KCl + KCN it gives a high [CN$^-$]

Page 29

1 Add NaHCO$_3$ and look for colourless gas that gives a white ppt. with Ca(OH)$_2$(aq)

2 Reflux with warm oxidising agent e.g. Cr$_2$O$_7^{2-}$(aq)/H$^+$

3 CH$_2$CH$_2$CH$_2$CH$_2$OH + 2[O] $\rightarrow$ CH$_2$CH$_2$CH$_2$COOH + H$_2$O

4 Hydrolysis of a nitrile

5 CH$_2$CH$_2$CH$_2$COOH + CH$_3$OH $\rightarrow$ CH$_2$CH$_2$CH$_2$COOCH$_3$ + H$_2$O

6 Methyl butanoate + water; conc. H$_2$SO$_4$

Page 31

1 Flavourings/solvents/plasticisers

2 Saponification; soap production

3 Propane,1,2,3-triol (or glycerol)

4 Butylpropanoate, butanol, sodium propanoate.

5 K_a close to one but NaOH reacts with carboxylic acid pulling equilibrium to RHS

6 Methyl methanoate

Page 33

1 C$_n$H$_{(2n+1)}$CO or

(NB. Z is not part of the acyl group)

2 Ethanoic acid

3 Ammonia

4 CH$_3$COCl + CH$_3$OH $\rightarrow$ CH$_3$COOCH$_3$ + HCl methyl ethanoate; hydrogen chloride

5 Cheaper; easier to control; less corrosive ('Easier to say' is not really a good reason!)

Page 35

1

2 Benzene's bonds are neither single nor double bonds but something in between. (C — C delocalised e$^-$.)

3 C$_6$H$_6$ does not have 3 $\times$ C=C bonds. Delocalised electrons are in a large, stable orbital

4 Electrophilic substitution

Page 37

1 2H$_2$SO$_4$ + HNO$_3$ $\rightarrow$ 2HSO$_4^-$ + H$_3$O$^+$ + NO$_2^+$

2

3 Aluminium/iron(III) halide e.g. AlCl$_3$

4 To polarise the electrophile so that it can draw electrons out of the π ring.

Page 39

1 The lone pair on the N atom

2 CH$_3$CH$_2$Br + CH$_3$CH$_2$NH$_2$ $\rightarrow$ CH$_3$CH$_2$NH CH$_2$CH$_3$ + HBr

3 Further substitution of H in amine

4 Addition-elimination

5 Haloalkane + ammonia; reduction of nitrile

6 Reduction of nitrile as it gives only one organic product.

Page 41

1 R–CH(NH$_2$)–COOH

2 CH$_3$–CH(NH$_2$)–COOH $\rightarrow$ R–CH(NH$_3^+$)–COO$^-$

3 Proteins

4 Amide or peptide

5 hydrogen bonds, ionic bonds, disulphide bridges

6 hydrolysis

Page 43

1 A C=C bond

2 Non-polar and saturated, non-biodegradable

3 $-[CO-CH_2-COOCH_2-CH_2-O]_n-$

4 Hexanedioic acid + 1,6–diaminohexane

5 It is hydrolysed to its monomers

Page 45

1 Free radicals, electrophiles, nucleophiles

2 $\bullet Cl$,$\bullet CH_3$ etc; HBr, HCl NO_2^+ etc; $:OH^-$, $:NH_3$ $:CN^-$ etc

3 Free radical substitution, electrophilic addition, electrophilic substitution, nucleophilic substitution, elimination, addition–elimination

4 a Free radical substitution b electrophilic addition c addition– elimination

5 Haloalkane – nitrile – acid

6 $C_6H_6 + CH_3COCl$ or $(CH_3CO)_2O$ with $AlCl_3$ – $CH_3COC_6H_5$ – reduce with H_2/Ni or $NaBH_4$

Page 47

1 Peak with highest m/z ignoring satellite peak

2 $M^{+\bullet} \rightarrow \bullet X + Y^+$

3 $m/z = 15$ $-CH_3^+$ / $29 = CH_3CH_2^+$ / $43 = CH_3CO^+$ / $57 = CH_3CH_2CO^+$

Page 49

1 Functional groups

2 a But-2-ene has peak at 1300 cm^{-1} (C=C) no peak at 1680–1750 cm^{-1}; butanone has no peak at 1100–1300 cm^{-1}(C=C) but has a strong peak at 1680–1750 cm^{-1}(C=O). b Ethanol has a peak at 3350 cm^{-1} but not 1680–1750 cm^{-1}; ethanoic acid has a broad peak from 2800–3350 cm^{-1} plus 1680–1750 cm^{-1}.

3 Compare its spectrum with propanone's and look for an exact match.

4 Water vapour, in your breath, has O–H bonds

Page 51

1 Tetramethyl silane Si(CH$_3$)$_4$

2 Large single peak to right of most others, volatile and inert so easily removed.

3 No. of non-equivalent protons, environment, relative number of protons producing peak and number of non-equivalent H neighbours.

Page 53

1 Number of different proton environments

2 If a group of protons have n non-equivalent neighbours their absorbance will split into n + 1 peaks.

3 So that there are no protons/H atoms in the solvent. If present their signals would swamp out those of the sample.

Module 5 Thermodynamics and further inorganic chemistry

Page 57

1 The heat change when one mole of compound is formed from its elements in their standard states.

2 $3C(s) + 3H_2(g) + \frac{1}{2}O_2(g) \rightarrow CH_3CH_2CHO(l)$

3 The heat change when one mole of gaseous ions dissolve in a large volume of water, to form aqueous ions.

4 $Ca^{2+}(g) + 2Cl^-(g) \rightarrow Ca^{2+}(aq) + 2Cl^-(aq)$

Page 59

1 An energy cycle containing all of the stages involved in the formation of an ionic compound from its elements.

2 Lattice enthalpy/enthalpy of hydration of potassium ions (g); enthalpy of hydration of bromide ions (g).

3 The average amount of heat energy needed to break a covalent bond in the gaseous state, taken over a range of compounds.

4 Mean bond enthalpies are average values and are not specific. Actual bond enthalpies depend on their environment. Energy cycles use practical values that are specific to the substances involved.

Page 61

1 $\Delta S = \Sigma S_{products} - \Sigma S_{reactants}$

2 $\Delta G = \Delta H - T\Delta S$

3 $\Delta G \leqslant 0$

4 Any balanced equation showing an increase in the number of moles of gas.

5 Both +ve

6 No, ΔG is +ve (+1.13 kJmol^{-1})

Page 63

1

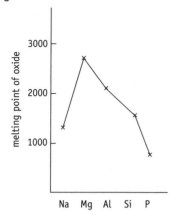

2 Na_2O, MgO ionic bond strength increase as M^{n+} gets smaller; ($\propto n$) Al_2O_3 structure differs from Na_2O and MgO so it has a lower m.p. SiO_2 macromolecular so high as many covalent bonds have to be broken; P_2O_5 molecular so only weak van der Waals forces to be overcome, so low boiling points.

3 $NaCl$, $MgCl_2$, are both ionic $AlCl_3$ or Al_2Cl_6, $SiCl_4$, PCl_5 are covalent.

4 $2Na + 2H_2O \rightarrow 2NaOH + H_2$

5 Mg very slow with water but rapid when hot (i.e. steam) and gives $Mg(s) + H_2O(g) \rightarrow MgO(s) + H_2(g)$ Na reacts vigorously with cold water making sodium hydroxide.

Page 66

1 Hydration involves the addition of water molecules intact but in hydrolysis the water molecules are split.

2 a $AlCl_3 + 6H_2O \rightarrow [Al(H_2O)_6]^{3+} + 3Cl^-$; **b** $SiCl_4 + 4H_2O \rightarrow [Si(OH)_4] + 4HCl$; **c** $PCl_5 + 4H_2O \rightarrow H_3PO_4 + 5HCl$

3 a 7; **b** 3/4; **c** 0

4 Ionic chlorides give neutral solutions. Covalent chlorides give acidic solutions.

5

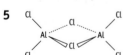

6 It reacts with acids and bases.

Page 69

1 Electron gain

2 a +5; **b** +5; **c** +4; **d** +3; **e** +3; **f** +6

3 $2S_2O_3^{2-} \rightarrow S_4O_6^{2-} + 2e^-$

Page 71

1 Hydrogen 0.00 V

2 Easier to set up and use / safer.

3 Electrode potentials are dependent on temperature, concentration and possibly pressure. Standards are 298K, 1 M solutions of ions, 100 kPa.

4 A list of redox half-equation arranged in order of their standard electrode potentials.

5 Iron(III) reduced to iron(II) and iodide oxidised to iodine.

$Fe^{3+}(aq) + I^-(aq) \rightarrow Fe^{2+}(aq) + \frac{1}{2}I_2(aq)$

Page 73

1 A metal atom or ion surrounded by a number of ligands.

2 A species that donates a lone pair of electrons to a metal atom or ion.

3 The number of lone pairs accepted or number of co-ordinate bonds.

4 Co-ordinate, the ligand provides both electrons to form the bond.

Page 75

1

$$\left[\begin{array}{c} H_2O \\ H_3N \cdots \underset{\underset{H_2O}{\overset{H_3N}{|}}}{\overset{|}{Cu}} \cdots NH_3 \\ NH_3 \end{array}\right]^{2+}$$

2 Change of ligand, co-ordination number or oxidation state of metal.

3 $[Co(H_2O)_6]^{2+} + 4Cl^- \rightleftharpoons [CoCl_4]^{2-} + 6H_2O$

Page 77

1 +2

2 Increase L to R

3 React with zinc + acid in a lightly stoppered flask

4 Manganate(VII) and dichromate(VI)

5 $5Fe^{2+} + MnO_4^- + 8H^+ \rightarrow 5Fe^{3+} + Mn^{2+} + 4H_2O$

6 $6Fe^{2+} + Cr_2O_7^{2-} + 14H^+ \rightarrow 6Fe^{3+} + 2Cr^{3+} + 7H_2O$

6 Cu(II) has an incomplete 3d sub-shell but in Cu(I) 3d is full.

Page 79

1 Colourless to pink

2 Sulphuric acid

3 It is a gradual change so there is no recognisable endpoint.

Page 81

1 The catalyst is in the same phase as the reactants + any suitable example

2 Adsorption, reaction, desorption

3 Adsorption is too strong so products are not released

4 Ethene + oxygen → epoxyethane

Page 83

1 A Lewis acid — accepts a pair of electrons e.g. the transition metal. A Lewis base — donates a lone pair of electrons E.g. ligand

2 $[Fe(H_2O)_6]^{2+}$ $[Cu(H_2O)_6]^{2+}$ $[Co(H_2O)_6]^{2+}$ $[Fe(H_2O)_6]^{3+}$ $[Cr(H_2O)_6]^{3+}$ $[Al(H_2O)_6]^{3+}$ $[V(H_2O)_6]^{3+}$

3 Hydrolysis, ligand exchange or substitution, redox

4 M^{3+} ions are smaller and more highly charged so have much higher surface charge densities. They polarise

the O–H bonds in water ligands much more so they break more easily resulting in more acidic properties.

Page 86

1 Chromium, cobalt and copper(II)

2 $2[M(H_2O)_6]^{3+} + 3CO_3^{2-} \rightarrow 2[M(H_2O)_3(OH)_3] + 3H_2O + 3CO_2$

Page 88

1 $[CoCl_4]^{2-} + 6H_2O \rightarrow [Co(H_2O)_6]^{2+} + 4Cl^-$

2 There is an increase in the number of species so there is likely to be more disorder.

Answers to end of module exam style questions

(Tick shows mark allocation.)

Module 4: Further physical and organic chemistry

Pages 54 and 55

1 a Solution buffered so $[H^+]$ constant ✓; **b** Both reactants negatively charged ✓ – repel each other ✓; **c** 1st ✓ – In exp 1 & 2 $[MnO_4^-]$ increased × 2.5 and so did rate ✓; **d** 1st ✓ in exp 2 & 3 both concentrations increased x 3 and rate x 9 so 2nd order overall ✓: 1st for $[MnO_4^-]$ so must be 1st for $[C_2O_4^{2-}]$ ✓; **e** rate = $k[MnO_4^-][C_2O_4^{2-}][H^+]$ ✓; **f** 3rd ✓; **g** rate = $k[Fe^{2+}]$ ✓ Zero order for MnO_4^- & $C_2O_4^{2-}$ as is exp 1 & 3 both concentrations are increased, whilst keeping $[Fe^{2+}]$ constant, but rate unchanged ✓. 1st order for Fe^{2+} as $[Fe^{2+}]$ doubled in exp 1 & 2 and rate also doubled – changes to $[MnO_4^-]$ & $[C_2O_4^{2-}]$ not relevant as order for both is zero ✓; **h** catalyst ✓ – affects rate but not a reactant ✓.

2 a 1 mol O_2 reacts with 2 mol SO_3 to give 2 mol SO_3 ✓ / for equal mol O_2 and SO_3 ✓ 1 mol O_2 must have reacted to give 2 mol SO_3 / leaving 2 mol O_2 and 1 mol SO_2 ✓ / total of 5 mol ✓/mole fraction = n/n_{total} ✓ so for SO_2 = 1/5 = 0.2 ✓; for O_2 and SO_3 = 2/5 = 0.4 ✓; **b** partial pressure = mole fraction × total pressure ✓; **c** pSO_2 = 0.2 × 1000 = 200 kPa ✓, pO_2 = 0.4 × 1000 = 400 kPa ✓; **d** $K_p = (pSO_3)^2/(pSO_2)^2(pO_2)$ ✓ **e** K_p = 0.01 ✓ kPa^{-1} units. ✓/correct substitution of values ✓.

3 a

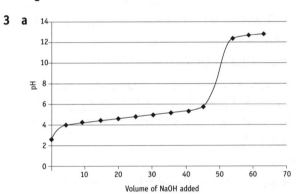

Scale and label ✓✓, plot ✓, line ✓.

b 50.0 cm^3 ✓; **c** vol NaOH × 2 vol acid ✓ ∴ acid concn 1.5M ✓; **d** phenolphthalein ✓; **e** Sudden pH rise from just over 7 to 11 ✓ covers range in which phenolphthalein changes colour ✓; **f** and **g** (in either order) At the start pH = 2.35 / can use $K_a = [H^+]^2/[HA]$ ✓ $[H^+]$ ✓ 4.47 × 10^{-3} $[HA]$ = 1.5 ✓ K_a = 1.33 × 10^{-5} mol dm^{-3} ✓ or pH when 25.00 cm^3 ✓. NaOH added, when acid is ½ neutralised ✓ so pH = pK_a ✓ = 4.87, K_a = 1.35 × 10^{-5} ✓ mol dm^{-3} ✓. Answers are slightly different as only 3 sig. fig.

h

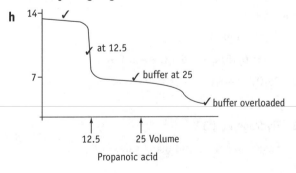

Propanoic acid

Module 5 Thermodynamics and further inorganic chemistry

Page 89 and 90

1 a $\Delta H^{\ominus} = \Sigma \, \Delta H^{\ominus}_{f\,products} - \Sigma \, \Delta H^{\ominus}_{f\,reactants}$ ✓

$\Delta S^{\ominus} = \Sigma \, S^{\ominus}_{products} - \Sigma \, S^{\ominus}_{reactants}$ ✓

$\Delta G^{\ominus} = \Delta H^{\ominus} - T \, \Delta S^{\ominus}$ ✓

$\Delta H = +31.0 \text{ kJmol}^{-1}$ ✓; $\Delta S = -21.2 \text{ JK}^{-1}\text{mol}^{-1}$ ✓;
∴ $\Delta G = +37.3 \text{kJmol}^{-1}$ ✓

b not feasible ✓ as ΔG is +ve ✓;

c ΔH is +ve and ΔS is –ve ✓ so ΔG is always +ve ✓;

d (Marks for correct step and sign.)

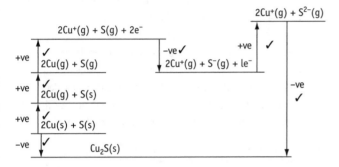

2 $[Cu(H_2O)_6]^{2+}$ **(A)** ✓ $+ 2OH^- \rightarrow [Cu(H_2O)_4(OH)_2]$ **(B)** ✓ $+ 2H_2O$ ✓

$[Cu(H_2O)_6]^{2+}$**(A)** $+ 4NH_3 \rightarrow [Cu(NH_3)_4(H_2O)_2]$**(C)** ✓ $+ 4H_2O$ ✓

$Ba^{2+}(aq) + SO_4^{2-}(aq) \rightarrow BaSO_4(s)$ ✓

b $CuSO_4$ ✓

3 a $MnO_4^- + 8H^+ + 5e^- \rightarrow Mn^{2+} + 4H_2O$ ✓;

b (i) $Fe^{2+} \rightarrow Fe^{3+} + e^-$ ✓ (ii) $C_2O_4^{2-} \rightarrow 2CO_2 + 2e^-$ ✓;

c $5Fe^{2+} + 5C_2O_4^{2-} + 3MnO_4^- + 24H^+ \rightarrow 5Fe^{3+} + 10CO_2 + 3Mn^{2+} + 12H_2O$ ✓5:3 ✓ equation correct

d No. moles of MnO_4^- ✓ 5:3 ✓
No. moles of $Fe^{3+}/Cr_2O_4^{2-}$ ✓
molarity ✓
0.0395 M or (0.0474 M)

4 a It has a full 3d sub-shell ✓; **b** The catalyst is in the same phase as the reactants ✓; **c** any suitable example ✓ ✓; **d** variable oxidation states ✓; **e** adsorption ✓ / reaction ✓ / desorption ✓; **f** Ag adsorbs too weakly ✓, W adsorbs too strongly ✓; **g** Large surface area ✓ of catalyst so more available to reactants ✓.

5 a 3186 ✓ ✓ –3231 ✓ ✓ = –45 kJmol⁻¹ ✓; **b** for knowing how to get value ✓ –45.3 kJmol⁻¹✓; **c** mean bond ✓ enthalpy very similar to actual bond enthalpies ✓; **d** $\Delta S = \Sigma S_{product} - \Sigma S_{reactant}$✓; –126 JK⁻¹mol⁻¹ ✓; **e** $\Delta G = \Delta H - T\Delta S$ ✓ –7.45 kJmol⁻¹✓; **f** 357K ✓ calculation ($\Delta G = 0$ so $T = \Delta H/\Delta S$) ✓ ✓; **g** the conditions used are not standard ✓.

6 a $6Fe^{2+} + Cr_2O_7^{2-} + 14H^+ \rightarrow 6Fe^{3+} + 2Cr^{3+} + 7H_2O$ ✓

b Dichromate(VI) cannot ✓ oxidise Cl^- to ½Cl_2 but manganate(VII) can ✓; **c** 127 ✓ calculation 5x ✓; **d** likely to be $FeCl_2$ ✓ so Cl^- is the likely anion ✓.

Module 6 Synoptic questions

Page 91 and 92

1 a $S_2O_8^{2-} + 2I^- \rightarrow 2SO_4^{2-}$ ✓ $S_2O_8^{2-} + 2e^- \rightarrow 2SO_4^-$ ✓ $2I^- \rightarrow I_2 + 2e^-$ ✓; slow as both ✓ ions are negatively charged so they repel ✓ each other.

b Fe^{2+} is oxidised to Fe^{3+} ✓ by $S_2O_8^{2-}$ and the Fe^{3+} ✓ is reduced back to Fe^{2+} by I^- so it acts as a catalyst ✓. This is possible as the redox potential for Fe^{3+}/Fe^{2+} is below that for peroxdisulphate ✓, but higher that the value for iodine/iodide ✓.

c Mg^{2+} cannot be reduced by I^- ✓ as its $E^{\ominus}$ is higher (more negative) ✓, so it cannot catalyse the reaction. Reaction slower as $[S_2O_8^{2-}]$ and $[I^-]$ decrease due to dilution ✓.

d The Fe^{3+} would catalyse the reaction ✓ as before but there would also be some oxidation of Cl^- if the I^- runs out before $S_2O_8^{2-}$ ✓.

2 Base ✓ lone pair donated to H^+ ✓ plus curly arrow ✓ to show this; nucleophile ✓ lone pair donated to electron deficient C atom ✓, e.g. $NH_3 + CH_3CH_2Br$ ✓ plus mechanism ✓; ligand ✓ lone pair donated to metal ion, e.g. $Co^{2+} + NH_3 \, [Co(NH_3)_6]^{2+}$ ✓ plus mechanism showing the donation of one lone pair ✓.

3 a Reaction 1 is exothermic ✓ so a lower T gives higher yield ✓ but a slower rate ✓.
∴Compromise needed to get acceptable yield and rate ✓. 3 mol gas react to give 1 ✓. A decrease in gaseous moles is favoured by a high pressure ✓ so the pressure used is likely to be high ✓ but it is more costly to build and run a plant at higher pressure ✓. A point is reached when increase in cost is not justified by increased yield ✓. A catalyst is likely to be used to increase rate without adversely affecting yield ✓. Reaction 2 is endothermic ✓ so yield increases as T increases ✓ so yield and rate can be increased by increasing the T ✓. Cost sets a

limit on T used but it is likely to be quite high ✓. A catalyst is still likely to be used to improve rate as high T = high fuel cost ✓. 2 mol of gas react to give 3 ✓. An increase in moles is favoured by lower pressure ✓ so pressure is likely to be low ✓. This will decrease rate ✓ and it will be costly to run at pressures much below atmospheric so the pressure is likely to be close to atmospheric ✓, i.e. a little above to pump the gases through the plant.

b For a reaction to be energetically possible it must have a negative free energy change ✓. To occur, ΔG must be negative and it must be possible to overcome the activation energy ✓. $\Delta G = \Delta H - T\Delta S$ If ΔH and ΔS are negative ΔG will be negative as long as $\Delta H > T\Delta S$ ✓. This will be so when T is small ✓. If ΔH and ΔS are positive ΔG ✓ will be negative as long as $\Delta H < T\Delta S$. This will be when T is high.

c Possible answer (many possibilities)

   i strong acid – fully ionised –solution should be dilute ✓

   $H_2SO_4(l) + H_2O(l) \rightarrow HSO_4^-(aq) + H_3O^+(aq)$ ✓

   pH of 1.00 M solution ✓ is below zero ✓ showing that all of the acid has ionised (and some of the HSO_4^- has)

   ii A catalyst

   conc. H_2SO_4 ✓ in formation of ester/dehydration of alcohol ✓

   $CH_3CH_2OH \rightarrow CH_2=CH_2 + H_2O$ ✓

   Acid protonates the OH of alcohol, ✓ followed by loss of H_2O. H^+ is then regenerated.

   iii An oxidising agent. Hot and conc. ✓

   e.g. $2NaBr + 2H_2SO_4 \rightarrow Na_2SO_4 + 2H_2O + SO_2 + Br_2$ ✓

   (solid NaBr) ✓$Br^- \rightarrow 1/2\ Br_2$ ✓

4 Label each solution A, B, C, D or E for ease ✓.

Add sodium hydroxide, drop by drop ✓, to each until there is no further change ✓. Three give no reaction: water, sodium carbonate and potassium sulphate; two give white ppt ✓: calcium chloride and aluminium sulphate ✓. One ppt. redissolves in excess aluminium sulphate ✓, so **aluminium sulphate and calcium chloride** ✓ have been identified. Add aluminium sulphate solution to a fresh sample of the three that did not react with sodium hydroxide. Two do not ✓ react but one, sodium carbonate, gives a white ppt ✓. plus a colourless gas ✓, so **sodium carbonate** has now been identified. If calcium chloride ✓ is added to the water and potassium sulphate that remain to be identified there is no

reaction with water ✓ but a white ppt. will form ✓ with **potassium sulphate**. The remaining solution must be **water** ✓.

5 How it works: Stages... vaporisation ✓, ionisation ✓ by electron gun ✓ to remove electrons ✓, acceleration ✓ by electrical field ✓, deflection ✓ by magnetic field ✓ and detection by charged plate ✓. Signal used to produce spectrum ✓. Angle of deflection depends on charge:mass ratio ✓ and magnetic field strength ✓. Field strength increased steadily to focus ions onto detector ✓ in order of increasing mass:charge ratio ✓. **a** If an element is used, the number of peaks = number of isotopes ✓ and relative heights give relative abundance ✓ so A_r can be calculated ✓. **b** If molecules are used the M_r ✓ is given by the peak with the highest m/z ratio ✓ – ignoring the satellite peak. **c** High resolution can distinguish between compounds with same integral M_r ✓ then compound can be identified ✓, peaks produced by fragments ✓and their relative heights ✓ may make it possible to determine structure ✓.

6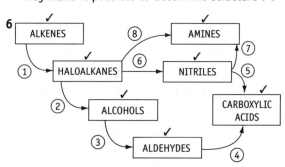

1. electrophilic addition ✓
2. nucleophilic substitution ✓
3. oxidation ✓
4. oxidation ✓
5. hydrolysis ✓
6. nucleophilic substitution ✓
7. reduction ✓
8. nucleophilic substitution ✓

7 Free radical $CH_4 + Cl_2$ ✓. Large excess of CH_4 ✓ leads to CH_3Cl being the major product. Explained by competition ✓: Electrophilic addition to an alkene ✓. The relative amounts of products determined by the relative stabilities of the carbocation intermediates ✓. This cannot be changed ✓ by altering the conditions. Substitution vs elimination ✓ in reaction between KOH and haloalkane. Warm, dilute aqueous KOH favours substitution ✓ whilst hot, concentrated, alcoholic KOH ✓ favours elimination. Or any other suitable examples.

8 Marks given for correct method, correct use of data collected, calculation. e.g.

1 Weigh accurately ✓ (i.e. to 0.001 g) 5–10g of salt (m) ✓

2 Dissolve in ~ 50 cm^3, dilute H_2SO_4 ✓

3 Add to a 250 cm^3 volumetric flask ✓

4 Add washings to flask ✓

5 Make up to exactly 250 cm^3 ✓ with water ✓

6 Rinse ppt with Fe(II) solution and discard washings ✓

7 Pipette 25.00 cm^3 into a clean conical flask ✓

8 Add ~15–20 cm^3 dil H_2SO_4 ✓

9 Rinse burette with 0.02M $KMnO_4$, ✓ discard washings ✓ then fill burette ✓ with $KMnO_4$ and zero/note the volume at start ✓

10 Add $KMnO_4$ (aq) to Fe(II) ✓ in flask with constant swirling ✓ until the solution first become pink ✓

11 Rinse sides of flask with water ✓ and if still pink ✓ note volume on burette ✓

12 Work out volume of $KMnO_4$ (aq) added ✓

13 Repeat to obtain concordant results ✓ i.e. so they are within 0.10 cm^3 ✓ of each other when the $KMnO_4$ (aq) is added drop by drop ✓ near the end point ✓

14 Use the average ✓ of the concordant results ✓ to calculate the number of moles of $KMnO_4$ added per titration ✓

15 Multiply this figure by 5 to get mole of Fe(II) in 25.00 cm^3

16 x 10 for moles of Fe(II) in 250.0 cm^3 ✓

17 Use $M_r = M/n$ to find a value for M_r ✓

[method may change slightly if potassium or sodium dichromate used]

Page 93

Module 6 Objective questions

1 C	11 C	21 C	31 A
2 B	12 C	22 B	32 B
3 D	13 D	23 D	33 B
4 C	14 A	24 D	34 A
5 A	15 C	25 A	35 D
6 D	16 B	26 C	36 C
7 A	17 C	27 D	37 D
8 B	18 D	28 A	38 B
9 D	19 B	29 C	39 C
10 B	20 A	30 B	40 D

Index

Camille Pissarro
Art Gallery of New South Wales
19 November 2005–19 February 2006
National Gallery of Victoria
4 March–28 May 2006

Curator and editor: Terence Maloon
Copy editor: Anna Macdonald
Catalogue compilation: Erica Drew
Catalogue design: Analiese Cairis
Curatorial assistance: Peter Raissis
Essayists: Terence Maloon, Richard Shiff, Joachim Pissarro,
Peter Raissis, Claire Durand-Ruel Snollaerts
Production: Cara Hickman
Catalogue transparencies: Edwina Brennan
Prepress: Spitting Image, Sydney
Printing: Australian Book Connection

Published by Art Gallery of New South Wales
Art Gallery Road, The Domain, NSW 2000, Australia
www.artgallery.nsw.gov.au

© 2005 Art Gallery of New South Wales

Cover
Camille Pissarro
cat. 22 *A cowherd on the route du Chou, Pontoise*, 1874
(detail), oil on canvas, 54.9 x 92.1, The Metropolitan
Museum of Art, gift of Edna H. Sachs, 1956

Cataloguing-in-publication data
Maloon, Terence.
Camille Pissarro / curator Terence Maloon; curatorial
assistance Peter Raissis; with additional essays by Richard
Shiff, Joachim Pissarro, Claire Durand-Ruel Snollaerts.

Includes bibliographical references.
ISBN 0 7347 6379 4

1. Pissarro, Camille, 1830–1903 – Exhibitions. I. Raissis,
Peter. II. Shiff, Richard. III. Pissarro, Joachim. IV.
Snollaerts, Claire Durand-Ruel. V. Art Gallery of New
South Wales. VI. National Gallery of Victoria. VII. Title.

Distribution
Art Gallery of New South Wales
Art Gallery Road, The Domain, NSW 2000 Australia
Yale University Press
32 Temple Street, New Haven, CT 06520, USA
Thames & Hudson
181A High Holborn, London WC1V 7QX, England

• All measurements are given in centimetres (height by
  width), unless otherwise stated.
• Titles of works are those given by the lenders, the most
  recent catalogue raisonné, or English translations thereof.

Abbreviations and published references

C   Jean Cailac, "The prints of Camille Pissarro.
     A supplement to the catalogue by Loys Delteil",
     The Print Collector's Quarterly, 19, 1932, pp. 75–85.

D   Loys Delteil, *Le peintre-graveur illustré:
     Pissarro, Sisley, Renoir*, Paris, 1923.

LM   Jean Leymarie and Michel Melot, *The Graphic
     Work of the Impressionists*, London, 1972.

PDRS   Joachim Pissarro and Claire Durand-Ruel Snollaerts,
     Pissarro: Critical Catalogue of Paintings, catalogue
     raisonné, Skira/Wildenstein Institute, Paris, 2005.

PV   Ludovic-Rodo Pissarro and Lionello Venturi,
     C. Pissarro, Paris, 1939.

pp. 2–3: cat. 13 *Effect of fog at Creil*, 1873 (detail), private collection, Switzerland

Camille Pissarro

Art Gallery of New South Wales
19 November 2005–19 February 2006

National Gallery of Victoria
4 March–28 May 2006

Lenders to the exhibition

Australia
Art Gallery of New South Wales, Sydney
National Gallery of Victoria, Melbourne
Queensland Art Gallery, Brisbane

Canada
Art Gallery of Ontario, Toronto
National Gallery of Canada, Ottawa

France
Bibliothèque Nationale de France, Paris
Musée de la Chartreuse, Douai
Musée des Beaux-Arts de Valenciennes
Musée d'Orsay, Paris
Musées de Pontoise:
    Musée Camille Pissarro, Musée Tavet-Delacour

Germany
Hamburger Kunsthalle, Hamburg

Switzerland
Kunsthaus Zürich
Kunstmuseum Bern
Kunstmuseum Sankt Gallen

United Kingdom
The British Museum, London
Fitzwilliam Museum, Cambridge
National Museums & Galleries of Wales, Cardiff
TATE, London
The National Gallery, London
The National Gallery of Scotland, Edinburgh
The Whitworth Art Gallery,
    The University of Manchester

United States of America
The Art Institute of Chicago, Illinois
The Baltimore Museum of Art, Maryland
Cincinnati Art Museum, Ohio
Columbus Museum of Art, Ohio
Corcoran Gallery of Art, Washington, DC
Dallas Museum of Art, Texas
Fine Arts Museums of San Francisco, California
Honolulu Academy of Arts, Hawaii
Indianapolis Museum of Art, Indiana
The J. Paul Getty Museum, Los Angeles, California
Kimbell Art Museum, Fort Worth, Texas
Los Angeles County Museum of Art, California
Memphis Brooks Museum of Art, Tennessee
The Metropolitan Museum of Art, New York
Museum of Fine Arts, Boston, Massachusetts
Museum of Fine Arts, Springfield, Massachusetts
National Gallery of Art, Washington, DC
Philadelphia Museum of Art, Pennsylvania
Princeton University Art Museum, New Jersey
Sterling and Francine Clark Art Institute,
    Williamstown, Massachusetts
Toledo Museum of Art, Ohio
Wadsworth Atheneum Museum of Art,
    Hartford, Connecticut
Yale University Art Gallery, New Haven,
    Connecticut

Private collectors in Australia,
Switzerland, the UK and the USA

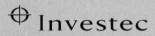

Principal sponsor Sydney

PRESIDENT'S
COUNCIL

Exhibition programme sponsor

Airline sponsor

RESMED

FOUNDATION

Major sponsor

Investec Bank (Australia) Limited is delighted to be associated with the *Camille Pissarro* exhibition. Values of passion, focus and creativity drive our business and stand us apart in investment banking. In Pissarro, we have more than found a kindred spirit and source of inspiration.

This tremendous and extensive exhibition showcases why Pissarro is revered as a liberating and complex artist, connected with a defining movement and yet distinctive and influential in his own right. His technical innovation, visual beauty and willingness to share his knowledge and expertise truly set him apart.

We congratulate the Art Gallery of New South Wales for bringing so many wonderful and important Pissarro works to Australia for the first time. Our particular thanks go to the passion and tenacity of senior curator Terence Maloon, who invested three years and plenty of hard work in making this captivating exhibition a reality.

On behalf of Investec, I invite you to enjoy the enlightening and out-of-the-ordinary experience that is the *Camille Pissarro* exhibition.

Brian Schwartz

Chief Executive Officer
Investec Bank (Australia) Limited

pp. 10–11: cat. 44 *The path to Le Chou, Pontoise*, 1878 (detail), Musée de la Chartreuse, Douai

Contents

Foreword

The virtues of devotion and probity resonate in the work of Camille Pissarro. It is not just because his paintings are honest and true but because they are strong, heartfelt and have a quietly resounding air of independence. It is such qualities that may have contributed to a view that Pissarro was a less colourful and eye-catching member of the Impressionist movement, and yet he was the only one of the group to participate in all eight of the epoch-making Impressionist exhibitions held in Paris between 1874 and 1886. Pissarro was a constant beat at the heart of the Impressionist movement as participant, artist, teacher, adviser and friend. He was not only regarded as the unifying figure of the Impressionist group, he was revered by the artists he mentored, including Cézanne, Gauguin, Seurat, Signac and Matisse.

The initial response of the critics and the public to the first Impressionist exhibition was by no means friendly. One of the works included was Monet's *Impression: soleil levant*, and it was this title that prompted the journalist Louis Leroy in the April 1874 edition of *Le Charivari* to dub the whole group "Impressionists". Certainly coined in the spirit of derision, the artists nonetheless accepted the term that was to become synonymous with the birth of modern western art.

Pissarro was born in 1830 on the island of St Thomas in the Danish West Indies to a prosperous French-Jewish merchant family.

At 20, convinced of his vocation as an artist, Pissarro and a Danish artist friend travelled and worked in Venezuela; he set up his first studio in Caracas. He must have felt something of an outsider when he finally settled in Paris in 1855, aged 25, despite his earlier schooling in France.

In Paris, Pissarro met Monet and Cézanne at the Atelier Suisse. He exhibited in several Salon exhibitions, although in the spirit of independence he was also represented in Napoleon III's Salon des Refusés, created to quell the protests of the increasing number of artists rejected from the official Salon. By 1873, he and his Impressionist colleagues had abandoned the Salon, a move which gave birth in 1874 to the first of the Impressionist exhibitions.

There is in Pissarro's work a subtle social and political sensibility, one determined by scruple and consideration. It is evident not only in the sheer authenticity of his close inspection and rendering of air, clouds, trees, foliage, buildings and the fleeting passage of light but equally in his obvious empathy for the human condition, for workers, peasants and people in general – whether the lone toiler in the fields or the bustling crowds on the boulevard. This is a quality that distinguishes him from his colleagues, Cézanne, Monet, Manet, Sisley and the other Impressionists. Pissarro's view and commitment are clearly stated in his comment, "our modern philosophy, which is absolutely social, anti-authoritarian and anti-mystical – a robust art based on sensation". Pissarro, in his extraordinary and constant correspondence, speaks much of "sensation" and declares his sensibility to the world burgeoning around him. The idea of modernity and progress clearly engaged his imagination. In 1897, he wrote, "I am delighted to be able to paint these Paris streets that people have come to call ugly, but which are so silvery, so luminous and vital. This is completely modern."

In spite of – or perhaps because of – his determined, independent stance and genuine social conscience, Pissarro was a kind and self-effacing man. His painting was his paramount concern. Cézanne described him as "a man to consult and something like *le bon Dieu*"; Mary Cassatt famously noted that "he could have taught stones to draw correctly". There was a quality of faithfulness about him and his work, which is demonstrated in his extensive and devoted correspondence and in his assiduous, accumulating brushwork. Interviewed by a newspaper towards the end of his life, Pissarro explained the essentials of his picture-making: "The great problem to solve is to bring everything, even the smallest details, into the harmony of the whole." As curator Terence Maloon notes, "… time and again, Pissarro links the morphology of clouds to trees and bushes … imparts diffuse, weightless, aerial qualities to the

phenomena of the earth and solid, textural and tactile qualities to events in the sky".

Ambitious in scale, this exhibition reveals a majestic, tireless and profound figure in the story of Impressionism, and we are greatly indebted to the many institutional and private lenders around the world who have so generously made works available.

Peasants' houses, Éragny, 1887 was the first Impressionist painting to enter the Gallery's collection, in 1935. The National Gallery of Victoria's splendid *Boulevard Montmartre, morning, cloudy weather*, 1897 and *The banks of the Viosne at Osny, grey weather, winter*, 1883 were acquired through the Felton bequest in 1905 and 1927 respectively. Together, these works provided the foundations and inspiration for this exhibition. We are delighted to be working again with the National Gallery of Victoria (where the exhibition will also be shown) and I extend my special thanks to its Director, Gerard Vaughan, and to Ted Gott and Tony Ellwood.

Absolutely central to the realisation of this exhibition has been our curator, Terence Maloon, whose unstinting drive and meticulous devotion to the project lie at the heart of its success. On this occasion, he has received valuable guidance and advice from Joachim Pissarro, the artist's great-grandson, and Claire Durand-Ruel Snollaerts who have produced the Wildenstein Institute's Pissarro

catalogue raisonné. Our curator of European works on paper, Peter Raissis, has rendered a crucial role with regard to the prints and drawings included in the exhibition. To all the contributors to the catalogue – Richard Shiff, Joachim Pissarro, Claire Durand-Ruel Snollaerts, Terence Maloon and Peter Raissis – I express my thanks.

An exhibition of this complexity and magnitude involves the entire Gallery, and in extending my appreciation to all I must make particular mention of Anne Flanagan, Erica Drew and Charlotte Davy of our Exhibitions department; Analiese Cairis for the catalogue design and exhibition material; Ursula Prunster, Brian Ladd and the Public Programmes department; and Belinda Hanrahan and her media and marketing team.

Finally, we record our sincere thanks to our sponsors for their support of the Gallery and this splendid exhibition: the principal sponsor, Investec; the Gallery's major exhibitions programme sponsor, the Gallery's President's Council under the chairmanship of David Gonski, President of our Board of Trustees; Qantas; the ResMed Foundation; media sponsors, the *Sydney Morning Herald* and JC Decaux; and supporting sponsors, the City of Sydney and the Sofitel Wentworth Hotel.

Edmund Capon
Director
Art Gallery of New South Wales

Pissarro and
the picturesque

Terence Maloon

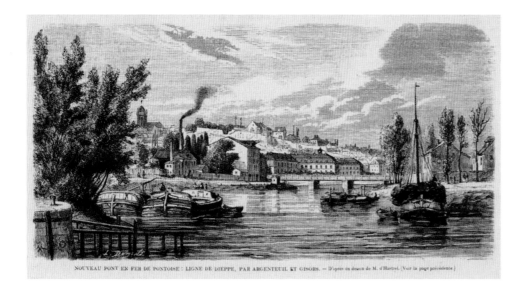

fig. 1
*New iron bridge of Pontoise:
railway line of Dieppe,
via Argenteuil and Gisors*
(after a drawing by Mr Hastrel)
1863
engraving
Private collection

Pontoise and the picturesque

Towards the end of 1863, a spate of articles appeared in the Paris newspapers reporting the progress of a new railway line that would link Paris to Dieppe in the north-west. The tracks already extended to the towns of Argenteuil, Saint-Ouen-l'Aumône, Pontoise and Gisors, connecting the Gare Saint-Lazare in Paris to the territories that later would become immortalised as "Pissarro country".

Ease of access to Paris was one of the reasons why Camille Pissarro decided to settle his young family in Pontoise in January 1866. He was preceded there by a homeopathic doctor from Paris, who was later to become an amateur painter and printmaker, art collector and the subject of two portraits by van Gogh – the legendary Dr Paul Gachet, no less. Dr Gachet had bought a house on the eastern outskirts of Pontoise in a semi-rural district known as L'Hermitage. Pissarro, at that time a father of two small children and a steadfast believer in homeopathic medicine, rented a house near him.

For about seventeen years, on and off, the town and its surrounding countryside provided a source of motifs for Pissarro's paintings, drawings and prints – "Pontoise, from every point of view, suits me," he declared to Claude Monet.[1] Indeed, it came to epitomise his idea of the picturesque.

An avant-garde painter's life is notoriously difficult, and Pissarro's was more so than most. During the three decades that followed his move to Pontoise, he experienced chronic problems in selling paintings and supporting a large household. Between 1863 and 1884, his wife Julie bore him eight children, three of whom died: one in infancy, another aged nine, the third as a young adult. In 1883, bleak economic prospects forced Pissarro to search for a house to rent in a remoter, cheaper location. While house hunting, the paintable features of the places he inspected were very much in the forefront of his considerations – he was searching for "a town able to replace Pontoise", he explained.[2]

The pictorial inadequacies of the places he visited were itemised in the letters he wrote to family and friends. He described how he had spent a day in Meaux, wandering around in sweltering heat:

> At first sight, the town itself has little character, the market is hideous, built [in iron] like the railway stations, less interesting than Les Halles in Paris which have character because of their proportions … The cathedral is superb. In places the surroundings are very good, [there are] interesting views that include the cathedral … [but] I fear it doesn't compete with Pontoise.[3]

He went to L'Isle-Adam to see a house with a big garden – but "I found the country hideous, a vast plain with little hills in the distance; long, long streets; sad, sad walls; stupid, stupid bourgeois houses! How could a painter stay there? I would be obliged to travel constantly, so you see the inconvenience."[4] L'Isle-Adam was unpicturesque; it was no Pontoise.

An article about the Paris–Dieppe railroad appeared in December 1863 in the periodical *L'Illustration*. It was accompanied by an engraving of the new iron railway bridge across the river Oise (see fig. 1), which showed a factory on the far bank of the river, a plume of dark smoke issuing from the larger of two smokestacks. The topical interest of the picture derived from the bridge and the smokestack – the first signs of the Industrial Revolution to make their mark on Pontoise. Best known for its export of veal and cabbages to the tables of Paris and for its populace of agricultural smallholders, Pontoise was still a quaint, rustic backwater. However, soon after he had settled there, Pissarro painted and drew the same motif portrayed in *L'Illustration*, focusing on the iron bridge and the smokestacks (see fig. 2). His ideas of the picturesque often had a timely, self-consciously contemporary orientation.

As further evidence of this, there is a letter that Pissarro wrote to the critic and collector Théodore Duret about some landscapes he had seen during his travels – landscapes replete with "old towers, churches, old houses with a romantic atmosphere".

fig. 2
Camille Pissarro
Banks of the Oise at Pontoise
1867
oil on canvas, 45.7 x 71.5
Denver Art Museum, gift of the
Barnett and Annalee Newman
Foundation in honour of
Annalee G. Newman

He speculated that it would be "very curious to do these with our modern eye", rendered in "*la peinture mate et claire*"[5] (the matte, bright painting style of the Impressionists). Pissarro was constantly modifying his technique – from the broad slabs of palette-knifed impasto that characterised his mid-1860s style to the intricately textured, tapestry-like paintings he developed in the early 1880s – so that he could add freshness and novelty to picturesque motifs, even to stale clichés of Old France. Equally, his constantly changing techniques recorded very up-to-date and topical phenomena, such as iron bridges, factories and telegraph lines.[6] In the book, *Pissarro and Pontoise*, the art historian Richard Brettell emphasised the scarcity of industrial sites in this part of France in the 1860s.[7] During the Second Empire, the wealth of France was still predominantly agricultural; the period has even been characterised as a "golden age of the peasantry".[8]

Pissarro found many diverse aspects of his surroundings good to paint – that is, he found them "picturesque", according to the primary meaning given to the word in French and English. Picturesque means "worthy of being painted; that which attracts attention, charms or amuses by an original aspect" (*Le Petit Robert*); "like, fit to be the subject of a striking picture" (*Oxford English Dictionary*); "visually charming or quaint, as resembling or suitable for a picture" (*Macquarie Dictionary*); "One says of a physiognomy, a garment, a site that they are picturesque when their well-pronounced beauty or character makes them worthy or at least susceptible to being represented in painting" (*Littré*).

The English painter Walter Sickert, one of Pissarro's warmest admirers, left a beautiful description of the vernacular features of the neighbourhood of L'Hermitage that found their way into Pissarro's paintings and came to represent a completely novel poetics of place, a distinctive variety of the picturesque:

It is still easy for the traveller, who only passes through Pontoise
– the city in the *Vexin Français* that reminded St Louis of

Jerusalem – to see the nourishment that the little orchards, bristling with bunches of leaves in the sun, the wood of old palings, the modest houses built of soft stone called *moellon*, diapered with black by smoke and with white by the droppings of pigeons, tinged with green by lichen, the linen hung out to dry by homely women in faded blue cotton gowns, must have been to Pissarro's talent.[9]

It was not only Pontoise that Pissarro immortalised. During the course of his long career, many other places were also unforgettably portrayed: La Varenne-St-Hilaire, Louveciennes, Éragny-sur-Epte, Dieppe, Montfoucault, Osny, Rouen, Le Havre, Paris and London.

Papa Corot

Without doubt, the most important influence to bear on Pissarro's notions of the picturesque was the art of Camille Corot. To be exact, there was a sub-category of Corot's immense production of landscape paintings that assumed crucial importance for Pissarro. Pissarro, like his fellow Impressionists, was very selective and biased in his appraisal of Corot. He tended to discount the hundreds of "classic", historical and allegorical landscapes that had secured Corot's reputation in the Salons and won his success with the bourgeois collectors of the time. In 1996, a gigantic exhibition of Corot's works, held in the Grand Palais, Paris[10], celebrated the bicentenary of his birth. *Corot 1796–1875* made a restitution of the "official" Corot and impelled a reconsideration of his complex and unwieldy achievement – rather to the detriment of the received idea that Corot was the forerunner of Impressionism. It called into review the perception of him as inveterate *pleinairiste*, a role model for the faux-naïf, anti-Academic artist, a dedicated and rather subversive exponent of picturesque landscape.

Pissarro first encountered Corot's work at the Universal Exhibition of 1855. Some forty years later, he recalled his astonishment at seeing one painting in particular – one among the six Corots on display at that exhibition.[11] The lesson he drew from it was:

> One can make such beautiful things with so little. Motifs that are too beautiful end up looking theatrical – think of Switzerland. Old Corot made lovely things at Gisors: two willows, a bit of water, a bridge – like the picture in the Universal Exhibition, what a masterpiece! Happy are those who see beautiful things in modest places, where others see nothing. Everything is beautiful, the whole thing is knowing how to interpret.[12]

Such was Corot's genius that he could imbue a minimally interesting motif with the rarest artistic qualities. It is instructive to see how early in his career Pissarro began to experiment with a sort of "zero degree" of the picturesque, essaying to "make beautiful things with so little" (see *Landscape*, c.1865, cat. 3 and *The Marne at La Varenne-St-Hilaire*, c.1864, cat. 4). It is also important to realise how inexplicable these paintings would have seemed to Pissarro's contemporaries. Many critics assumed his paintings lacked a subject and were therefore devoid of any real significance.

"One sees that Mr Pissaro [sic] is not banal through an inability to be picturesque," a sympathetic critic protested in 1866, struggling to justify the paradox he had observed between the powerful impact of one of Pissarro's paintings and its ostensibly uninteresting subject.[13] Even one of his staunchest supporters couldn't resist a joke at his expense: "He often comes to paint insignificant sites, where nature herself makes so little of a picture, that he paints a landscape without making a picture."[14]

Pissarro's great criterion in art was unity. Unity and harmony should be the ultimate goal of a painter, he believed: "For us, the search for unity is the goal towards which every intelligent artist should aspire, and even with great faults it is more intelligent, more artistic to do this than to remain bogged in romanticism."[15] Corot's work had proven the efficacy of this ideal: his paintings transcended the slightness and banality of their motifs because they resolved into a perfect unity.

Consequently, unity was the keynote of the picturesque. A work became picturesque by dint of being harmonious and unified. Unity, Pissarro explained, was what "the human mind gives to vision […] That's where our impressions, initially scattered, are coordinated."[16] The instinct for unity was personal and individual, because the terms that made up a work of art were freely chosen and could be freely changed. This leeway of choice always revealed something distinctive about the sensibility and temperament of the artist.

Each painter interpreted nature in a significantly different manner, Corot had explained to Pissarro: "We do not see in the same way; you see green and I see grey and 'blonde'. But this is no reason for you not to work at values, for that is the basis of everything, and in whatever way one may feel and express oneself, one cannot do good painting without it."[17] By "values", Corot meant tonal relationships, but in a more important sense "values" were the constituents of unity. The contribution that a patch of tone made to the integrity and harmony of the whole was what made it valuable.

For several years, from 1856 onwards, Pissarro visited Corot for informal instruction and criticism in his studio at 56, rue Paradis-Poissonière in the 10th arrondissement of Paris. Despite the sporadic and casual nature of their encounters, Pissarro obtained permission to describe himself as a "pupil of Corot" in his early submissions to the Salon exhibitions in Paris.[18] Corot initiated Pissarro into a means of painting based on carefully differentiated, delicately harmonised intervals of tone. In fact, Pissarro became the most direct conduit between Corot's method of plein-air painting and the method of the Impressionists. In principle, these were one and the same thing – a way of "painting in masses" and seeking to record an instantaneous effect: "I always try to see the effect at once

… In the same way I work on all the parts of my picture at once, gradually improving each one till I have found the complete effect," Corot explained.[19] "Work at the same time upon sky, water, branches, ground, keeping everything going on an equal basis and unceasingly rework until you have got it," Pissarro advised a young painter who had come to him for instruction in 1896–97, virtually paraphrasing his master Corot. "Paint generously and unhesitatingly, for it is best not to lose the first impression," he added.[20]

From very early in his career, Corot took care to preserve the freshness of colour and vivacity of handling of his small studies (*études*), which were usually painted outdoors. According to the conventional criteria of the time, these *études* were unfinished and could not be exhibited in public – they were inadmissable in the Salon exhibitions, for reasons given by the art historian Albert Boime: "Because the *étude* lacked the imaginative and premeditative look of the atelier, the Academy could not accept it as a true work of art."[21] Nonetheless, Corot was justifiably proud of them, as he confided in his notebook:

> I have observed that everything done at first go is franker, better formed, and I have known how to take advantage of many accidents – whereas, when it is reworked, the primitive, harmonious colour is lost.[22]

Indeed, much of the impact of Corot's *études* on the history of modern art stems from his daring and original ways of "taking advantage of accidents". In so doing, he deftly sidestepped the clichés of Academic composition, whereas his studio landscapes generally *were* those clichés, with a veil of lovely tonality and charming handling thrown over them.

One of Corot's most important lessons to Pissarro and the Impressionists was to encourage them to maintain an open mind and to use their intuitive judgment vis-à-vis how, when and why a pictorial composition "worked". A constant refrain in Pissarro's advice to his sons and to all the aspiring painters who came to learn from him was to avoid preconceived methods, systems, formulae. Reality had to be taken by surprise: "Everything is beautiful, the whole thing is knowing how to interpret."[23]

At some point in the late 1860s or early 1870s, Pissarro became fascinated with Japanese prints. Their influence reinforced the lessons of Corot's *études* in encouraging unconventional, experimental gambits of composition. Pissarro's compositions sometimes involved an extreme simplicity or a great intricacy, and sometimes hinged on a provocatively "obvious" symmetry or an audacious asymmetry (see *La Sente de Justice, Pontoise*, 1872, cat. 12, and *Still-life: apples and pears in a round basket*, 1872, cat. 14). Pissarro's response to an exhibition of Japanese prints in Paris in 1883 is on record: "It is marvellous," he wrote to Lucien. "This is what I see in the art of this astonishing people: nothing that leaps to the eye, a calm, a grandeur, an extraordinary unity, a rather subdued radiance which is brilliant nonetheless. What sobriety and what taste!"[24] One may deduce from these remarks how Pissarro sought to make the qualities of Japanese art the virtues of his own paintings, drawings and prints.

The aesthetic refinement and intellectual brilliance of the Japanese printmakers were evident to western artists through the formal qualities of their pictures – through the variety and ingenuity of their compositions. If a westerner discounted the superficial exoticism of Japanese imagery, it was easy to recognise that the printmakers had portrayed the small incidents of daily life, that they were interested in landscape and that they subscribed to an analogous cult of the picturesque. They obviously believed that "everything is beautiful, the whole thing is knowing how to interpret", affirming the creed of Pissarro's generation of painters.

Pissarro likened the unity of a work of art to a "pivot". An artist resolved and synthesised a work by intuition, by feeling, as if everything he added or subtracted affected the poise of the work on an imaginary pivot. As he explained to Lucien:

You are right, all in all, to follow your feeling in painting. Indeed, the different manifestations of art do not contradict one another, there is always the same pivot […] What you say about the modesty of your intention is a bit like Papa Corot used to say to us: "I have only a little flute, but I try to play the right note."[25]

Accords

Pissarro's notion of the pivot seems to explain a great deal about his unusual attitude as a painter – how, in the spirit of experiment, he would frequently alter his technique, his painting implements, the colours of his palette and the character of his mark-making. If he believed that the "pivot" of the work of art was there for all time and for all comers, and if he believed that unity and harmony could be attained in an infinite variety of ways, and that the end (unity) always justified the means, Pissarro would have had little or no compunction in altering, time and again, the material constituents of his paintings.

Some may object that the evidence of Pissarro's aesthetic theory cited here is scattered through letters, interviews and the reminiscences of various acquaintances spanning several decades. Didn't his ideas develop, didn't they change over the years? There were changes in certain respects, no doubt, but it is reasonable to assume that many of Pissarro's ideas remained constant, and that the most important were already established in the mid-1860s. Consider, for example, the explanation he gave to a newspaper reporter in Le Havre during the last months of his life:

I see only spots [*taches*]. When I begin a painting, the first thing I try to fix is the accord. Between the sky, this land and this water there is necessarily a relationship. It can only be a relationship of accords, and therein lies the great difficulty of painting. What

interests me less and less in my art is the material side of painting (the lines). The great problem to solve is to bring everything, even the smallest details, into the harmony of the whole – that is to say, into the accord.[26]

The notion of a painting successively built up of "accords" until it achieves a totality of "accord" is fundamental to Pissarro, and can already be intimated in works of the 1860s, such as *The Marne at La Varenne-St-Hilaire* (cat. 4), where a consistently dense paint surface is rippled with sinuous tracks of the brush, creating a corresponding reflection in the sky and the worn pathway, establishing their accord. The boldly simplified zones of *Banks of the Marne in winter*, 1866 (cat. 7) ring out in thunderous accord. Time and again, Pissarro links the morphology of clouds to trees and bushes (*A cowherd on the route du Chou, Pontoise*, 1874, cat. 22, and *The highway*, 1880, cat. 47). He loves to situate a white horse in a landscape (in reflection of the clouds) or a field worker dressed in washed-out blue or grey (in reflection of the sky). He often imparts diffuse, weightless, aerial qualities to the phenomena of the earth and solid, textural, tactile qualities to events in the sky.

His predilection for linking the foreground and the background, up and down, left and right, figure and field – saturating the composition with accords – had the effect of breaking down conventional distinctions and hierarchies, seeming to compress and "level" his compositions. This feature greatly disconcerted the early viewers of Impressionism, as evidenced by Louis Leroy's famous rebuke of a painting by Pissarro in the first Impressionist exhibition: "It has neither head nor tail, top nor bottom, foreground nor background."[27] The same idiosyncratic features that Leroy observed in 1874 persist in a masterpiece of 1900, *The wash house at Bazincourt* (fig. 7), a glorious tissue of accords.

It seems that the word "accord" is particular to Pissarro. It does not appear to have had universal currency among the Impressionists – it was not conspicuous in Monet's vocabulary, for example (it is

absent from both Steven Z. Levine's useful index of key terms for Monet, and from the very comprehensive index of Virginia Spate's monograph).[28] Yet the concept of "accord" was important for Cézanne (who used it as a synonym for a favourite word of his, "rapport"), and it also occurs in relation to Corot. There was a legend of Corot perpetuated by the art critic Gustave Geffroy:

> All the time he didn't spend in front of his easel, no matter if he was alone or in conversation, [Corot] was preoccupied with the thought of accords almost in spite of himself, of the harmonies between things that he saw everywhere, that he had reproduced yesterday and was going to reproduce again.[29]

We can add to this the invaluable testimony of Pissarro's eldest son Lucien, who, for the benefit of his youngest brother, Paul-Émile, remembered the ferment of ideas, the constant experimentation with technique and the profound friendship that united Pissarro and Cézanne, which the young Lucien had witnessed during the years 1872 to 1874:

> At the time, Cézanne lived in Auvers [near Pontoise] and every day he walked the three kilometres to come and work with Papa. He had endless theories. I remember one day they had special palette knives sent from Paris in order to make palette-knifed studies […] As usual, they didn't continue painting with palette knives for long, and tried using a divided touch. I think this was the time when Cézanne began to paint with his vertical divisions and Papa adopted long brushes and painted in tiny little commas. To say who influenced whom is impossible … An amateur painter from Pontoise often came to see them and there were endless discussions. I remember a snatch of a phrase which assumes great significance today: "But Mister So-and-So, we don't paint a figure, we make Accords!"[30]

The picturesque in question

Lucien Pissarro didn't say who uttered these words. It may have been Cézanne, or Pissarro, or both in chorus – such was their complicity. What were the theories he mentions? What did they endlessly discuss? The speculative bent of their conversations must have been established in the early days of their friendship, for we obtain momentary glimpses in some of Cézanne's letters which evidently refer to ongoing discussions with Pissarro. In 1866, Cézanne wrote to Pissarro: "You are perfectly right to talk about grey. That alone reigns in nature, yet there is a frightful difficulty in capturing it."[31]

During the period of their regular companionship in plein-air painting sessions, they must have raised the question of whether or not volumetric modelling was an indispensable convention of painting, because in 1876 Cézanne alluded to it in a letter to Pissarro from Provence:

> The sun here is so terrific that it seems to me that objects rise up in silhouette, not only in black and white, but in blue, red, brown, violet. I may be wrong, but it seems to me the opposite of modelling.[32]

These stray remarks alert us to some of Pissarro's abiding concerns. He was attracted to the colourism of paintings by Turner, Monet and Renoir ("the rare bird whose plumage is resplendent with all the lovely colours of the rainbow"[33], as he put it in a letter to Lucien), yet he was also powerfully attracted to the tonalism of Corot and Whistler – so much so, that he was able to entertain the contrary idea that "grey alone reigns in nature". Right until the end of his life, the "values" in Pissarro's paintings oscillated between spectral colours and grisaille ("Let us work hard and try to make dazzling greys,"[34] he urged Lucien), as if his ultimate ambition were to efface the difference between them.

Similarly, the concern for modelling comes and goes in Pissarro's painting. It seems to go into abeyance in the 1870s during the

fig. 7
Camille Pissarro
The wash house at Bazincourt
1900
oil on canvas, 65.5 x 81
Musée d'Orsay, Paris, bequest
(02 Legacy), Antonin Personnaz

heyday of Impressionism, yet it becomes important again in his figure paintings from circa 1880 – works he produced during the so-called "crisis of Impressionism", more or less at the same time as Renoir's recoil from Impressionism and the momentous coming of age of Cézanne's art. Renoir said he was trying to remedy the lack of modelling and solidity, as well as the neglect of drawing that was sanctioned in Impressionist painting, and Pissarro seems to have set out to make similar amends in a series of monumental figure paintings from 1880 to 1883 (*Woman washing dishes*, 1882, cat. 48, *The little country maid*, 1882, cat. 49, and *Peasants resting*, 1881, cat. 50).

Creating an illusion of volume in a painting requires creating the complementary illusion of deep space, which is necessary to establish the roundness and the bulk of an object or a figure. In *Woman washing dishes*, the figure acts as a pivot between a plunging spatial illusion on the left and a compressed planar arrangement on the right. One of the defining characteristics of Impressionism is that the space around and between things is always treated as if it were as substantial, full and pictorially significant as the things themselves. Space is never empty. So how does an Impressionist reinstate an illusion of volume? In such works as *Peasants resting*, Pissarro refuses to let the surroundings recede, to empty out and become subsidiary to the figures. He wants to endow his figures with volume and mass, yet here he denies them the void that would give them an added dimension.

The typical outcome of such a contradiction of intent is an inconsistent, conflicting, dysfunctional pictorial space – analogous to the pictorial space we see in many of Renoir's and Cézanne's paintings of the same period.[35] All three artists lent their very considerable ingenuity and skills to exploring and resolving the pictorial challenges that were exposed – as did Vincent van Gogh who, very soon after he came onto the scene in Paris a few years later, found his own distinctive way of reconciling an intensely expressive plasticity ("modelling") with the congested spatial

field of Impressionism. It is more than likely that van Gogh looked very attentively at *Peasants resting* and the related works produced by Pissarro at that time.

Cézanne found an ingenious compromise to the problem of modelling, as he recounted to the painter Émile Bernard a few months after Pissarro's death. The fact that his remarks seem just as pertinent to Pissarro's paintings as to his own may well be due to their origin in past conversations. "There's no line, there's no modelling, there are only contrasts," Cézanne explained. "It isn't black and white [that is, the tonal differences] that create the contrasts, but sensations of colour. Modelling arises from the exact rapport of tones. When they are harmoniously juxtaposed and when they are all there, the picture models itself. One shouldn't say *modelling*, one should say *modulating*."[36]

Typically, Pissarro's painting is all modulation, inviting the viewer to follow a perpetual transit of colour and tone. This aspect of his painting happens to coincide with one of the English writer William Gilpin's definitions of the picturesque, for it represents a situation where the eye is led to "glide up and down among endless transitions".[37] The Reverend William Gilpin (1724–1804) was the leading theorist of picturesque landscape in England at the end of the eighteenth century. Irregularity and roughness were keynotes of the picturesque landscape, he stated:

Turn the lawn into a piece of broken ground: plant rugged oaks instead of flowering shrubs; break the edges of the walk: give it the rudeness of a road; mark it with wheel-tracks; and scatter around a few stones, and brushwood; in a word, instead of making the whole *smooth*, make it *rough*; and you make it also *picturesque*.[38]

This description summons many of Pissarro landscapes to mind (for example, *The fence*, 1872, cat. 10). A degree of roughness in the execution of a landscape painting was not only permissible, Gilpin thought, it was desirable because it eliminated detail, forcing the viewer to step back and assess the scene as a whole. "The province of the picturesque eye is to *survey nature*; not to *anatomize matter*," he wrote. "It throws its glances around in the broad-cast style. It comprehends an extensive tract at each sweep. It examines *parts*, but never descends to *particles*."[39]

Despite Gilpin's fame in the English-speaking world, French theorists of landscape painting generally denied him and other English writers on the picturesque even a passing mention in their treatises.[40] There is no evidence that Pissarro was acquainted with these writings either, although he was influenced by the post-Gilpin picturesque tradition in English painting through the works of Turner and Constable, which he had studied in London museums in 1870. Despite Constable's and (to a lesser extent) Turner's impact on painters of the Fontainebleau school in France, the British landscapists were effectively segregated as foreign and seditious by the purists of the French Academy. For these Academicians, the most honourable type of landscape painting was *le paysage classique*, also known as *le paysage historique*, based on seventeenth century French prototypes. In effect, *le paysage historique* was a pastiche of the idealised landscapes of Nicolas Poussin, Gaspard Dughet and Claude Lorrain. Between 1816 and 1863, the Academy held an annual Prix de Rome competition for *le paysage historique*, and it is possible that the drawing in the Yale University Art Museum (*Landscape with trees, two figures on a road and mountains in the background*, c.1855, cat. 1) is a student work prepared by Pissarro for this competition.[41]

The pictorial and the picturesque

According to the Academicians' doctrine, a painting was a meticulously crafted invention that was created in a studio. Consequently, a landscape painted outdoors, in direct contemplation of the motif, was patently *not* an invention. A landscape study could furnish the raw material for a work of art, but it could not be a work of art in its own right, since it was a *copy*. "Wherefore should I wish for a copy?" asked the Permanent Secretary of the Institut des Beaux-Arts, Antoine Quatremère de Quincy (1755–1847). "What need have I of the appearances of things whose reality I am wholly indifferent to? What worth can I attach to the image, when I hold its model in contempt?"[42] The business of the artist was to address the ideal, not to copy the real. And so it is easy to see how ideas associated with the picturesque could remain contentious in France throughout much of Pissarro's lifetime.

For the Academicians, as for most nineteenth century connoisseurs of the old masters, the ideal condition of a landscape painting was signalled by the colour brown. "Gaspar Poussin's green landscapes have no charms for me," an English connoisseur confessed in his diary. "The fact seems to be that the delightful green of nature cannot be represented in a picture. Our own [John] Glover has perhaps made the greatest possible exertions to surmount the difficulty, and give with fidelity the real colours of nature; but I believe the beauty of his pictures is in an inverse ratio to their fidelity; and that nature must be stripped of her green livery, and dressed in the browns of the painters, or confined to her own autumnal tints in order to be transferred to canvas."[43]

Apropos of this predilection for brown, there is an amusing story of a disagreement that broke out between John Constable and his patron, Sir George Beaumont. Sir George had "recommended the colour of an old Cremona fiddle for the prevailing tone of everything,

and this Constable answered by laying an old fiddle on the green lawn before the house".[44] In fact, before the advent of the Impressionists, very few European painters had made extensive use of green in their landscapes. Pissarro's *The road to Ennery, near Pontoise*, 1874 (cat. 17) and *A cowherd on the route du Chou, Pontoise*, 1874 (cat. 22) must have seemed less pleasantly tranquil and more confronting to contemporary viewers because of the unsparing use of green – green being the colour of non-ideality, the colour of realism.

As *Road to Port-Marly*, c.1860–67 (cat. 2) and *The Marne at La Varenne-St-Hilaire*, c.1864 (cat. 4) reveal, Pissarro's attachment to the picturesque long predated his 1870 voyage to London, so we must be wary of overemphasising his English connections and overestimating the importance of his exposure to Turner and Constable. Nonetheless, one of the works he painted in London, the Kimbell Museum's *Near Sydenham Hill*, 1871 (cat. 9), is based on a cliché of picturesque composition, using two flanking trees to frame a distant prospect. The contrast between the robust browns and blacks of the tree trunks and branches and the dissolving pastel tints edging towards the horizon defines the picture's depth of field. Pissarro deliberately chose this compositional schema to remind the viewer of paintings already seen, of picturesque landscapes which appeal to the mind's eye, inviting an imaginary journey into the interior distance. Isn't that one of the defining features of the picturesque?

Likewise, the Hamburger Kunsthalle's *Resting beneath the trees, Pontoise*, 1878 (cat. 43) is based on an archetype of picturesque composition. The compositional schema is reminiscent of Constable's *Dedham Vale*, Turner's *Crossing the brook* and any number of paintings, in fact, which have Claude Lorrain's *Hagar and the angel* (or something very like it) as their prototype. These two Pissarro canvases are "rough" and "rugged" beyond anything the Reverend Gilpin might have countenanced in a landscape painting – so much so that the idyllic prospect frays before our eyes, unravelling into

gritty opacity. Any viewer who wants imaginatively *to enter* the scenes Pissarro portrayed is brought up short by the resistant materiality of the paint surface.

Why did Pissarro sabotage such a seductive illusion – an illusion that, technically speaking, was so easily within his grasp? Why had he neglected to smoothe and blend the edges of his brushstrokes, which could have affirmed and enhanced the picturesque charm of the painting? Why had he painted so hurriedly and left off so abruptly? During Pissarro's lifetime, the public's response to his work was plagued by questions such as these, and one may well understand the critics' perplexity (which is amply documented in the survey on pages 225–46), even if one no longer shares their way of looking at his paintings.

As most critics acknowledged at the time, Pissarro's apparent faults and deficiencies were not due to his lack of talent or ability nor, as one of them put it, to his "inability to be picturesque".[45] Yet he represented a maddening paradox for them: from a distance his paintings extended the promise of picturesque illusion, but on closer inspection they disintegrated into crude markings, or into the conspicuous thickness and heaviness of the impasto. How was one to explain the raison d'être of such pictures? Were they part of a deliberate campaign of demoralisation and disenchantment, waged by a nihilist aiming to bring down all that was fine, noble, beautiful and intelligible?

Consider the violent response of American reviewers to the first Impressionist exhibition in New York, in 1886: "Communism incarnate, with the red flag and the Phrygian cap of lawless violence boldly displayed, is the art of the French Impressionist," declared one critic.[46] "The Paris Impressionists […] are pressing a deliberate campaign against legitimate art," wrote another. "The landscapes after [Monet], Pissaro [sic] and Guillaumin, are simply insolent in the crudity and rudeness of their work […] Impressionists, at least this group, are simply expositors of the social and moral cultus of

Zola-ism, of sensualism and voluptuousness, or sheer atheism […] The dreariness and unspirituality of their landscapes is thus accounted for."[47]

Strange to say, these responses were quite valid. It was perfectly reasonable to ascribe the assertive materiality and defective illusionism of Impressionist painting to "materialism", "atheism" and "Zola-ism" – that is, to the artists' refutation of the religiosity, idealism and transcendental mood-making that had all but saturated western painting until that time.[48] American critics were especially sensitive to this aspect of Impressionism, because it struck at something very dear to their hearts. The great shibboleth of American landscape painting was the "sublime", which was evoked by the treatment of light, space and scale. The sublime was "the feeling of infinity"[49], and Impressionist paintings appeared entirely bereft of it.

Seen in a wider context, Pissarro's idiosyncratic treatment of the picturesque takes on additional significance. For its most trenchant analysis, we have to wait for the publication of "The Subject in Painting", a text by Paul Signac which appeared in 1935.[50] This was, in effect, Signac's last testament – he died the same year, aged seventy-two. Despite the thirty-year difference in their age, Signac had been Pissarro's closest painter friend during the period of their mutual involvement with Neo-Impressionism (1885–89). True to the spirit of his mentor, Signac subsequently became a benevolent guardian of younger artists affiliated to the avant-garde, and he remained a militant of the modernist persuasion to the end of his days.

Signac argued that there were two fundamental aspects to the art of painting: the *picturesque* and the *pictorial*. The picturesque involved everything concerned with the motif, the overt subject or the occasion of a painting – whether it be circumstantial, geographical, literary, historical, polemical or sentimental. In contrast, the pictorial comprised "all the aesthetic and formal forces of organisation,

harmony and materials".[51] In the nineteenth century, because of the hegemony of the Salon and the Academy in France, painting had been dominated by subject matter (that is, by the picturesque). "Even today," Signac added, "our society, in refusing culture to the masses, imposes on them this enslavement to the subject."[52]

Signac described the overriding "cause" of modernism as follows: the Impressionists, their predecessors, their allies and their followers had all worked to bring discredit to the picturesque in order to promote the ascendancy of the pictorial. This might seem a rash claim, but Signac, writing in 1935, was justifying the radicalism of Pierre Bonnard and Henri Matisse – the contemporaries he most admired.

Could Pissarro have recognised his aims in this assertion? That would be most unlikely. There is no detectable ambivalence, no negative connotation in the way he used the word "picturesque" in his correspondence. Picturesque always indicated something affirmative, attractive, desired. Pissarro assumed that the picturesque and the pictorial had to work in unison. It was inconceivable, then, that there could be an art based on "organisation, harmony and materials" alone. Equally, it was inconceivable for Pissarro that everything Signac had assimilated into his notion of the picturesque could be "discredited", alienated and exiled from the art of painting.

Nonetheless, as soon as we acknowledge the terms of Signac's dichotomy, we notice how the claims exerted by the motif in Pissarro's paintings were constantly being challenged by the claims of composition, textures, colours, tones and "accords". Because the viewer was never allowed to overlook the pictorial aspects of Pissarro's paintings, this gave a subversive spin to his treatment of the picturesque.

Photography and the picturesque

Camille Pissarro was born in 1830, four years after Nicéphore Nièpce took the first ever landscape photograph from his window, in an exposure that lasted for several hours. By the end of Pissarro's life, a craze for picture postcards had inundated almost every household in France with a glut of landscape imagery (see pages 220–23). The exposure time required for a photograph had shrunk from hours to minutes, then to a split second. The era of the snapshot had arrived. In the interim, photography changed people's idea of the picturesque from something worthy of being painted to something worthy of being photographed.

Nineteenth century artists and critics regarded the photograph as the archetype of realism: "The principal character of the realist painters is to take their subjects and their inspiration from ordinary life, and to paint what they see, how they see it, without the preconception of convention, without distinguishing the noble and the ignoble. They take the daguerreotype and the photograph as the archetype of art," a critic wrote in 1852.[53] Highly experimental composer that he was, one of the novelties of Pissarro's paintings was that they emulated the casual, artless look of photographs. His strategy of "cropping" images was inspired not only by Japanese art but by photography. He was chastised in 1874 by a reviewer of the first Impressionist exhibition for violating a sacred rule of composition: "He has committed the grave error of painting cast shadows over his field, projected from trees situated beyond the frame – trees which the spectator can only surmise, not being able to see them."[54] Those extraneous shadows were a typical feature of photographs, not of paintings, yet Pissarro repeatedly made use of them (see *The highway*, 1880, cat. 47, *Woman washing dishes*, 1882, cat. 48, and *Peasants' houses, Éragny*, 1887, cat. 52).

When they decided to hold the first and the sixth of their group exhibitions (in 1874 and 1881) at 35, boulevard des Capucines, in premises formerly occupied by the most prominent professional photographers in Paris, Gustave Le Gray and Félix Nadar, did the Impressionists want to impress the public with their rapport with photography? Awareness of photography was a defining feature of what Pissarro called "our modern eye".[55] His notion of the "pivot"[56] is very relevant when we consider what happens when a photographer looks into a camera, lining up the image, finding its equilibrium in the viewfinder. A photographer's approach to composition is very different from a painter's or draughtsman's: he or she recognises or discovers the composition rather than contriving, willing and manhandling it into being. Nonetheless, "there is always the same pivot", as Pissarro acknowledged: the balance, harmony and unity of a photograph depends on the same aesthetic instinct that a painter exercises. The painters of the nineteenth century who sought to capture fleeting effects and take reality by surprise, who wanted to discover design in accident and order in chaos, were receptive to everything that photography could teach them.[57]

The die-hard conservatives in the Paris art world liked to contrast their own ideal of the work of art as a purely intellectual creation with what they assumed was simply the mindless copying of nature in Impressionist painting, something they often likened to photography. They were convinced that the Impressionists were vastly less imaginative, intelligent and skilful than their Academic counterparts. Pissarro could have countered their criticisms by saying he was not just copying but interpreting, transposing, synthesising – he was composing all the time, from the first brushstroke to the last.

Yet the questions persist. How faithful was Pissarro to his motifs? How literal are his paintings? Did he change and rearrange things? Bonnard thought Pissarro took considerable liberties: "The painters who were able to approach the motif directly are very rare – and

fig. 10
cat. 89

Camille Pissarro
Rue de l'Epicerie, Rouen
1898
The Metropolitan Museum
of Art, New York

fig. 11

John Rewald
Rue de l'Epicerie
1939
silver gelatin print
John Rewald Papers,
National Gallery of Art,
Washington, DC, Gallery Archives

fig. 12 (opposite, left)
cat. 95

Camille Pissarro
*The stone bridge in Rouen,
dull weather*
1896
National Gallery of Canada,
Ottawa

fig. 13 (opposite, right)

John Rewald
Pont Corneille in Rouen
1939
silver gelatin print
John Rewald Papers,
National Gallery of Art,
Washington, DC, Gallery Archives

those who were able to extricate themselves [from the spell of the motif] did so through a very personal means of defence [...] The Impressionists went to the motif, but they were better defended than the others against the object itself because of their procedures, their manner of painting. This is even more apparent in Pissarro's case: he arranged things; system plays a greater part in his painting."[58] Elsewhere, Richard Brettell has given clear proofs of Pissarro editing, simplifying and rearranging his motifs.[59]

Yet there are photographs which correspond extremely closely to Pissarro's paintings, where it is evident that his alterations were very subtle. Here and there, he would eliminate a redundant element, freshen a colour, raise or lower the height of a building, revise the spacing, change the colour of a roof, redesign or realign a grid of windows, lessen the tonal contrasts, knit the textures together, flatten out the perspective, and so on, while remaining quite faithful to the general character of the scene. In 1939, John Rewald photographed several motifs that Pissarro had painted in Rouen (see fig. 11 and 13), and when we compare the photographs with the paintings, the similarity is far more impressive than any signs of "arranging" or any evidence of a "method" being used as a "defence against the object", as Bonnard had alleged.

However, in spite of the appealing picturesqueness and the steadfast fidelity of Pissarro's paintings, it is their pictorial qualities that really matter: the success of the composition, the beauty of the colour, texture and handling, the conviction and integrity of their aesthetic purpose. Surprisingly, some of Pissarro's most "literal" paintings are among his most intensely original and prophetic. Consider the planar compositions of *Rue de l'Épicerie, Rouen*, 1898 (fig. 10, cat. 89) or the Toledo Museum of Art's *The roofs of old Rouen* (PV973): both seem to point to Picasso's and Braque's Cubism a decade later. And, in the intricate choreography of the Art Gallery of Ontario's *Pont Boïeldieu in Rouen, damp weather*, 1896 (cat. 96), a multitude of off-kilter elements finds a perfect inner balance – a forerunner, surely, to André Derain's 1906 paintings of the port of London.

Towards the end of the nineteenth century, many of the places where Pissarro painted were revisited by a legion of photographers, some of whom set up their tripods very close to where he had once positioned his easel. Did the photographers know that painters had frequented those places? Were any of them acquainted with Pissarro? Some of their photographs were even reminiscent of his most provocatively "insignificant" and desolate pictures. There were also "staged" pictures of agricultural workers (see, for example, the postcard on page 222). Had they seen similar images in Pissarro's paintings? The ubiquity of picture postcards in France ultimately devalued and discredited the currency of the picturesque in painting, but Pissarro's place in the history of art no longer relied on this.

His posterity was assured, as Georges Lecomte recognised as early as 1892 when he wrote: "The story of his endeavours is tantamount to the complete history of Impressionism, up to its most recent manifestations; they even seem to preface the art of tomorrow."[60] In addition, as we have seen, the story of his endeavours was also intricately bound up with the changing fortunes of the picturesque.

1 Pissarro, letter to Monet, 18 September 1882, in Janine Bailly-Herzberg (ed.), *Correspondance de Camille Pissarro*, vol. 1, Presses Universitaires de France, Paris, 1980; vols 2–5, Éditions Valhermeil, Saint-Ouen-l'Aumône, 1986–96; vol. 1, p. 165.

2 Pissarro, letter to Lucien Pissarro, 4 June 1883, *Correspondance*, ibid., vol. 1, p. 214.

3 ibid., pp. 214–15.

4 Pissarro, letter to Lucien Pissarro, 14 December 1883, *Correspondance*, ibid., vol. 1, p. 261.

5 Pissarro, letter to Théodore Duret, 28 September 1882, *Correspondance*, ibid., vol. 1, p. 167.

6 Telegraph lines appear in Pissarro's *La Route au bord du chemin de fer, effet de neige*, 1873 (PV205), National Museum of Western Art, Tokyo. Richard Brettell cites the critic Jules-Antoine Castagnary's condemnation (in 1857) of a landscape by Jeanron which showed telegraph poles – see Richard Brettell, *Pissarro and Pontoise*, Yale University Press, New Haven, 1990, p. 76.

7 Brettell, ibid.

8 Christophe Charle, *Histoire sociale de la France au XIXe siècle*, Éditions du Seuil, Paris, 1991, p. 87 and pp. 141–79.

9 Walter R. Sickert, *A Free House! Or The Artist as Craftsman*, Macmillan, London, 1947, p. 141.

10 See exhibition catalogue, *Corot 1796–1875*, Grand Palais, Paris, 1996.

11 For descriptions of Corot's paintings in the Exposition Universelle, see Ralph E. Shikes and Paula Harper, *Pissarro – His Life and Work*, Quartet Books, London, 1980, pp. 39–40. However, Shikes and Harper mistranslate Pissarro's words in the quotation below (footnote 12), pluralising his *tableau* and *chef-d'œuvre* into "paintings" and "masterpieces" – Pissarro was referring to a single painting, not several.

12 Pissarro, letter to Lucien Pissarro, 26 July 1893, *Correspondance*, op. cit. (footnote 1), vol. 3, p. 351.

13 Jean Rousseau, "Le Salon de 1866 – IV", *L'Univers Illustré*, Paris, 14 July 1866, p. 448.

14 Théodore Duret, "Le salon – Les naturalistes", *L'électeur libre*, Paris, 12 May 1870, p. 61.

15 Pissarro, letter to Lucien Pissarro, 1 October 1888, *Correspondance*, op. cit. (footnote 1), vol. 2, p. 254.

16 Pissarro interview by Paul Gsell (1892), quoted by John House, "Camille Pissarro's idea of unity", in Christopher Lloyd, *Studies on Camille Pissarro*,

Routledge & Kegan Paul, London, 1986, p. 17. House's essay explores several phases of Pissarro's career where his idea of unity seemed to alter or became more developed and explicit.

17 Corot's advice to Pissarro quoted in John Rewald, *The History of Impressionism*, Secker & Warburg, London, 1980, p. 101, originally cited in Arsène Alexandre, *Claude Monet*, Bernheim-Jeune, Paris, 1921, p. 36.

18 Sisley also was accepted into the Salon of 1867 as "*élève de Corot*" – see Lionello Venturi, *Les Archives de l'Impressionnisme*, Durand-Ruel Éditeurs, Paris–New York, 1939, 2 vols; vol. 1, p. 28.

19 Corot, as recorded by the painter Mme Aviat, in *Corot raconté par lui-même et par ses amis*, Pierre Cailler Éditeur, Vésenaz-Genève, 1946, p. 97.

20 Notes of Louis Le Bail recording Pissarro's instruction, in John Rewald, *The History of Impressionism*, op. cit. (footnote 17), p. 458.

21 Albert Boime, *The Academy and French Painting in the Nineteenth Century*, Phaidon, London, 1971, p. 150.

22 Corot in a notebook, c.1825, quoted in *Corot raconté par lui-même*, op. cit. (footnote 19), p. 87.

23 See above, footnote 12.

24 Pissarro, letter to Lucien Pissarro, 24 April 1883, *Correspondance*, op. cit. (footnote 1), vol. 1, pp. 198–99. Cf. Pissarro's remarks about Japan to Duret ten years earlier – "a bold and very revolutionary country in art", letter to Théodore Duret, 2 February 1873, *Correspondance*, ibid., vol. 1, pp. 78–79.

25 Pissarro, letter to Lucien Pissarro, 23 February 1894, *Correspondance*, ibid., vol. 3, p. 434.

26 Robert de la Villehervé, "Choses du Havre: les dernières semaines du peintre Camille Pissarro", *Havre-Eclair*, 25 September 1904, p. 2.

27 Louis Leroy, "L'Exposition des impressionnistes", *Le Charivari*, Paris, 25 April 1874, p. 79.

28 "Index of Key Terms" in Steven Z. Levine, *Monet and His Critics*, Garland Publishing, New York, 1976, pp. 463–466; Virginia Spate, *The Colour of Time – Claude Monet*, Thames & Hudson, London, 1992.

29 Gustave Geffroy, *Corot*, Éditions Nilsson, Paris, 1924, p. 15.

30 Letter from Lucien Pissarro to Paul-Émile Pissarro, undated, 1912, in Anne Thorold (ed.), *Artists, Writers, Politics: Camille Pissarro and His Friends*, Ashmolean Museum, Oxford, 1980, pp. 8–9.

31 Paul Cézanne, letter to Pissarro, 23 October 1866, in John Rewald (ed.): *Paul Cézanne – Correspondance*, Bernard Grasset, Paris, 1978, p. 125.

32 Paul Cézanne, letter to Pissarro, 2 July 1876, ibid., p. 152.

33 Pissarro, letter to Lucien Pissarro, first fortnight of February 1883, *Correspondance*, op. cit. (footnote 1), vol. 1, p. 174.

34 "*Travaillons ferme et tâchons de faire des gris épatants!*" – letter to Lucien Pissarro, 31 January 1896, *Correspondance*, ibid., vol. 4, p. 159.

35 In effect, it is the embryonic pictorial space of Modernism that we encounter in these paintings, which artists such as Matisse, Picasso and Braque would subsequently make their own.

36 Émile Bernard, "Paul Cézanne" (originally published in *L'Occident*, July 1904), in P.M. Doran (ed.), *Conversations avec Cézanne*, Macula, Paris, 1978, p. 36.

37 Gilpin quoted in Joshua C. Taylor (ed.), *Nineteenth-Century Theories of Art*, University of California Press, Berkeley and Los Angeles, 1989, p. 49.

38 William Gilpin, "Essay on Picturesque Beauty" (1792) in Joshua C. Taylor (ed.), ibid., p. 51.

39 ibid., p. 58.

40 For example, Charles-Jacques-François Lecarpentier, *Essai sur le paysage* (originally published in 1817), Rumeur des Ages, La Rochelle, 2002; and Jean-Baptiste Deperthes, *Théorie du paysage* (originally published in 1818), Rumeur des Ages, La Rochelle, 2002. Gilpin's "Three Essays on the Picturesque" had been published in a French translation in 1799, but it was assumed to be an attack on "*le beau idéal*" and so it was less than enthusiastically received. The most comprehensive expression of the picturesque in France was Baron Taylor's 21-volume collection of lithographs, *Voyages pittoresques et romantiques dans l'ancienne France*, published between 1821 and 1878.

41 In 1855–56, Pissarro was enrolled in classes held at the private ateliers of François-Édouard Picot (1786–1868) and Isidore Dagnan (1794–1873), both of whom were teachers at the École des Beaux-Arts. Dagnan was a respected landscapist. See Ralph E. Shikes and Paula Harper, *Pissarro – His Life and Work*, Quartet Books, London, 1980, p. 43. On *le paysage historique* and its alternative, *le paysage campagnard*, see Albert Boime, *The Academy & French Painting in the Nineteenth Century*, Phaidon, London, 1971, pp. 133–48. It is instructive to compare the illustrations

of paintings by Prix de Rome contestants in Boime's book (plates 121–124) with the Yale drawing.

42 Antoine Quatremère de Quincy, *An Essay on the Nature, the End and the Means of Imitation in the Fine Arts*, trans. J.C. Kent, Smith, Elder & Co., London, 1837, excerpted in Charles Harrison, Paul Wood and Jason Gaiger (eds), *Art in Theory 1815–1900*, Blackwell, London, 1998, pp. 123–24.

43 Henry Matthews quoted in C.R. Leslie, *Memoirs of the Life of John Constable* (originally published in 1843), Phaidon, London, 1951, p. 80, footnote 2.

44 ibid., p. 114.

45 Jean Rousseau, "Le Salon de 1866 – IV", *L'Univers Illustré*, Paris, 14 July 1866, p. 448. The relevant section of Rousseau's review can be found on p. 228.

46 Anonymous, "The Impressionists", *Art Age*, New York, March 1886.

47 *The Churchman*, New York, 12 June 1886. For a survey of the American responses to Impressionism, see H. Huth, "Impressionism comes to America", *Gazette de Beaux-Arts*, Paris, April 1946, pp. 226–52.

48 On Pissarro's "vehement opposition to any form of religion", see Joachim Pissarro, "Pissarro's Memory", in *Camille Pissarro – Impressionist Innovator*, The Israel Museum, Jerusalem, 1995, p. 19 passim.

49 "*Que le sublime en art, défini par ses effets, boulverse notre âme [..] tantôt en nous procurant, par la peinture, le sentiment de l'infini ...*" – Charles Blanc, *Grammaire des arts du dessin* (originally published in 1867), École nationale supérieure des beaux-arts, Paris, 2000, p. 90.

50 Paul Signac, "Le Sujet en peinture", (originally published in *L'Encyclopédie française*, 1935) in *D'Eugène Delacroix au néo-impressionnisme* (ed. Françoise Cachin), Hermann, Paris, 1978, pp. 174–90.

51 ibid., p. 174.

52 ibid., p. 175.

53 Claude Vignon, "Variétés – Salon de 1852", excerpted in Jean-Paul Bouillon, Nicole Debreuil-Blondin, et al. (eds), *La Promenade du critique influent*, Hazan, Paris, 1990, p. 30.

54 Jules-Antoine Castagnary, "Exposition du boulevard des Capucines: Les Impressionnistes", *Le Siècle*, Paris, 29 April 1874, p. 3.

55 See above, footnote 5.

56 See above, footnote 25.

57 See Gary Tinterow, "Realist Landscape", in Gary Tinterow and Henri Loyrette, *Origins of Impressionism*, The Metropolitan Museum of Art, New York, 1994, p. 73.

58 Angèle Lamotte, "Le Bouquet de Roses, Propos de Pierre Bonnard recueillis en 1943", *Verve*, vol. v, nos 17 & 18, Paris, August 1947, n.p.

59 Richard Brettell, op. cit. (footnote 6), pp. 5–8.

60 Georges Lecomte, *Exposition Camille Pissarro*, Galeries Durand-Ruel, Paris, February 1892, n.p.

The restless worker

Richard Shiff

fig. 14
Paul Gauguin
The quarries of Le Chou near Pontoise
1882
oil on canvas, 59.4 x 73
National Gallery of Canada, Ottawa,
purchased with the assistance of
a gift from H.S. Southam, Ottawa,
1947 (no.4842)

Restless experiment

Having struggled to recover what I lost,
while not losing what I was able to learn ...

Camille Pissarro, March 1896[1]

If you examine Pissarro's œuvre in its entirety,
despite its fluctuations, [you find] an essentially
intuitive art of consistent distinction.

Paul Gauguin, September 1902[2]

By Camille Pissarro's self-effacing estimation, his enterprise was
a series of intuitions, experiments and risks – some beneficial,
others not. He could only hope that on balance the results would
keep profiting his art. With the allusion to "fluctuations", Paul
Gauguin's account similarly suggests unevenness in the pattern
of Pissarro's creativity.

Yet Gauguin's locution *de belle race* ("of consistent distinction")
seems to imply that, no matter what might be happening at
particular moments, Pissarro would sustain the highest quality
in his work; it would emerge as if it were genetic or inborn.
Quality was not at issue in his case, at least not in the eyes of a
partisan admirer like Gauguin, who had himself been criticised
for unevenness. He, too, experimented with an unusually wide
range of techniques and media – gaining, losing, gaining again.

Viewers should not expect "perfection" from Gauguin, Odilon Redon
remarked, for "the powerful stream of his [diverse] production
prevents him from achieving it".[3] To similar effect, Maurice Denis
referred to an excess of artistic energy – Gauguin's "inexhaustible
imagination".[4] Such an attribute was double-edged: it enhanced
the artist's creativity while keeping it unruly. Gauguin accepted
this feature as his own, and perceived it in Pissarro, as well

– "an excessive and unfailing drive".[5] He made this statement
in 1902, two decades after the period of his close alliance with
Pissarro. Long after his primitivist and symbolist art had diverged
from the Impressionism of his old master, Gauguin considered
Pissarro as a force with which future artists would have to reckon.

Like Gauguin, Pissarro remained forever restless. His creativity
meandered, turning in every direction for inspiration. Some called
him imitative, whereas to others he appeared unanchored and
indiscriminate.[6] Whatever else he may have been, he was pure
enlightenment to Gauguin. He tutored and encouraged his younger
colleague, especially in 1879 and during the early 1880s, although
never without reservation concerning Gauguin's character and
values.[7] Instinctively aggressive, Gauguin seems to have competed
with Pissarro, even while learning from him. Their paintings show
this, for Gauguin exaggerated Pissarro's Impressionist technique of
applying paint additively, in discrete strokes, each stroke presumably
corresponding to a bit of sensation, each conveying a quality of
immediacy.[8] In Gauguin's version, the facture sometimes appears
more assertive, but usually becomes more regular, evidently self-
consciously so and therefore less spontaneous.

In a number of paintings created in the company of Pissarro, such
as *The quarries of Le Chou near Pontoise,* 1882 (fig. 14), Gauguin
stressed the inherent rhythms of the clustered and patterned
brushmarks, so that a decorative arabesque appears to flow through
the natural scene. When he painted the same rising pathway seen
in Pissarro's *Steep road at Osny,* 1883 (cat. 46), he turned his linear
marks in the direction of the road, converting the motif into a single
bold curve, whereas many of the strokes within Pissarro's complexly
textured form were transversal. The simplified technique of
Gauguin's version (*Landscape at Osny*, 1883; Ny Carlsberg Glyptotek,
Copenhagen) flattened what remained of conventional illusionistic
space; this was his wilful alteration of Pissarro's model, which was
already radical enough.[9] In fact, at the time of their close association,
Pissarro worried over his own "crude, rough-cut execution":

"I would like to have a smoother application, nevertheless retaining the same qualities of wildness [*qualités sauvages*], while eliminating the distracting harshness that allows my canvases to be viewed only when lit from the front"[10] – thus, minimising the visible roughness of the surface.

As soon as Gauguin acquired the Impressionist effects he perceived in Pissarro's works, he made them more pronounced. Yet Pissarro's own experimental variations on Impressionist technique – including his remarkable etchings of 1879 (see *Rainy effect*, cat. 29, *Twilight with haystacks*, cat. 31, *Horizontal landscape*, cat. 37) – were in many ways more extreme. It seems that one "wild" and restless soul was leading the other to experiment in turn, with Gauguin readily accepting the challenge but Pissarro retaining the lead.[11] Pissarro's initial efforts to render the sensation of nature emphasised the materiality of his viscous paint, whether applied by brush or palette knife. Such a direct "touch" may have been wild and primitive, but it immediately freed Pissarro of the tasteful lines and shapes that structured conventional depiction. Despite the linear branching of its pictured trees, his *Near Sydenham Hill*, 1871 (cat. 9) reveals no linear edges or contours in the depiction itself, nor are its strokes distinct enough to be viewed in isolation, as if each were charged with an identifiable emotion. Impressionist naturalism had many precedents, but it avoided the institutionalised residues of the era's preeminent models: Ingres' classical delineation (too standardised) and Delacroix's romantic gesture (too theatrical).[12] Pissarro's technique relied instead on layering colours over and beside each other so that the surface became both earthy (thick with pigment) and atmospheric (chromatically bright yet subtly nuanced, like rainbow light). This way of painting may raise a contradiction in logic, but not in sensation: Pissarro shows that a densely stroked surface of clotted paint can generate the sensation of air, light and transparency. His touch is as solid and weighty as the air it renders is vaporous. A sympathetic writer later described the situation this way: "The brightness of his palette envelops objects in atmosphere.

... He paints the smell of the earth."[13] Ambient air can be thick with odour, temperature and flow, and this, as well as the position and ponderation of things, is the effect Pissarro conveyed. Gauguin was impressed.

The extraordinary sensory presence of *Near Sydenham Hill* and other works of its type did not prevent the restless painter from exploring alternatives. Throughout his career, he experimented with mixing spectral colours to create a grey light that would shine with the chromatic force of the same colours in their saturated purity. This project led to two very different kinds of image: *Piette's house, Montfoucault, in the snow*, 1874 (cat. 23) and *The artist's palette with a landscape*, c.1878 (cat. 51). For *Piette's house*, Pissarro created a surface of greyish tones brushed over a brownish ground, in which traces of the component primary colours remain visible, often as no more than the track of a single bristle of the brush that must have been mixing luminous "greys" from reds, yellows and blues. As if to indicate his method, Pissarro signed the picture in brilliant pinkish violet, overlaid with a more neutral greenish blue – presumably two of the hues submerged within the chromatic greys. For *The artist's palette*, he allowed his six component colours – white, yellow, orange-red, violet-red, blue, green – to remain in place, unblended, as they would be on a working palette. Within the palette, he demonstrated how an Impressionist picture could be created with the same limited number of colours applied as separate strokes juxtaposed and also blended, to produce a full range of harmonised hue and value. In 1890, Lucien Pissarro drew his father holding a palette comprising six colours, perhaps the same spectral hues (including white) from which all effects of Impressionist light could be formed, whether "grey" or fully chromatic.[14]

Pissarro's cross-generational adaptation of George Seurat's Neo-Impressionist technique may have been the riskiest of his experiments. He met Seurat in October 1885, and for the next four years or so approached the much younger painter's pointillism through his own kind of stroke-by-stroke separation of colour

fig. 15
Georges Seurat
Study for *Le Bec du Hoc, Grandcamp*
1885
oil on panel, 15.6 x 24.5
National Gallery of Australia, Canberra,
purchased with proceeds from
The Great Impressionists exhibition, 1984

(often called divisionism, although both Impressionists and Neo-Impressionists were "divisionists"). *Apple picking at Éragny-sur-Epte*, 1888 (cat. 54) and *L'Ile Lacroix, Rouen (effect of fog)*, 1888 (cat. 57), although they appear different from each other, are fully mature examples of Pissarro's Neo-Impressionism. But the modest *Ploughing at Éragny*, c.1886 (cat. 53) may be still more experimental, for here Pissarro adopted an alien application of paint, along with an unusual format and type of support. A revealing comparison could be made with any of Seurat's small panel paintings of the period, such as *Study for 'Le Bec du Hoc, Grandcamp'*, 1885 (fig. 15). If Seurat became a figure on whom Pissarro's restless aesthetic curiosity could settle, it never settled entirely comfortably. To Seurat's disciplined structural application of tiny strokes of contrasting hue, Pissarro brought impetuous movement – perhaps he became to Seurat what Gauguin had been to him. By Seurat's standards, Pissarro introduced too many secondary mixtures of colours and allowed the play of his brushwork to become too gestural; his marks flow across the diminutive surface of *Ploughing at Éragny* in large, bold waves. The disparity between the two techniques corresponds to Pissarro's explanation of his abandonment of Neo-Impressionism proper: having tested out the pointillist method in many guises, he determined that it made it "impossible to pursue fleeting sensations, in order to give [to the image] life and movement".[15] Restless as ever, he seemed to reverse his proposition only two days later: "We are sick with the fever of painting quickly, which hardly suits [serious] art."[16]

Usually, the more numerous and the more densely applied the painter's marks, the more labour is involved in articulating the surface. Seurat impressed his peers with his capacity for work, and so did Pissarro. *Ploughing at Éragny* exhibits the intensity of painter's work, while it also illustrates field workers' labour – perhaps *this* painter's most common theme at the time. With *Ploughing at Éragny* and analogous images, the work of art encounters the work of industry and society. An analogous confrontation occurs in many

of Pissarro's designs for fans – a challenging format for composition (in this respect, like painting on a palette) – in which scenes of agrarian, peasant labour have been set into a shape connoting urban, bourgeois leisure. For *Peasant women placing pea-sticks in the ground*, 1890 (fig. 16), Pissarro arranged the figures so that their performance of work becomes a rhythmic pictorial movement. He created a pictorial beauty corresponding to the beauty of a type of vigorous work, revealed in its potential to offer more sensory pleasure than many forms of leisure.[17]

Restless work

What a poet you are, my dear Pissarro, and what a worker [ouvrier]!

Octave Mirbeau, October 1891[18]

Camille Pissarro has been a revolutionary through the revitalised working methods [renouvellements ouvriers] with which he has endowed painting.

Octave Mirbeau, February 1892[19]

Pissarro was sixty-one when he received these two compliments from the critic Octave Mirbeau, who had become his friend. Despite his years, he was working as hard as ever. Unlike Cézanne, nearly a decade younger, Pissarro does not seem to have thought of himself as old, nor did his admirers: "[He has] unchanging spiritual youth [and] the look of an ancestor who remained a young man."[20] This extremely active painter was nevertheless destined to be typecast as a past master by a younger generation of self-proclaimed symbolists. When this happened – particularly with the publication of Albert Aurier's "Le symbolisme en peinture" in March 1891 – Pissarro's

fig. 16
Camille Pissarro
*Peasant women placing
pea-sticks in the ground*
1890
gouache and chalk, 39 x 60.2
Ashmolean Museum, Oxford

special sense of the "work" of painting became central to the redeeming social value his advocates perceived in his art.

In "Le symbolisme en peinture", Aurier used the example of Gauguin to identify a practice of painting analogous to the new symbolism in literature. By 1891, of course, Gauguin's art had evolved, and Aurier set his procedures and aims against those of Pissarro and Claude Monet, representatives of an older Impressionist generation. He implied that Impressionists resembled naturalist novelists rather than symbolist poets; despite the personalised and often quite idiosyncratic nature of their observations, they sought nothing more than the materiality of things, their "reality". Impressionism, Aurier argued, amounted to realism – albeit realism in subjective form, "with all the deformations of a rapid, subjective synthesis". Subjectivity was more of a primary element in the new symbolism rather than an inescapable by-product, so Aurier gave Pissarro and Monet little credit for manifesting it themselves. It was as if these Impressionists had somehow trivialised subjectivity by recording it naively in their paintings; to Aurier, they seemed to translate and transform their sensory impressions without revealing idealist, mystical or imaginative content of any sort. Gauguin departed from such mundane realism. He took less interest in representing the character of an object or a scene, considering instead that the things of material life were signs of things "beyond": thoughts, dreams, transcendental ideas, emotions of every kind.[21]

Clearly, Aurier sought to differentiate between two modes of painting, neither of them truly conventional. In a grander scheme, both might be regarded as emotionally expressive, subjectively "poetic", ideologically challenging and sufficiently innovative in their technical operations to confound the typical bourgeois collector.[22] Given our residual tendency to separate Impressionist from post-Impressionist styles and meanings, it appears that Aurier's strategy succeeded. Despite the brilliance of his polemic, however, he was no more consistent from essay to essay than other critics of his day, caught up in shifting patterns of alliance and competition within the dense Parisian network of artists, writers, dealers and collectors. Although he emphasised the limitations of Impressionism in 1891, just a year later he would virtually exempt Monet, lauding his "divine radiances".[23] And only a year previously, in March 1890, he had been admiring Pissarro's "poetry", his capacity to discover character, thought and emotion in both the objects to be represented and their eventual pictorial forms.[24] On that occasion, reviewing works of the late 1880s, Aurier stressed Pissarro's interest in Neo-Impressionist technique and intense effects of colour. What the painter actually thought in return is unknown, but the critic's remarks elicited his grateful acknowledgment, as was customary.[25]

At one moment, in 1890, Pissarro is the imaginative poet; at the next moment, in 1891, he becomes a mere imitator of mundane reality. Needless to say, Aurier's writing irritated and frustrated Pissarro, if only because it seemed to waver in its evaluation, reducing the Impressionist position to a caricature. This was an opinion worthy of a conservative traditionalist, not a reformer or renovator, as Aurier, with his philosophical allusions to Neoplatonism, was pretending to be.

Aurier may have functioned in Pissarro's mind as the most prominent among a number of critics (surrogates for Gauguin) whom, he felt, represented Gauguin's position, exaggerating the originality of the "symbolist" achievement, just as Gauguin did. In January 1891, Pissarro nevertheless contributed to the campaign to gather critical support for Gauguin's Tahitian project, lobbying to assure the participation of Mirbeau.[26] Only three months later, however, around the time he read Aurier's essay, he was referring to Gauguin as a "schemer" who would "crush whoever stands in [his] way", an artist well aware of the reactionary social movement towards religiosity, superstition and mysticism, and quite willing to take advantage of it.[27] So, Pissarro's complaint against Gauguin – and, eventually, against Aurier, too – associated intellectual sophistry and personal corruption with generalised social decay and political failure.

fig. 17
Camille Pissarro
Stevedores
illustration for *La Plume* 5
1 May 1893
Research Library, The Getty Research
Institute, Los Angeles

Pissarro had a copy of Aurier's "Le symbolisme en peinture" which he annotated, demonstrating how perturbed he was by the critic's logic of (false) distinctions.[28] His marginal notes appear in the form of ironic gestures of shock and disbelief, actually a rather typical way for him to express himself verbally. At times, his commentary pushes Aurier's points to absurdity, seriously undermining the argument. He suggests, for example, that it should take more to be the vaunted "ideist" than the obfuscations offered by Aurier's words or Gauguin's images: "You aren't ideist just because you're ridiculously obscure, inharmonious, [and] formless on purpose."[29] (Aurier used *ideist* as a neologism in avoidance of the expected term *idealist*, which would have implied conventional academic practice. Ideas were essential, primary, generative, whereas an idealist's ideal was a secondary concept, derivative and somewhat artificial. Set against *idealist*, *ideist* had its advantages, but the word also verged on preciosity.)

Pissarro's objection to Aurier and symbolism was twofold.[30] First, he could see no fundamental difference between the artistic qualities of the old Impressionism and the new symbolism; in their content, both were, or should have become, poetic and idealist – or "ideist", to use the symbolist's term. Both art forms should also be individualistic yet fully rational in their technique, so that their imagery would be accessible. Pissarro sent his annotated Aurier to his son Lucien, who certainly agreed. In fact, Lucien discovered a point worth disputing that his father had missed: there could be no difference between the definitive quality that Aurier ascribed to symbolist practice – *émotivité* (emotivity, emotionalisation) – and what the Impressionists had always called *sensation*, an impassioned, personalised experience of sensory perception.[31] Second, Pissarro believed that there was a political failing in Aurier's symbolism, just as there was in Gauguin himself. His final marginal note announces that art "must respond to social ideas and not to the return of mysticism".[32] On one hand, he was countering Aurier by denying symbolists had any exclusive claim to visual poetry or idealisation;

on the other hand, he was challenging their grasp of political needs and realities, and even their commitment to a reputable, progressive set of values. He labelled Aurier and his symbolist cohort as counter-revolutionary, however unwittingly they may have fallen into the role (which perhaps made it that much worse).[33]

Pissarro's own politics were anarchist, corresponding to a programme advocating decentralised social organisation and industry with the aim of assuring beneficial working conditions, leisure time and the greatest opportunity for aesthetic development in the individual.[34] In one of its less utopian social roles, art represented the potential for individual autonomy within an urbanised society regarded as increasingly depersonalised and alienating. Creative artists, at least to the extent that they avoided venality, would become emblematic figures of the self-fulfilled individual. The anarchist project could hardly be fostered by a symbolist art that seemed socially and intellectually elitist or esoteric, for such an art would never speak to the "masses" whom Pissarro and other anarchists championed – rural peasants, urban labourers and also, perhaps, all who identified with the simple values of a life of work, whether or not they belonged economically to the so-called working classes.[35] Living out his life at the margins where expanding towns abutted rural hamlets (such as Osny, where he and Gauguin painted during the early 1880s), Pissarro considered himself a worker. His work was art – very often the art of representing workers and work (see fig. 16 and 17, as well as *Woman emptying a wheelbarrow*, 1880, cat. 41, *Sower*, 1896, cat. 72, *Woman sewing*, 1895, cat. 98). From Aurier's essay on Gauguin, Pissarro surmised that symbolist art might develop out of musings and fantasy alone, with hardly any physical work involved. Take Aurier's position "to its absurdity, and it is no longer necessary to draw or paint!"[36] he commented.

The two broad complaints Pissarro lodged against Aurier's symbolism – that it involved no sincere artistic "work" and that it neither reflected nor furthered proper social values – found expression before and after the fact of "Le symbolisme en peinture" in writings by two critic friends who shared his anarchism. One was Mirbeau, a well-established novelist and journalist, and the other was Georges Lecomte, an emerging literary figure, destined to be Pissarro's biographer. In December 1891, a number of months after "Le symbolisme en peinture" was published, Mirbeau encountered Aurier on the street and reported back to Pissarro: "I explained to him that Camille Pissarro was a great *idealist*."[37] To which, as to everything else Mirbeau asserted, Aurier placidly concurred, apparently unable to explain or defend his contrary views: "Smiling stupidly, he agrees with everything ... People like this, Pissarro, will make me end up liking Boileau [the epitome of traditionalism]." With the same sarcasm, Mirbeau suggested that so much "Gauguinesque foolishness" made him long for Ingres' drawings.[38]

Around 1891, as if in response to Aurier's "symbolist" claims, Mirbeau and Lecomte appear to have adjusted their critical vocabulary to accommodate the symbolist stress on an internal, imaginative, "poetic" source for artistic expression (hence, Mirbeau's insistence on the "idealism" of Pissarro's art). Lecomte, for example, in his assessment of how art had evolved by 1892, argued that Impressionists had liberated themselves from empirical attempts to record precise appearances; they had become painters of "imaginative vision (*le rêve*)". It followed that symbolists were "less innovative than they think" – so much for the differences Aurier presumed.[39] Another of Lecomte's statements may have been directed at Aurier's claim that Pissarro engaged in imitative, materialistic "realism". The critic applied a quasi-Baudelairean theory, entailing a dual commitment to the particular and the universal – a proposal that might satisfy old Impressionists and young symbolists alike. Pissarro's talent, he argued, was not simply "realist"; instead, by fixing "transitory effects" according to his "personal intervention", he devoted himself to the "work of philosophical synthesis". In the end, he attained a "quality of Permanent Beauty" – the capitalisation, Lecomte's, competes with Aurier's philosophical pretensions.[40]

Perhaps the most rewarding feature of Lecomte's writings on Pissarro is the sense of artistic transformation and process, the movement from spontaneous yet selective observation to representation and thematised idea, all facilitated by the painter's technical mastery. Lecomte attended well to details of formal organisation: "Pissarro modifies lines according to the direction of his ornamental motif ... An arching tree, casting a curved shadow on the ground, will inaugurate an arabesque completed by the bent back of a peasant woman." The writer is describing the painter's pictorial elements and devices, source of the visual poetry of an anarchist harmony alive in a healthy relationship between worker and nature, labour and earth (compare *Apple picking at Éragny-sur-Epte*, 1888, cat. 54). Lecomte developed an imaginative literary conceit, following the lead of the pictorial: "The robust health of the peasant women [with] their fresh skin tones shines forth in this decor of sunlight and greenery."[41] It remains to ask if all this "health" derives from direct observation of peasants in their environment, or from the coordination of an artist's temperament, desire and cultivated skill, or from the critic's crafted metaphor.

The answer must take into account the fact that Lecomte had adopted aspects of contemporary psychological and aesthetic theory, including the belief in merging subject and object in experience, whether that of a rural painter, an urban writer or a peasant in the fields. A statement by the aesthetician Gabriel Séailles is typical of the thinking of the era: "Art is nothing but the image of an imaginative vision [*l'image d'un rêve*] reflected in an appearance it has created and which reproduces it ... It unites material substance and immaterial thought."[42] Intellectuals of Lecomte's time conceived of representation operating reciprocally; a pictorial image reconfirmed and strengthened the mentality that brought it to life. Under such conditions, art becomes self-fulfilling, the material and gestural realisation of the artist's being – the manifestation of his or her thoughts, desires, emotions and imaginative phantasms. These are the qualities which Pissarro would isolate as *sensation* and *émotion*, and which Aurier – using a vocabulary that masked external material factors – might identify with *désir, idée* or *rêve*. Lecomte and Mirbeau did not hesitate to turn Aurier's own language towards the defence and praise of Pissarro. And much of this language could be found in the writings of mutually admired sources, such as Stéphane Mallarmé: "To *suggest* [the object], there you have the dream [*rêve*] ... gradually to evoke an object in order to show a state of mind."[43]

As the manifestation and satisfaction of desire, emotion, imagination and other psychological forces, artistic creation becomes inherently rewarding, a kind of unalienated labour to be repeated without end, and very much identified with life itself. But such artistic activity easily becomes obsessive. It did so, notoriously, in the case of Vincent van Gogh – at least in Aurier's admiring account, and in Mirbeau's, as well, when he claimed that the painter's overwhelming compulsion to create led him to work himself to death.[44] If van Gogh represented art taken to excess, Pissarro, known as a tireless worker, represented the healthy and sound version of the same restless desire to create – or, as both Mirbeau and Pissarro might have said, to work. Work was, or at least should be (as anarchists knew), creative and self-affirming: "Work [*travail*] is a marvellous regulator of both mental and physical health."[45] The quotation belongs to Pissarro, as recollected by Mirbeau after the painter's death, in testimony to a most productive life. Van Gogh died young; the more regulated Pissarro survived to full maturity.

Mirbeau published a carefully crafted essay on Pissarro in January 1891, shortly before Aurier's "Le symbolisme en peinture" appeared. Although Mirbeau seems to have been somewhat displeased with it, it formed the gist of two subsequent publications about his friend.[46] It introduced Pissarro as an anarchist (without using the word) concerned with social equality and "harmony", one who manifested in his own being a "moral harmony" that united him with his "work [*oeuvre*]". Pissarro was no vain elitist, nor did he pretend to aesthetic mystification: "He firmly believes that the painter 'exists within [ordinary] humanity', just as the poet, the farmer, the doctor, the blacksmith [and other workers] ... Paintings should no longer constitute inaccessible treasures or sacred religious fetishes, but contributions to society valued like all other productions of human enterprise [*travail*] or genius."[47]

Discussing the paintings themselves, Mirbeau tended to refer to pictorial qualities as opposed to specifics of subject matter. His descriptive language emphasised the colour, light, rhythm and general liveliness of Pissarro's paintings, contrasting them with the relatively dark, stiff, stage-set qualities of his Barbizon predecessors, Théodore Rousseau and Jean-François Millet (an unfair comparison, perhaps motivated by the social and religious conservatism attributed to the Barbizon masters).[48] The reader surmises that the living harmony of Pissarro's forms corresponds to the ideal social harmony anarchists imagine – their *rêve* or dream, so to speak. Formal harmonies represent this ideal as much or more than any other aspect of painting.[49] Accordingly, Mirbeau developed analogies of human and pictorial organicism. Because of the way Pissarro painted them, it was as if his images of nature became living, breathing organisms, possessing vital fluids. One of Mirbeau's crucial words was *sève* (literally, the sap of a plant), by which he meant an energy and a flow in man and nature. This word was applied similarly in Aurier's 1890 essay on Pissarro, in appreciation of a landscape: "We sense the silent rising of saps, the delightedly animated life of cells."[50] Criticism acknowledged Pissarro's paintings as alive, animated, vivid.

Noting thematic analogies between Millet and Pissarro but differences in the degree of integrative harmony, Mirbeau concluded his account with another striking organicist metaphor: "The earth is for [Millet] merely a backdrop against which he places his figures in relief ... In [Pissarro's] works, man is ... fused with the earth, where he appears solely in his function as human plant [*plante humaine*]."[51] Millet, it seems, set his figures against their ground, causing them to remain apart, whereas Pissarro "planted" his figures, as if he were realising an agrarian ideal as much as creating pictorial harmony. Mirbeau's phrasing also evokes the argot, *une belle plante*, an exemplary specimen of human life: Pissarro's artistic labours kept him healthy and (as colloquial English puts it) preserved this "fine specimen".[52]

Social work

Oh! to have a fine, healthy practice of art, like old Corot, like Claude Monet, like Camille Pissarro! Is it not also about fantasy [*rêve*], their painting? ... They are such profound workers [*ouvriers*]!

"Lucien", in Octave Mirbeau's *Dans le ciel*, serialised September 1892–May 1893[53]

Alluding to Pissarro's "human plant", Mirbeau also referred to "broad [formal] generalisations".[54] The one was as healthy a sign as the other, and an analogy was implicit: the free-flowing form that derives from an artist's creative work is like the form of a life fulfilled through productive partnership with the environment. Pissarro echoed this notion in a remark to Mirbeau in April 1892. The occasion was the painter's response to a challenge in anarchist Peter Kropotkin's book, *La conquête du pain*: "The love of the *land* and of what grows on it is not acquired by making studies with the brush; it is acquired only in service to the land. And without loving it, how can it be painted?"[55] To which Pissarro replied: "Let us be artists first and we will have the ability to experience everything, even a landscape, without being a peasant."[56] He was recommending that artists attend to the harmonious structures a healthy act of painting would reveal through its own working process, as valid as labour on the land. Such individualised expression – an artist's involvement with a natural site or an inventive theme – would

have inherent value. The development of a normative image of natural and social harmony might inspire committed anarchists, but only as an inessential supplement to the work of painting. For Pissarro and his like-minded critics, any sincere artistic creation functioned directly as political and spiritual affirmation, whereas the symbolist pursuit of an art of religious mysticism was either reactionary or simply misguided.

Mirbeau's fictional character "Lucien", a painter, not only appreciated Pissarro's kind of art, he expressed comparable anarchist views as he fulminated against the State for denying the right "to think, to imagine, to create, even to feel".[57] Whether or not representing workers, Pissarro directed his art at personal feeling. He told his son Lucien, also a painter: "You must have persistence, a strong will and liberated sensations – relieved of everything other than their own feeling."[58]

Dans le ciel belongs to 1892, and so does Pissarro's *The conversation (Two young peasant women)* (fig. 18). The art historian John House provides a moving description of it: "Two female figures, one in reverie, one resting from work, face each other, each characterised by what lies beyond her – a shady bank [also fruit trees] beyond the *rêveuse* [the daydreamer], a staked field beyond the woman who has been working; the forms of the landscape enfold them and define their spheres of activity, presenting a harmonious coexistence between women and nature."[59] With this organic harmony comes a harmonious social circuit of (ungendered) labour and leisure, also expressed by *The conversation*: a harmony between the human functions of physical work and mental imagination, between body and mind, between these convergent sources of sensation. Just as the two states coexist in Pissarro's depiction, harmonious in their composition, so they become integrated in the life of every individual in a society dedicated to cultivating feeling – aesthetic, emotional, intellectual – as successfully as it cultivates plants.[60] Pissarro's pensive woman, with head resting on hand and appearing to gaze inwards, contrasts with the more upright figure who leans

fig. 18

Camille Pissarro
The conversation
(Two young peasant women)
1892
oil on canvas, 89.5 x 116.5
The Metropolitan Museum of Art,
gift of Mr and Mrs Charles Wrightsman,
1973 (1973.311.5)

against her implement of work and seems, by comparison, to turn her attention outwards. Through the pictorial device of the women's "chat" or conversational exchange, Pissarro suggests the possibility of bridging their difference, undoing the separation between rural labour and a labourer's imaginative, mental life (the dream, or *rêve*).

Work in the future anarchist society was to yield not only material sustenance but the leisure to satisfy artistic, philosophical and intellectual needs previously met only in the life of the economically privileged. The thematics of *The conversation* accorded not only with Mirbeau's *Dans le ciel* but with Kropotkin's *La conquête du pain*: "The five to seven hours a day available to each [worker] after having devoted several hours to the production of [material] necessities will amply suffice to satisfy all needs for luxury, [which] would be fulfilled by art."[61] Since artists' customary work integrated physical execution (Pissarro's *renouvellements ouvriers*) and mental projection (*le rêve*), it became an effort that was its own reward. Enjoyable work was already luxury, a rival to leisure.

Pissarro, Mirbeau and Lecomte might have been in unqualified agreement with Aurier on one point: "symbolist" painters such as van Gogh and Gauguin, reaching beyond reason and common understanding, shocked the bourgeois with their excesses. In van Gogh's case, as Aurier wrote, the result was "often sublime, sometimes grotesque, [and] always revealing something all but pathological", which healthy – that is, normal – individuals would never have experienced.[62] Pissarro's art, by contrast, expressed its worker's thoughts and feelings through application of a certain kind of pictorial science – that is to say, through restless experimentation, perhaps more intuition than logic. Yet his liberating individualism, manifested by an imaginative visual poetry as much as by technical innovations, could appear as incomprehensible as van Gogh's pathology to the bourgeois capitalist collector whose dulled senses accommodated only the easiest satisfactions. Gauguin, after all, had characterised Pissarro's artistic will as "excessive".[63] Despite their shock factor, Pissarro's intense colours and compellingly rhythmical

forms derived not from sickness and instability but from the health and sanity of an integrated personality. As Lecomte put it, "Powers of reason ceaselessly support [his] exquisite, forceful instinct."[64] Similarly, Mirbeau identified Pissarro as "a robust artist – healthy".[65]

Recall Pissarro's own statement on his condition: "Work is a marvellous regulator of both mental and physical health." Although marketed like a commodity, the work of painting became therapeutic for both artist and viewer of Pissarro's generation, and this function has remained associated with artistic production to the present time. Pissarro's anarchist utopia, however, has never been attained; with few exceptions, workers have not become worker–artists. So, there remains a question as to whether or not the work of art (in the double, metonymic sense of "work", as both activity and product) lost too much of its critical and transformative capacity when it became associated with leisure of the bourgeois sort.[66] For better or worse, the therapeutic, leisure value of art has been repeatedly acknowledged as a social benefit, certainly by the psychoanalytic profession, although art historians often seem to regard this connection as demeaning, perhaps because it undermines elitist distinctions. Yet this, in turn, is precisely why Pissarro painted – to eliminate hierarchy and privilege.

1 Pissarro, draft of a letter to Henry Van de Velde, 27 March 1891, in Janine Bailly-Herzberg, ed., *Correspondance de Camille Pissarro*, 5 vols. (Paris: Presses Universitaires de France [vol. 1], Valhermeil [vols. 2–5], 1980–91), 4:179; hereafter, *CCP*. The present essay extends the arguments of several related publications: see Richard Shiff, "The Work of Painting: Camille Pissarro and Symbolism", *Apollo* 136 (November 1992): 307–10; "To Move the Eyes: Impressionism, Symbolism, and Well-being, c.1891", in Richard Hobbs, ed., *Impressions of French Modernity: Art and Literature in France 1850–1900* (Manchester: Manchester University Press, 1998), 190–210; "The Primitive of Everyone Else's Way", in Guillermo Solana, ed., *Gauguin and the Origins of Symbolism* (Madrid: Fundación Colección Thyssen-Bornemisza, 2004), 64–79. I thank Charlotte Cousins and Adrian Kohn for essential aid in research. All translations are mine.

2 Paul Gauguin, *Racontars de rapin* (Paris: Falaize, 1951), 35. Gauguin dated this manuscript September 1902.

3 Odilon Redon, statement in Charles Morice, ed., "Quelques opinions sur Paul Gauguin", *Mercure de France* 48 (November 1903): 428–29. Redon was offering a summary opinion of Gauguin, who died in May 1903; Pissarro died several months later, in November 1903.

4 Maurice Denis, "L'influence de Paul Gauguin" (1903), *Théories 1890–1910: Du Symbolisme et de Gauguin vers un nouvel ordre classique* (Paris: Rouart et Watelin, 1920), 168.

5 Gauguin, *Racontars de rapin*, 35.

6 In 1881, Paul Mantz labelled Pissarro a "mannerist" whose professions of aesthetic liberation and fidelity to nature were deceptive or self-deceptive (Mantz, "Exposition des oeuvres des artistes indépendants" [23 April 1881], in Ruth Berson, ed., *The New Painting, Impressionism 1874–1886: Documentation*, 2 vols. (San Francisco: The Fine Arts Museums of San Francisco, 1996], 1:357). Among critics who commented on the series of Impressionist exhibitions, Mantz (1821–95) was one of the few who were older than Pissarro. Given the range of his perspective, he referred to Camille Corot of the 1830s and 1840s as "the greatest Impressionist", whose practice defined "the golden age of Impressionism". Accordingly, Mantz regarded those who followed Corot's teaching during the 1870s – this would include Pissarro and Gauguin – as decadent imitators rather than healthy experimenters (Mantz, "L'exposition des peintres impressionnistes" [22 April 1877], *The New Painting, Impressionism 1874–1886: Documentation*, 1:166). In fact, at the Salons of 1864 and 1865, Pissarro identified

himself as a pupil of Corot and often sought the older painter's advice during that decade (see John Rewald, *The History of Impressionism* [New York: Museum of Modern Art, 1961], 101). His contemporaneous works (cat. 3, 4, 5 and 7) have a Corot-like sense of colour and tonality. The association of Corot's style with later Impressionism remained strong. Nearing death, art dealer Paul Durand-Ruel, who represented both Corot and Pissarro, remarked that he "always imagined [Paradise] with the gentle harmoniousness of a landscape by Corot or Camille Pissarro" (see Georges Lecomte, *Ma traversée* [Paris: Robert Laffont, 1949], 139).

7 See Pissarro, letter to his son Lucien, 31 October 1883, *CCP*, 1:245–246.

8 On Impressionist sensation, see Richard Shiff, "Sensation, Movement, Cézanne", in Terence Maloon, ed., *Classic Cézanne* (Sydney: Art Gallery of New South Wales, 1998), 13–27.

9 On Pissarro and Gauguin at Osny, see Anne-Birgitte Fonsmark, *Manet, Gauguin, Rodin … Chefs-d'oeuvre de la Ny Carlsberg Glyptotek de Copenhague* (Paris: Réunion des Musées Nationaux, 1995), 122–25.

10 Pissarro, letter to Lucien, 4 May 1883, *CCP*, 1:202.

11 This seems to have been Gauguin's mode of operation – during the 1880s, he also learnt from Paul Cézanne, Edgar Degas and others. He was recognised for his ability to appropriate an alien style, suppressing his own aesthetic personality (see Charles Morice, "Salons et Salonnets", *Mercure de France* 10 [January 1894]: 68).

12 On the context for precedents to Impressionism, see Richard Shiff, "Natural, Personal, Pictorial: Corot and the Painter's Mark", in Andreas Burmester, Christoph Heilmann and Michael F. Zimmermann, eds., *Barbizon: Malerei der Natur – Natur der Malerei* (Munich: Klinkhardt & Biermann, 1999), 120–38.

13 Octave Mirbeau, "L'Exposition Internationale de la rue de Sèze (II)" (1887), *Combats esthétiques*, ed. Pierre Michel and Jean-François Nivet, 2 vols. (Paris: Séguier, 1993), 1:335.

14 Lucien Pissarro's drawing of his father at work illustrated a biographical account; see Georges Lecomte, "Camille Pissarro", *Les hommes d'aujourd'hui* 8, no 366 (March 1890), n.p. For documentation regarding Pissarro's choices of pigment colours, see Richard Shiff, *Cézanne and the End of Impressionism* (Chicago: University of Chicago Press, 1984), 204–07, 299–300; Anthea Callen, *The Art of Impressionism: Painting technique and the making of modernity* (New Haven: Yale University Press, 2000), 144–51.

15 Pissarro, letter to Henry Van de Velde, 27 March 1896, *CCP*, 4:180. This is the context in which Pissarro referred to regaining what he had lost (see the first epigraph, p. 34). Virtually the same explanation appears in 1890 in Lecomte's biographical study ("Camille Pissarro", *Les hommes d'aujourd'hui*, n.p.), presumably informed by Pissarro himself. See also: Pissarro, letter to Lucien, 6 September 1888, *CCP*, 2:251 ("The [Neo-Impressionist] point [is] more monotonous than simplified"); letter to Lucien, 20 February 1889, *CCP*, 2:266 ("The working method [*travail d'exécution*] does not seem to me quick enough and does not respond immediately enough to sensation").

16 Pissarro, letter to Esther Isaacson, his cousin's daughter, 29 March 1896, *CCP*, 4:183.

17 Compare Kathleen Adler, "*Objets de luxe* or propaganda? Camille Pissarro's Fans", *Apollo* 136 (November 1992): 304.

18 Octave Mirbeau, letter to Pissarro, 18 October 1891, in *Correspondance avec Camille Pissarro*, eds. Pierre Michel and Jean-François Nivet (Tusson: Le Lérot, 1990), 51; hereafter, *CMP*.

19 Octave Mirbeau, "Camille Pissarro" (1 February 1892), *Des artistes*, 2 vols. (Paris: Flammarion, 1922–24), 1:147.

20 Georges Lecomte, "M. Camille Pissarro", *La Plume* 3 (1 September 1891): 302. Soon afterwards, Félix Vallotton stated that Pissarro's art had "never been more vibrant, stronger or younger" (Vallotton, "L'exposition Pissarro", *Gazette de Lausanne*, 7 March 1892, quoted by Terence Maloon and Claire Durand-Ruel Snollaerts, "Pissarro and his critics" in this catalogue, pp. 242–43).

21 Albert Aurier, "Le Symbolisme en peinture: Paul Gauguin", *Mercure de France* 3 (March 1891): 157–58, 162–64.

22 On this cluster of issues, see John House, "Camille Pissarro's Idea of Unity", in Christopher Lloyd, ed., *Studies on Camille Pissarro* (London: Routledge & Kegan Paul, 1986), 15–34; Richard Thomson, *Camille Pissarro: Impressionism, Landscape and Rural Labour* (London: South Bank Centre, 1990); Paul Smith, "'Parbleu': Pissarro and the Political Colour of an Original Vision", *Art History* 15 (June 1992): 223–47. See also Shiff, *Cézanne and the End of Impressionism*, 1–52, 187–89; "Review Article", *Art Bulletin* 66 (December 1984): 681–90.

23 Albert Aurier, "Claude Monet", *Mercure de France* 5 (April 1892): 304.

24 G. Albert Aurier, "Camille Pissaro [sic]", *Revue indépendante* 14 (March 1890), 506–07.

25 See Patricia Townley Mathews, *Aurier's Symbolist*

Art Criticism and Theory (Ann Arbor: UMI, 1986), 114.

26 The campaign also involved Stéphane Mallarmé; see Charles Morice, *Paul Gauguin* (Paris: H. Floury, 1919), 81–82. Mirbeau responded sympathetically: "Imagination (*le rêve*) leads [Gauguin] into majestic contours, towards spiritual synthesis, to eloquent, profound expression" (Mirbeau, "Paul Gauguin" [16 February 1891], *Des artistes*, 1:126).

27 Pissarro, letters to Lucien, 20 April and 13 May 1891, *CCP*, 3:66, 81–82.

28 See Belinda Thomson, "Camille Pissarro and Symbolism: some thoughts prompted by the recent discovery of an annotated article", *Burlington Magazine* 124 (January 1982): 14–23. The document is in the collection of the Ashmolean Museum, Oxford. I thank Joachim Pissarro for supplying a complete copy of Pissarro's annotations.

29 Marginal note, Pissarro's copy of Aurier, "Le Symbolisme en peinture", 159.

30 Compare Pissarro, letters to Lucien, 13 and 20 April 1891, *CCP*, 3:63, 66.

31 Lucien Pissarro, letter to his father, [early] May 1891, in Anne Thorold, ed., *The Letters of Lucien to Camille Pissarro, 1883–1903* (Cambridge: Cambridge University Press, 1993), 217; see also, B. Thomson, "Camille Pissarro and Symbolism", 19.

32 Compare Pissarro, letter to Lucien, 13 April 1891, in which he distinguishes his modern "anarchist philosophy" from the retrogressive "mysticism" to which the younger generation was attracted (*CCP*, 3:63).

33 Pissarro, letter to Lucien, 8 July 1891, *CCP*, 3:103.

34 On the anarchist convictions of Pissarro and his critical supporters, see Eugenia W. Herbert, *The Artist and Social Reform: France and Belgium, 1885–1898* (New Haven: Yale University Press, 1961); Reg Carr, *Anarchism in France: The Case of Octave Mirbeau* (Manchester: Manchester University Press, 1977); Richard Thomson, "Camille Pissarro, 'Turpitudes Sociales' and the Universal Exhibition of 1889", *Arts Magazine* 56 (April 1982): 82–88; R. Thomson, *Camille Pissarro*, 11, 59–65; Ralph E. Shikes, "Pissarro's Political Philosophy and His Art", in Lloyd, *Studies on Camille Pissarro*, 35–54; T.J. Clark, *Farewell to an Idea* (New Haven: Yale University Press, 1999), 55–137. Pissarro contributed images of workers to a number of anarchist essays and projects: see his *Stevedores* [see fig. 17, p. 38] in André Veidaux, "Philosophie de l'anarchie", *La Plume* 5 (1 May 1893): 189; and his letter to Mirbeau, 30 September 1892, *CCP*, 3:261. Writers attached politically charged labels,

such as *anarchiste* and *intransigéant*, to artists, applying these as loose metaphors. Cézanne, hardly an anarchist, was said to make "anarchist's painting" (Théodore Duret, *Histoire des peintres impressionnistes* [Paris: H. Floury, 1922], 146).

35 On art inaccessible to the masses, see Jean Grave, *La société future* (Paris: Stock, 1895), 360.

36 Marginal note, Pissarro's copy of Aurier, "Le Symbolisme en peinture", 159.

37 Mirbeau, letter to Pissarro, 14 December 1891, *CMP*, 75 (original emphasis). For Aurier's use of "idealist", see "Le symbolisme en peinture", 158.

38 Mirbeau, letter to Pissarro, 16 December 1891, *CMP*, 78. See also Pissarro's reply to Mirbeau, 17 December 1891, *CCP*, 3:169.

39 Georges Lecomte, *L'Art impressionniste d'après la collection privée de M. Durand-Ruel* (Paris: Chamerot et Renouard, 1892), 260–61.

40 Lecomte, "M. Camille Pissarro", *La Plume*, 302. The relevance to Aurier was first suggested by B. Thomson, "Camille Pissarro and Symbolism", 23.

41 Lecomte, *L'Art impressionniste*, 79–80.

42 Gabriel Séailles, "L'Origine et les destinées de l'art", *Revue philosophique* 22 (1886), 347.

43 Stéphane Mallarmé, statement of 1891, in Jules Huret, *Enquête sur l'évolution littéraire* (Paris: Charpentier, 1913), 60 (original emphasis).

44 Albert Aurier, "Les Isolés: Vincent van Gogh", *Mercure de France* 1 (January 1890), 24–29; Mirbeau, "Vincent van Gogh" (31 March 1891), "Vincent van Gogh" (17 March 1901), *Des artistes*, 1:130–37, 2:141–48.

45 Mirbeau, "Vincent van Gogh" (1901), "Préface au Catalogue des oeuvres de Camille Pissarro, exposés chez Durand-Ruel du 7 au 30 avril 1904", *Des artistes*, 2:146–47, 226.

46 Octave Mirbeau, "Camille Pissarro", *L'Art dans les deux mondes*, 10 janvier 1891, 83–84; "Camille Pissarro" (1892), "Préface" (1904), *Des artistes*, 1:145–53, 2:224–34. On Mirbeau's dissatisfaction, see his letter to Monet, 22 January 1891, in Octave Mirbeau, *Correspondance avec Claude Monet*, ed. Pierre Michel and Jean-François Nivet (Tusson: Le Lérot, 1990), 114.

47 Mirbeau, "Camille Pissarro", *L'Art dans les deux mondes*, 83.

48 Mirbeau, "Camille Pissarro", *L'Art dans les deux mondes*, 84. Pissarro was often irritated to be regarded as a follower of Millet; see R. Thomson, *Camille Pissarro*, 51.

49 Hence, Pissarro denied that a picture having "anarchist" content must be of workers: "All arts are anarchist when something is beautifully well done!" (letter to Mirbeau, 30 September 1892, *CCP*, 3:261).

50 Aurier, "Camille Pissaro [sic]", 513.

51 Mirbeau, "Camille Pissarro", *L'Art dans les deux mondes*, 84.

52 Along such lines, Mirbeau, an amateur horticulturist, would call van Gogh a "fine human flower": Mirbeau, "Vincent van Gogh" (1891), *Des artistes*, 1:134. Mirbeau also used *sève* in relation to van Gogh (1:136).

53 Octave Mirbeau, *Dans le ciel* (1892–93), ed. Pierre Michel and Jean-François Nivet (Caen: L'Echoppe, 1989), 130–131. In part, Mirbeau modelled the unstable, tormented "Lucien" on van Gogh.

54 Mirbeau, "Camille Pissarro", *L'Art dans les deux mondes*, 84.

55 Pierre Kropotkine, *La conquête du pain* (Paris: Tresse & Stock, 1892), 148 (original emphasis).

56 Pissarro, letter to Mirbeau, 21 April 1892, *CCP*, 3:217.

57 Mirbeau, *Dans le ciel*, 110. Pissarro appreciated Mirbeau's fiction; see his letter to Mirbeau, 4 December 1892, *CCP*, 3:283.

58 Pissarro, letter to Lucien, 9 September 1892, *CCP*, 3:256 (emphasis eliminated).

59 John House, "Camille Pissarro's *Seated Peasant Woman*: The Rhetoric of Inexpressiveness", in John Wilmerding, ed., *Essays in Honor of Paul Mellon* (Washington: National Gallery of Art, 1986), 163.

60 Hence, Pissarro would study the land but would "form [its] true poem" within the contemplative space of his working atelier: Pissarro, quoted in Paul Gsell, "La tradition artistique française: L'impressionnisme", *Revue politique et littéraire* 49 (26 March 1892): 404. Pissarro was unhappy with Gsell's use of his interview material; see his letter to Mirbeau, 21 April 1892, *CCP*, 3:217.

61 Kropotkine, *La conquête du pain*, 151. Compare Grave, *La société future*, 362, 368.

62 Aurier, "Les Isolés: Vincent van Gogh", 26–27.

63 Gauguin, *Racontars de rapin*, 35.

64 Lecomte, "M. Camille Pissarro", *La Plume*, 302.

65 Mirbeau, "Exposition de peinture (1, rue Laffitte)" (21 May 1886), *Combats esthétiques*, 1:278.

66 Compare Robert L. Herbert, *Impressionism: Art, Leisure, and Parisian Society* (New Haven: Yale University Press, 1988), 303–06.

Pissarro's Neo-Impressionism

Joachim Pissarro

fig. 19
Claude Monet
Poppy field in a hollow near Giverny
1885
oil on canvas, 61.5 x 81.3
Museum of Fine Arts, Boston,
Juliana Cheney Edwards Collection

"A new phase in the logical advancement of Impressionism"[1]
– with these words, Camille Pissarro in 1886 described the
beginning of a new art movement: Neo-Impressionism. He would
wholeheartedly embrace Neo-Impressionism, leaving behind the
principles of Impressionism he had espoused since 1874.

The sentence describing this new art movement appeared in a letter
by Pissarro to an American author, Celen Sabbrin. Since Pissarro
had never met the author, he did not realise he was writing to a
woman. He addressed her as "Monsieur Sabbrin", unaware that
Celen Sabbrin was the pseudonym of Helen Abbott (1857–1904),
who had studied piano in Paris in the late 1870s and went on to
become a biochemist and a physician in the US. She was interested
in parallels between science and art, a central concern of Neo-
Impressionism. Pissarro was intrigued by Sabbrin's article, "Science
and Philosophy in Art", in which she interwove metaphysical
interpretations of Claude Monet's paintings with rigorous formal
description. She saw in Monet's poppy field (*Poppy field in a
hollow near Giverny*, 1895, fig. 19) "the beginning of life's course",
imagining Monet's field as a metaphor for the free course of the soul.
"Nature throws no obstacle to her progress; there is no warning
hand to hold the soul from running to her own destruction; and
the indifference of nature to suffering or happiness is terrible to
contemplate,"[2] she wrote. Without transition, she moved to a careful
reading of the elements of the composition:

> The grassy bank is covered with many colored grasses, the
> different colors giving the effect of light and shadow. These
> different patches are formed like triangles. The entire picture
> can be looked upon as the interior of a geometrical solid.
> The poppy field is a parallelogram; diagonal lines run across
> it from one to the opposite corner, and these large triangles
> can in turn be divided into smaller ones.[3]

Alternating "science" and "philosophy" (or geometry and
metaphysics), Sabbrin concluded her analysis of Monet's painting:

It is significant that in one of Monet's highest expressions of
thought, the unbending principles of geometrical form are the
most clearly discernible. It may be claiming too much to say
that mathematical principles are the basis of all truth, but that
the two are nearly related must be acknowledged.[4]

Hitherto, little attention has been given to the impact of this text
on Pissarro. Evidently, he recognised an affinity with what he
and his colleagues were attempting: to inter-relate scientific and
philosophical concerns. Pissarro was so impressed by the text,
he decided to translate it so that his close friends and Neo-
Impressionist colleagues could read it. He sent a copy to Paul
Signac, and another to Félix Fénéon, who did not like it very
much. It is easy to see why it should have struck a chord with
Pissarro, as it reflected two of his central interests. Sabbrin not
only proposed to apply scientific principles to art, her philosophical
conclusions also converged with Pissarro's utopian and anarchistic
beliefs. When Sabbrin discerned in Monet's poppy field an elating
but frightening theme, Pissarro was surely impressed. "Nature
throws no obstacle to her progress; there is no warning hand
to hold the soul from running to her own destruction," Sabbrin
wrote, and perhaps Pissarro felt these lines were meant for him.
The references to nature and human suffering and the miseries
of social injustice were themes close to Pissarro's heart. Musing
on Monet's poppy field, Sabbrin wrote:

> The heart weakens and the soul is faint at what she sees.
> It is the end of the struggle of the human race; all work
> and thought have been of no avail; the fight is over and
> inorganic forces proclaim their victory. The scene is a
> striking reality. Nature is indifferent, and her aspects are
> meaningless, for what indications of the unavoidable end
> come from seeing that gay flowered field? It is a mockery,
> and that mind which has once felt the depth of the
> thoughts expressed in this painting can only seek safety
> in forgetfulness.[5]

Pissarro's enthusiasm for this text was conditioned by the fact that it was written about his close rival, Monet, whose talent Pissarro simultaneously acknowledged yet mistrusted, due to Monet's rebuttal of the "scientific gains" made by Neo-Impressionism. When Sabbrin proclaimed Monet "the philosopher of the Impressionist school", Pissarro may have felt this title was misattributed. Hence, he felt the need to draw the author's attention to the "new phase" in the "logical march of Impressionism" whose concerns were echoed in Sabbrin's article. Pissarro evidently expected that she might feel more empathy for his work and that of the "Néos" than for Monet's. Accompanying his *félicitations* to the author of "Science and Philosophy in Art", Pissarro sent a copy of Fénéon's article, "Les impressionnistes en 1886", published in *La Vogue* (13–20 June 1886), in order to show the kinship between her writing and the thoughts of Félix Fénéon, the theoretical mouthpiece of Neo-Impressionism. "I believe, Sir, that you will find a certain interest in it,"[6] he added.

As far as we know, Pissarro's attempt to stimulate Sabbrin's interest in Neo-Impressionism was unsuccessful, but his enthusiasm for the text is very revealing. This is the only publication he ever translated. It resonated with his interests in a utopian atheism, his conviction that the only way out of mankind's despair was in "seeking safety in forgetfulness" through art. Because of its involvement with science, he looked to Neo-Impressionism to bring art to a higher plane of awareness, engaging with the "inorganic forces" that had caused society's decay.

This is how Pissarro described the situation that prevailed at the beginning of the Neo-Impressionist adventure:

> From all sides I hear … that France is lost, that it's in decadence, that Germany gains ground, that artistic France is beaten by mathematics, that the future belongs to mechanics, engineers, big financiers, Germans and Americans – as if we could foresee the future's surprises. Indeed yes, France is sick, but from what? … That's the question. It's sick from transformation. And it could very well die from it, that's for sure …[7]

Pissarro believed that transformation was inevitable – at the same time, he could recognise negative aspects arising from this transformation. This ambivalence was an integral feature of his experiment with Neo-Impressionism, the first artistic movement dedicated to the reform of Impressionism.

Impressionism and Neo-Impressionism

Pissarro is the only artist who went from Impressionism to Neo-Impressionism, and his statements form links between the two movements. Here, for example, is the definition he gave of Impressionism in 1883:

> … really, Impressionism should only be a theory purely of observation, without forgetting the fantasy, the liberty, the grandeur, and in the end everything that makes an art great.[8]

And here, three years later, is his definition of Neo-Impressionism:

> To search for a modern synthesis through methods that are based in science, in the theory of colours discovered by Mr Chevreul, following the experiments of Maxwell and the measurements of N.O. Rood. To substitute optical mixture for the mixture of pigments – in other words, the decomposition of tones into their constituent elements, because optical mixture stirs up much more intense luminosities than the mixture of pigments. As for execution, we ignore it. It has scant importance – art has nothing to do with it, in our opinion. Originality only consists of the character of the drawing and the particular vision of each artist.[9]

When Pissarro described Neo-Impressionism as a new phase in the logical march of Impressionism, we should bear in mind the definitions he gave of each movement. His definition of Neo-Impressionism is very technical and down to earth, almost like a set of guidelines in an instruction manual. In contrast, his definition of Impressionism is stern and assertive, more personal. Pissarro's style of expression is usually slightly awkward and cumbersome, yet his arguments are cogently and clearly conveyed. Impressionism is a mixture of pure observation and freedom of improvisation, or fantasy. These notions might seem contradictory: if Impressionism is a pure theory of observation, what possible role could it offer to fantasy? On the face of it, Impressionism faithfully records the flux of appearances passing before the artist, but Pissarro insists that pure observation be combined with freedom, fantasy and everything that makes art great. His theory reconciles objectivity with subjectivity. This can be compared with the critic Jules-Antoine Castagnary's formulation, prompted by the first Impressionist exhibition, in 1874:

> They are impressionists in the sense that they render not the landscape but the sensation produced by the landscape.[10]

The argument is identical: Impressionism consists equally of rigorous observation and fantasy. The Impressionists stand halfway between realism and idealism: they do not merely "copy" or "imitate" nature, they paint what they see, *as they see it*. Their subjective sensations are an essential component of their art. Reality is not separable from the way it is seen. As Zola often repeated in reference to Manet's and Pissarro's works, a work of art was "a bit of nature seen through a temperament".[11]

The danger of this fragile equilibrium of precise observation and free and fanciful interpretation is that one side of the equation can always outweigh the other. Interestingly, Castagnary, in his first text written on the Impressionists, foresaw the problem that would bring dissension to the ranks of the Impressionists. Virtually predicting the Neo-Impressionist crisis twelve years later, Castagnary wrote in 1874: "In a few years, the artists who today are gathered in the boulevard des Capucines will have split up." The division would arise between those who could draw perfectly and those who gave free rein to their dreams and imagination (such as Cézanne):

> Those who follow the path of improving their drawing will discard Impressionism as an art that has become truly superficial for them. The others who neglect to ponder and learn will have pursued the impression to excess. The example of Mr Cézanne (*A modern Olympia*) can show them right away what result awaits them. From idealisation to idealisation, they will end up with this degree of untrammelled romanticism, where nature is only a pretext for dreams, and where the imagination becomes powerless to formulate anything other than subjective, personal fantasies – which, because they are out of control and unverifiable in reality, have no echo in common reason.[12]

This text beautifully anticipates Pissarro's 1883 definition of Impressionism – a theory of pure observation that doesn't lose sight of fantasy, freedom, grandeur or, in brief, anything that makes art great – and his definition of Neo-Impressionism in 1886. It also anticipates the way the Impressionists split up: Pissarro situated the Neo-Impressionist debate along the dividing line of accomplished drawing versus fantasy. Referring to a conversation he had with Albert Dubois-Pillet, Pissarro concurred with his colleague's prediction that even Monet would eventually have to join the Neo-Impressionists, given the "great purity of design" that Neo-Impressionism had remedied, which had been a deficiency in Impressionism.

When Pissarro joined the Neo-Impressionists, he became extremely critical of his Impressionist colleagues, especially Monet, who seemed the very incarnation of Impressionism. One day, Paul Durand-Ruel asked Pissarro (who happened to be passing his gallery and spoke good English) to explain the qualities that distinguished Monet's works to an American client, Mr Robertson, a director of the American Art Association. The ensuing dialogue was fascinating. Although he was no longer convinced of Monet's uncompromised talent, Pissarro agreed to encourage Robertson's appreciation of Monet's painting, "without any effort, knowing the extent of the talent of this artist, despite his errors".[13] Monet had not accepted the "truth" of the new, irresistible logic of the evolution in painting that led to Neo-Impressionism. Pissarro was caught in a difficult situation, having to defend Monet's art. He thought Monet should have gone further, whereas Robertson, who could not even begin to understand Monet's art, thought he had already gone too far. In the end, Pissarro found himself sympathising with the American who "didn't understand anything", conceding that he too saw "disorder" and "fantasy" in Monet's work:

> ... Robertson can't cope with this art. Where does this horror stem from? It is true that he knows nothing about it. However, this simple man ought to be alarmed by the disorder that springs from this romantic fantasy which, despite the talent of the artist, no longer accords with our era ...[14]

Here, Pissarro and Robertson ("who doesn't know anything at all about it") concur in their critique of Monet.

The logical march of Impressionism, announced in the letter to Sabbrin, was not universally recognised. If, from a logical point of view, there seemed a natural progression from Impressionism to Neo-Impressionism, Neo-Impressionism constituted a harsh critique of Impressionism. It proposed to reform the older movement by introducing a more scientific and systematic method, debunking the intuitive approach advocated by the Impressionists. Impressionism was soon derided as "disorderly", "dishevelled" and "romantic" – terms that Pissarro himself used to denigrate the Impressionists. As the young succeed the old, so the Neo-Impressionists blithely claimed to be the inheritors of Impressionism. If Neo-Impressionism was a *phase nouvelle* (a new phase) within Impressionism, and a consequence of *la marche logique de l'Impressionnisme* (the logical march of Impressionism), there was no going back.

The logical momentum was unstoppable, according to Pissarro. He wondered how long it would take for Monet to realise his mistake and join forces with the Neo-Impressionists. Neo-Impressionism became a dogma for Pissarro, who assumed its legitimacy was self-evident. He believed it was an inevitable step any Impressionist would have to take: one had to choose between working with history or against it. The fact that none of his former colleagues were in agreement with him did not divert him from his new convictions. When he opted to join ranks with Seurat and Signac, he placed himself outside Impressionism and, by the same token, alienated himself from his former comrades-in-arms with whom, twelve years earlier, he had struggled to establish the legitimacy of Impressionism. Pissarro is always remembered as the only Impressionist who participated in all eight Impressionist exhibitions (1874–86), yet he took part in the last Impressionist exhibition not as an Impressionist but as a Neo-Impressionist. A section of the exhibition was cordoned off, and there Camille Pissarro, Georges Seurat, Paul Signac and Pissarro's eldest son, Lucien Pissarro, established their independent republic. All four exhibited works that proclaimed an exception to the eighth Impressionist exhibition, purporting to show a way forward in *la marche logique de l'Impressionnisme*.

It became difficult to endorse Pissarro's new paintings, which looked so different from the art he had advocated ten or fifteen years before. Jean Ajalbert, a critic who reviewed the eighth Impressionist exhibition, declared:

> It is difficult to speak of Camille Pissarro ... What we have here is a fighter from way back, a master who continually grows and courageously adapts to new theories.[15]

Another reviewer praised "the wonderful versatility of Pissarro's talent".[16] Virtually every reviewer who commented on Pissarro's work in the last Impressionist exhibition noted his extraordinary capacity to change his art, revise his position and take on new challenges. Marcel Fouquier compared his work of 1886 with the time when Pissarro "painted squares of cabbages in a beam of light, employing the most violent violets of the whole school". He added, he "has had the curiosity to diversify … [and] has made extraordinary progress in his practice".[17]

Pissarro was twenty-nine years older than Seurat (1859–91) and thirty-three years older than Signac (1863–1935). Signac was exactly the same age as Lucien Pissarro (1863–1944). Once such a die-hard Impressionist as Pissarro had turned his back on Impressionism, it was apparent that Impressionism had no chance of surviving – hence, the new phase in the logical march of Impressionism actually signalled its demise.

Pissarro was not a simple artist. In the early 1880s, shortly before he became a Neo-Impressionist, he had produced large-scale figure paintings depicting different settings and contexts and using diverse media to convey different implications. All the while, he was preoccupied with various formal problems, repeatedly reflecting on problems of "execution". A few months before his conversion to Neo-Impressionism, Pissarro declared to Lucien: "I have at last found my execution, which has been bothering me for the past year."[18] But he completely transformed his execution a few months later. Neo-Impressionism did not solve all the problems of execution for Pissarro; at the end of his involvement with Neo-Impressionism, he was still concerned with questions of execution. When Theo van Gogh offered to organise an exhibition of his work in 1889, Pissarro realised he needed to woo back a number of collectors who had not seen his work for a while. This compelled him to reconsider his Neo-Impressionist phase, while the question of execution remained paramount:

> … it's certain that I need to attract a number of art lovers who haven't seen a show of my work for a long time, it's something that has to be done. It is important for the future and necessitates mental strain and considerable work. I need to retreat into

myself, like the monks of time past, and tranquilly, patiently build up the work. I'm in excellent conditions at this time and I'm not short of ideas. I'm working a lot in my head, thinking about execution.[19]

The question of execution was a thread traversing the whole Neo-Impressionist phase of Pissarro's work. At first, Neo-Impressionism seemed to offer some solutions, or formulae, yet Pissarro could not stand formulae and steered away from any such regulation.

His adherence to Neo-Impressionism was complicated by the fact that he had sought new solutions and new opportunities before Neo-Impressionism could offer them to him. In 1885, he described his situation to his dealer Paul Durand-Ruel as a "crisis":

> I have worked a lot in Gisors. Unfortunately, the changing weather made me sick with impatience. I still have not been able to finish, as I had wished, some rather curious, extremely difficult motifs. I am even more overwhelmed not to have finished those studies that I am in the process of changing, where I am impatiently hoping for some sort of result.[20]

The expectation of a result *was* satisfied in 1885, but a complete change during and after 1886 brought with it a different range of problems and solutions. More than any of his Impressionist colleagues, Pissarro went on radically questioning his painting techniques and methods – and this was the case throughout the 1880s, before, during and after Neo-Impressionism. When we survey these changes, we note that the emphasis shifts from an iconographical problem (what should the paintings represent?) to a more technical problem (how could the execution be improved?). Here, again, it seems as if the text by Sabbrin had an effect on him: the expansive fields in Pissarro's paintings prefigure, in some respects, abstraction. They seem to fit Sabbrin's description of wide expanses of "grassy banks" whose many-coloured grasses suggest interminable depth while "a sense of solemnity steals over the observer".

These transformations were followed by serious consequences. Pissarro severed ties with his old colleagues Monet and Degas. He who had held a pivotal role in the organisation of the Impressionist exhibitions, who was effectively the oldest Impressionist (two years older than Manet, four years older than Degas, and nine or ten years older than Cézanne, Monet, Renoir and Sisley), now exhibited with artists who were his son's age. His erstwhile friends and colleagues were bewildered; even today, we are challenged to explain Pissarro's recklessly idealistic initiative. The artist was then in his mid-fifties, with six children, striving to win recognition and needing to sell a sufficient number of works to pay his bills. However, he was utterly convinced that this new form of art answered the needs of history. Neo-Impressionism introduced rigorous discipline and science into the practice of painting, remedying what he was now inclined to see as the disorderly weakness of Impressionism. He became a zealot of Neo-Impressionism, just as he had been a zealot of Impressionism. Yet the Neo-Impressionist interlude would last only four years.

As Friedrich Schlegel wrote almost a century before the Impressionist era: "That which does not cancel itself out is worth nothing."[21] Is this not what Pissarro felt about his art, and about Impressionism in general? Impressionism had grown old and needed to "cancel itself out". Part of the intrinsic logic of creative innovation was that all idioms died and were replaced in new phases of the ineluctable march of progress. Such views of history (or of the history of art) have led to significant problems. This is the notion of historicism, of history which seems to obey its own internal forces. In this view, man can only submit to history; an artist cannot make history but merely accept its compulsions. It is significant that Pissarro wholeheartedly subscribed to this positivist, utopian sense of history as an irresistible progress. Where and how would it end? These are questions Pissarro did not – and probably could not – ask himself. Was every "new phase" of art equally acceptable? We know, for example, that Pissarro categorically rejected Gauguin's attempt to explore new artistic avenues (see Richard Shiff's essay,

"The restless worker", page 33). We also know that he had scant interest in the Nabis, who were proposing a "new phase", but of a different kind. His vision of the unstoppable march of history was not pluralistic – Pissarro was dogmatic in his position. On the other hand, he was unflaggingly self-critical and able to change his opinions, so that he could plunge into Neo-Impressionism but leave it four years later.

The necessity of self-transformation was at the core of Pissarro's conception of art. His adaptability and his strenuous self-criticism ensured that Impressionism never congealed into a formula (unlike Academic art, which the Impressionists fought at every level). He gave great scope to individual innovation and great weight to inter-subjective responses to other artists' creative transformations. This ensured that the creative process – the modes and methods of artistic production – received greater emphasis than the product. The end result, the work of art, mattered only insofar as it was bound to a vital process of exploration. These assumptions were put to the test in some of Pissarro's figure paintings. His figures, when engaged in their own reverie, have no message to deliver. They are utterly anti-allegorical. To ascribe to them an expression of concern for the plight of mankind or a relish for the idylls of the countryside may reveal more about the viewer than the artist or his figures.[22] They are simple. They are simply there, without effects. Their simplicity can be seen as unfathomable (see *Peasants resting*, 1881, fig. 22, cat. 50), even though the ramifications of Pissarro's representation may be more complicated. The figures appear autonomous: their presence, their thoughts are not explicable through any external factors or through a story we can identify. This autonomy of presence suggests a parallel with the autonomy of the brushstroke, which Pissarro strove to master. For a few years, the dot (*le point*) appeared to offer a solution. It was, in a sense, an atomised brushstroke. It could not be broken down into further units; it was the smallest unit of paint that could be committed to canvas. Put together into a tapestry or mosaic, the dots lost their

molecular identity, melting into each other and establishing, as Félix Fénéon put it, the "essence of painting":

> Step back a bit, and all these varicoloured spots melt into undulating, luminous masses; the brushwork vanishes, so to speak; the eye is no longer solicited by anything but the essence of the painting.[23]

The quest for purity and autonomy informed much of the Neo-Impressionist endeavour. Pissarro approved of such notions. The silent, self-reliant figures of the early 1880s (see cat. 48–50) form an analogy to his late 1880s abstractions of wide expanses of bright, multicoloured, multi-form surfaces made up of tiny marks of paint (see *Apple picking at Éragny-sur-Epte*, 1888, fig. 20, cat. 54).

Making the transition from Impressionism to Neo-Impressionism was not easy. While Pissarro prepared for his most important Neo-Impressionist show at Theo van Gogh's gallery, he described his state of anticipation, which "... demands considerable mental tension and a lot of work. I have to retire within myself, like the monks of the past, and quietly, patiently construct the work."[24] A patient execution is best appreciated by a patient observer. Pissarro's Neo-Impressionist art yields little to those in a hurry; there is scant aesthetic satisfaction in the short term. Pissarro drew attention to this at the beginning of his Neo-Impressionist phase in a remarkable letter to his son. He observed that two of his most successfully resolved paintings had been snubbed in London:

> Remember that I have a rustic, melancholic temperament, rough and wild of aspect. I can only please in the long term if there is a grain of indulgence in the observer, but for the passerby whose glance is too hasty, he sees only the surface – having no time, he moves on![25]

Because these paintings do not conceal their technique and make an ostentatious feature of fractured brushwork, they promote awareness of the dialectical exchange between the painter and the viewer. The astonishing investment of energy and the pulverising of colour into myriad touches on the canvas direct attention towards the conceptual and technical production of such paintings. Here, the finely divided touches are not called for or motivated by representational subject matter. The tiny units of colour function better in a representational sense in some parts of the painting than in others. They are more convincing when they represent freckled leaves, dappled grass or brilliantly coloured fruits (see *Apple picking at Éragny-sur-Epte*), rather than a gate, a path or tree trunks. In fact, Pissarro's Neo-Impressionist techniques evaded any kind of formula. His media included pen and ink on paper, watercolour and gouache. There are several occurrences of Pissarro reinterpreting an earlier motif in the Neo-Impressionist technique: *Brickyard at Éragny* (PV724) is a Neo-Impressionist replica of an identically titled work of 1884 (PV661) and *Apple picking at Éragny-sur-Epte* (PV695) is a variant of a painting (PV545) from 1881. In some of the Neo-Impressionist paintings, there are dense, thick, parallel, vertical strokes which are reminiscent of Cézanne's so-called "constructive stroke"[26], to which Pissarro had responded in a work of circa 1884, *Landscape at Osny* (fig. 21). Pissarro, like Signac, acquired four studies by Cézanne, which he described to Lucien:

> From a distance, you often perceive a completely different idea of things, but all the same when you are too close to something, you see nothing – as with a painting by Cézanne when it's right under your nose. Speaking of Cézanne, I have bought myself four of his studies, which are very curious.[27]

As we have noted, during the year when Pissarro met Signac and Seurat, he was endeavouring to organise his brushwork, using the kind of diagonal, parallel, blocky brushstrokes that evoked Cézanne, more than anyone. This preoccupation predisposed Pissarro to understand Signac, Seurat and their aims, enabling him to take full advantage of their techniques. As is evident in *Apple picking at Éragny-sur-Epte*, Pissarro did not paraphrase Cézanne, Seurat or Signac. He integrated some of their procedures into his own

fig. 23

Camille Pissarro
Woman in a field
(also called *Spring sunshine
in a field at Éragny*)
1887
oil on canvas, 45.4 x 65
Musée d'Orsay, Paris,
bequest of Antonin Personnaz

idiom, creating an altogether different result – more complex, provocative and heteroclite. *Apple picking at Éragny-sur-Epte* amalgamates chaos and order, Impressionism and Neo-Impressionism. In the foreground, the earth is composed of multifarious strokes (curves, dots, crosses, hieroglyphics). In other areas of the painting, there is different notation – such as criss-crossing, comma-like strokes, parallel diagonals – which attests to the fact that Pissarro's conception of Neo-Impressionism was anything but formulaic. He maintained a sense of jazziness and ceaseless improvisation. He was an avid reader of Chevreul's treatises concerning the separation of colours and of Ogden Rood's *Théorie scientifique des couleurs*, a crucial text which scientifically established the difference between light-colours and matter-colours, pointing out that the latter (for example, oil paints) could not match or copy light. Rood encouraged artists to consider complementary oppositions, such as purple/green instead of the conventional red/green.

Pissarro's experiments with colour contrasts were of enduring importance and outlasted his Neo-Impressionist phase. Even though he grew impatient with the rigorous methods of Neo-Impressionism, he retained luminous colour contrasts that were derived from his Neo-Impressionist experiments (see *Springtime, grey weather, Éragny*, 1895, cat. 61). As we have seen, Pissarro took liberties with the scientific rigour of Neo-Impressionist theory and aspired to introduce a sense of poetry into his works. This is evident in *L'Ile Lacroix, Rouen (effect of fog)*, 1888 (cat. 57). Ambivalence was a characteristic feature of Pissarro's art. He kept in constant tension the impulses of rigour and improvisation, system and spontaneity, science, poetry and philosophy. No wonder, then, that the text by Sabbrin he discovered in 1886 impressed him so much.

Pissarro drew close to Seurat for the reason that, as the art historian Robert Herbert has pointed out, Seurat's works implied the artistry of a poet and a technician.[28] We have seen how the fragmentation of brushwork (divisionism) made Pissarro's pictorial touch all the

more crucial to the complicated process of creation. We have also seen how his work led to clusters of variations, all linked because of certain types of repetition (see *Peasants' houses, Éragny*, 1887, cat. 52, and *Woman in a field,* also called *Spring sunshine in a field at Éragny*, 1887, fig. 23). We may discern in Pissarro's work what the eminent linguist Roman Jakobson identified as an essential component of poetic language: the tendency towards repetition.[29] However, Seurat's repetitions and variations of themes and motifs evolve towards the final elaboration of a unique masterwork, the outcome of a pyramid-like or hierarchical procedure. Pissarro, on the other hand, produces a sequence of related works which do not culminate in a masterpiece but form a more or less consistent chain. His works need to be seen next to each other. They corroborate and complement each other through their technical, compositional and representational relationships, forming a continuum which the observer can enjoy in its integrality.

There is a painting, *Woman chopping wood*, 1890 (PV757), which meticulously repeats an earlier composition of 1888 but goes beyond the Neo-Impressionist idiom, pointing towards other possibilities. The title of this painting inadvertently alludes to a French colloquialism. It suggests that Pissarro can make fire out of any kind of wood: *Pissarro fait feu de tout bois.*

1 *Correspondance de Camille Pissarro*, ed. Janine Bailly-Herzberg (Éditions du Valhermeil, Paris, 1986), vol. 2, p. 56; hereafter, *CCP*.

2 Celen Sabbrin (pseudonym for Helen Abbott), "Science and Philosophy in Art", 1886; quoted by Charles Stuckey, *Monet: A retrospective* (Hugh Lauter Levin Associates, New York, 1985), p. 126.

3 ibid.

4 ibid., pp. 126–27.

5 ibid., p. 127.

6 *CCP*, op. cit., vol. 2, p. 56; letter to Monsieur Celen Sabbrin, 18 June 1886.

7 *CCP*, op. cit., vol. 2, p. 30; letter to Lucien Pissarro, circa 3 March 1883.

8 *CCP*, op. cit., vol. 1, p. 178; letter to Lucien Pissarro, 28 February 1883.

9 *CCP*, op. cit., vol. 2, p. 75; letter to Paul Durand-Ruel, 6 November 1886.

10 Jules-Antoine Castagnary, "L'Exposition du boulevard des Capucines: Les Impressionnistes", *Le Siècle*, 29 April 1874; quoted by Richard Shiff, *Cézanne and the End of Impressionism* (The University of Chicago Press, Chicago, 1984), p. 2, and by Hélène Adhémar, *Centenaire de l'impressionnisme* (Éditions des Musées Nationaux, Paris, 1974), p. 265.

11 Émile Zola, "Proudhon et Courbet" (1866), "M.H. Taine, artiste" (1866), *Mes Haines, Mon Salon, Édouard Manet* (Paris, 1879), pp. 25, 229; see also Zola, "Mon Salon" (1866), *Ecrits*, pp. 59–60, and Zola's letter to Valabrègue, 1864, in *Correspondance 1858–71*, ed. Maurice Le Blond (Paris, 1927), p. 248; the above references are quoted by Shiff, op. cit., p. 88.

12 Castagnary, op. cit; quoted by Adhémar, op. cit., p. 265.

13 *CCP*, op. cit., vol. 2, p. 101.

14 ibid.

15 Jean Ajalbert, *La Revue moderne*, Marseille, 20 June 1886; quoted in Charles Moffett, et al., *The New Painting: Impressionism 1874–1886*, The Fine Arts Museums of San Francisco, San Francisco, 1986, p. 461.

16 Rodolphe Darzens, *La Pléiade*, May 1886; quoted by Charles Moffett, op. cit., p. 461.

17 Marcel Fouquier, *Le XIXe Siècle*, 16 May 1886; quoted by Charles Moffett, op. cit., p. 463.

18 *CCP*, op. cit., vol. 1, p. 342; letter to Lucien Pissarro, 21 August 1885.

19 *CCP*, op. cit., vol. 2, p. 303; letter to Georges Pissarro (Pissarro's second son), 7 October 1889.

20 *CCP*, op. cit., vol. 1, p. 336; letter to Paul Durand-Ruel, June/July 1885.

21 Friedrich Schlegel, *Literary Notebooks, 1797–1801*, London, 1957, p. 226; quoted by Tzvetan Todorov, *Theories of the Symbol*, trans. Catherine Porter (Cornell University Press, Ithaca, 1982), p. 170.

22 See John House, "Camille Pissarro's Idea of Unity", in *Studies on Camille Pissarro*, ed. Christopher Lloyd (Routledge & Kegan Paul, London, 1986), p. 30.

23 Félix Fénéon, "Neo-Impressionism", *L'Art moderne*, 1887; in Linda Nochlin, *Impressionism and Post-Impressionism 1874–1904* (Prentice-Hall, New Jersey, 1966), p. 110.

24 *CCP*, op. cit., vol. 2, p. 303; letter to Georges Pissarro, 7 October 1889.

25 *CCP*, op. cit., vol. 1, p. 252; letter to Lucien Pissarro, 20 November 1883.

26 Theodore Reff, "Cézanne's Constructive Stoke", *Art Quarterly* 25, no 3 (Autumn 1962), pp. 214–17. Reff was one of the first art historians to point out the importance of Cézanne in the formation of Neo-Impressionism: "The neo-impressionists, although far more consistent than Cézanne in their rationalization of both color and touch, may have derived from him a heightened awareness of the canvas as a surface to be covered with tiny strokes."

27 *CCP*, op. cit., vol. 1, pp. 293–94; letter to Lucien Pissarro, early March 1884.

28 Robert Herbert, "Introduction" in *Seurat* (Éditions de la Réunion des musées nationaux, Paris, 1991), p. 22. "If one wishes to find the complete artist, poet and technician simultaneously, we should redefine the 'science' of Seurat."

29 See Todorov, op. cit., p. 279: "One tendency of poetic language in particular engages Jakobson's attention: the tendency toward repetition." As Roman Jakobson pointed out in "Grammatical Parallelism and its Russian Facet", *Language*, 42 (1966), p. 399: "On every level of language, the essence of poetic artifice consists in recurrent returns."

Pissarro and printmaking: an introduction

Peter Raissis

For Camille Pissarro, printmaking was a vital aspect of his career, and he devoted himself to it with particular intensity. His often radically experimental approach resulted in prints of startling originality for their time. Working in the technique of etching primarily, but also lithography and to a lesser degree monotype, his innovative achievements rank him alongside Degas as the most significant printmaker of the Impressionists.

Consistently, his themes were derived from the eternal France of the countryside and the peasantry. Later, he found subjects for his prints in the modern life of the city, in images of bathing women and in portraiture. All were fundamentally an extension of his painting interests. The printed image was not so much a means of exploring alternative subject matter as one through which he could freely experiment with novel pictorial effects in a process of continuous research.

Pissarro was a prolific printmaker (he left behind more than two hundred plates), yet his explorations in the medium were essentially personal and non-commercial. He was attracted by both the inherent capriciousness of printmaking and its exacting procedures. Compulsively, he worked and reworked his plates, fashioning unusual printed landscapes of great atmospheric beauty, and he committed himself to the laborious and drawn-out process of elaborating his compositions over multiple states. Surprisingly, however, the number of impressions pulled during his lifetime is very small, and rarely were they issued in editions. Indeed, his reputation as a printmaker has been based largely on the posthumous editions commissioned by his family.[1]

It was commonly his practice to print only as many impressions as pleased him for his own requirements, and some of his most appealing images exist as unique impressions (cat. 31–34 and 37). He lavished much attention on printmaking, despite the fact that his endeavours were poorly remunerated. "As for etching," he wrote in 1880 to his friend, the critic Théodore Duret, "I have neither the time nor the means to make trial proofs … The need

to sell pushes me towards watercolour; etching is abandoned for the moment."[2] In spite of displaying non-conformity to the trends of the contemporary print world, Pissarro was nevertheless not indifferent to his work's saleability. He showed in 1889 and 1890 with the Société des Peintres-Graveurs at the Durand-Ruel gallery in Paris and sent prints to exhibitions and on consignment to dealers in Paris and abroad, in London, Brussels, Dresden and New York.[3] His prints were not expensive, but they never sold as well as he hoped: "What a pity there is no demand for my prints," he despaired. "I find this work as interesting as painting, which everybody does, and there are so few who achieve something in printmaking."[4]

Writing to his son Lucien in the early 1890s, he regretted that the public cared more for deluxe reproductions than original prints, and also observed that "those who call themselves pure art collectors have complete collections of Charles Jacque, of Buhot and, among these, things by Seymour Haden, Whistler and Legros slip in, but that is all."[5] In a similar vein, he explained to Lucien:

> Up till now I have felt around me nothing but indifference for work in black and white … As for the collectors, I assure you what they really like is Charles Jacque, Buhot, Bracquemond, or Legros when he imitates Rembrandt; the same goes for Seymour Haden. But engraving based on vivid sensations, no! And it is this that I try for when I do engravings![6]

Since he did not possess his own press until 1894, he had to rely on fellow artists – Degas and Charles Jacque and his son – who made their printing presses available to him during his brief visits to Paris. On other occasions, he sent his plates to the professional printers Salmon or Delâtre, although he disliked the latter's fashionable practice of *retroussage*, which involved passing muslin over the wiped plate to drag ink out of the indentations, thus making the printed lines softer. He complained to Lucien of plates having been spoilt by "that frightful Delâtre, he pours on too much sauce".[7]

Inking and wiping the etched plate, selecting inks and papers for printing, and deciding the number of impressions to produce should be the preserve of the artist alone. In this respect, Pissarro followed Whistler, who had famously reclaimed the printing of his own plates as a vital part of the creative process:

> Whistler makes drypoints mostly, and sometimes regular etchings, but the suppleness you find in them, the pithiness and delicacy which charm you derive from the inking which is done by Whistler himself; no professional printer could substitute for him … Now we would like to achieve suppleness before the printing.[8]

Etching Revival

When he first tried his hand at printmaking, in about 1863, Pissarro was in his early thirties. His decision to take up etching – and not another of the various techniques available to him at the time – was not unexpected, for he did so at a time that might be described as the height of the Etching Revival, when the medium carried distinctly avant-garde connotations. The far-reaching resurgence of etching witnessed in France during the Second Empire brought together leading contemporary artists and critics, fervent in their promotion of the print as an original work of art. Not since the time of Rembrandt, more than two hundred years earlier, had etching been adopted with such enthusiasm, and never had it been invested with so much meaning. In a famous essay first published in *Le Boulevard* in 1862, Baudelaire noted:

> … it would seem that we are about to witness a return to etching … this technique, so subtle and superb, so naive and profound, so gay and severe, a technique which can paradoxically accommodate the most varied qualities and which is so admirably expressive of the personal character of the artist has never enjoyed

a really great popularity among the vulgar … [it is] the crispest possible translation of the artist's personality.[9]

The crucial appeal of etching, as noted by Baudelaire, was its responsiveness to the vagaries of the artist's hand and the fact that it permitted him to draw spontaneously. An etching also looked distinctly autographic, in stark contrast to, for example, the regular and stylised lines of an engraving. It is for this reason that etching came to be regarded by many as superior to such commercially popular nineteenth century printing practices as lithography, wood engraving and steel engraving, all of which smacked of their lowly association with reproductive illustration. Etching was the medium par excellence for the *peintre-graveur* who directly executed his own designs and usually carried out or supervised the printing, as opposed to the engraver who copied the compositions of others for dissemination in large numbers to audiences far and wide.

The Etching Revival was given its official voice in 1862, when the print publisher and dealer Alfred Cadart and the printer Auguste Delâtre established the Société des Aquafortistes. Its aim was to promulgate the status of the print as an original work of individual creativity, capable of typifying some of the values accorded to painting and drawing. Certainly, Pissarro shared such attitudes, and over the ensuing years he would realise these ambitions in his own innovative prints.

He upheld the principle of the print as an authentic and personal statement and vehemently disapproved of *gillotage*[10], a process of mechanical reproduction used in the illustrated journals that flourished in the second half of the nineteenth century. Pissarro objected so strongly to the process because he felt it distorted the artist's handiwork: "*gillotage* is to engraving what false Turkish rugs are to real ones … At last I have the *Figaro* engravings printed in London. They didn't need to go so far to get them done so badly, it's horrible, horrible!!!"[11] Later, he disparaged it as "expeditious *gillotage*, paying such great services to commerce".[12]

In defending the importance of the etching as an original work of the artist's mind and hand, the Société des Aquafortistes was an unambiguous stand against the proliferation of uniform, mechanically reproduced images, of which photography was thought to be the most insidious. "The need to react against the positivism of the mirror-instrument," wrote Théophile Gautier in 1863, in his preface to the Society's inaugural publication, "has caused more than one painter to take up the etching needle … Indeed, no means is simpler, more direct, more personal than etching … The acid bites into the exposed parts of the metal and deepens the grooves, which exactly reproduce each line drawn by the artist. If the process succeeds the plate is ready. It can be printed. One can even entertain the idea of the master sparkling with life and spontaneity, without the mediation of any translation. Every etching is an original drawing … This Society has no other creed than individualism."[13]

Pissarro's first etching, *By the water's edge*, c.1863 (fig. 24), coincides in date with the early days of the short-lived Société des Aquafortistes. This gentle scene, with two figures sitting on a bank in a forest interior, is typical of the work published by the Society and shows the influence of the Barbizon school. In particular, with its luminous meshes of thinly etched lines carried neatly to the edges of the plate, it owes much in feeling and style to Corot's prints. Pissarro is listed as a member of the Society, but he neither exhibited with the Society nor published etchings through its annual albums. Yet, as Janine Bailly-Herzberg has shown, there is a direct link between Pissarro's origins as a printmaker and the Society; he made etchings annually until the Society disbanded in 1867 and did not resume printmaking until 1873.[14]

Subsequently, Pissarro's printmaking was nurtured in a close-knit group of artist–friends gathered around Dr Paul Gachet, renowned as the physician who treated Vincent van Gogh. Under the pseudonym van Ryssel, Gachet was a dedicated amateur etcher. In 1872, he bought a house in the village of Auvers-sur-Oise, close

to Pontoise, where Pissarro had returned the same year. He set up a printmaking studio in the attic of his house and, for a short period in 1873, he encouraged Pissarro, Cézanne and Armand Gillaumin to experiment; he printed their hastily drawn plates for them on his press.[15] The landscape prints of the so-called Auvers Group are characterised by sketchy, spirited line work and a deliberately unrefined execution, which was part of Gachet's aim for authenticity. Pissarro's *Portrait of Paul Cézanne*, 1874 (cat. 100), printed by Gachet, typifies this period of collaborative endeavour, and its carefully placed signature and dedication in pencil – "*à mon ami Luce*" – declare its uniqueness. It relates closely to a portrait painted at about the same time and, although it captures something of the new approach learnt at Auvers, its purely linear style is far from the radical redirection Pissarro's work would take in the next five years.

Pissarro and Degas: experimental printmaking

By far the most important phase of Pissarro's stylistic evolution as a printmaker occurred in 1879–80, during a period of intensely creative collaboration with Edgar Degas.[16] Pissarro and Degas made perhaps an unlikely pair; their social backgrounds were as diverse as their political beliefs or the subjects they chose to represent. But they were nevertheless united in their dedication to printmaking. By now, Pissarro was an accomplished etcher, but Degas was the more experienced printmaker. He initiated Pissarro into the mysteries of his unorthodox approaches and together they carried out bold experiments, exploring the medium's boundless creative potential. One senses the fervour of experimentation in the marginalia of *Woman on the road*, 1879 (cat. 27), where trials or imperfections are proudly admitted as parts of the print. In other works, such

fig. 24
Camille Pissarro
By the water's edge
c.1863
etching, 31.4 x 23.7
only state
Museum of Fine Arts Boston,
George Peabody Gardner Fund
(inv 63.323)

fig. 25
Edgar Degas
*Mary Cassatt at the Louvre:
the Etruscan Gallery*
1879–80
softground etching, drypoint,
aquatint and etching, 26.7 x 23.2
vii of 9 states
Museum of Fine Arts, Boston,
Katherine E. Bullard Fund
in memory of Francis Bullard, 1984

as *Path in the woods, Pontoise*, 1879 (cat. 25 and 26), technical exuberance almost obliterates the forest. The two foreground figures, as well as the houses and church in the distance, are virtually indistinguishable from the textural encrustations of the plate.

The striking feature of the prints of 1879 is their technical daring. They defied all received opinion about what etching could be; even calling them etchings seems hardly apt when they are compared with the hundreds of conventional scenes that poured off printing presses in Paris in the wake of the Etching Revival. According to Barbara Stern Shapiro, who has studied Pissarro's graphics in great detail, the 1879 prints are "a veritable cuisine of intaglio processes. Along with the little known softground, the novel use of aquatint, the dusting of resin grains, the salt grain method, as well as the drypoint and etched line work, they also incorporated printing imperfections and accidents with acid into the design of the print ... the intention to fashion impressionistic, painterly images without obvious contours or noticeable etched lines was brilliantly achieved."[17]

Before he met Degas, Pissarro was unfamiliar with aquatint – a tonal method of etching used to create graded shades between pale grey and black in imitation of the effects of a wash drawing. Degas knew how to manipulate aquatint to great effect and sent the following advice to Pissarro:

It is necessary for you to practise dusting the particles in order, for instance, to obtain a sky of uniform grey, smooth and fine. That is very difficult, if one is to believe Maître Bracquemond ... This is the method. Take a very smooth plate (it is essential, you understand). Degrease it thoroughly with whitening. Previously you will have prepared a solution of resin in very concentrated alcohol. This liquid ... evaporates and leaves the plate covered with a coating, more or less thick, of small particles of resin. In allowing it to bite, you obtain a network of lines, deeper or less deep, according as to whether you allowed it to bite more or less.[18]

In the same letter, Degas encouraged Pissarro to seek out copper plates instead of zinc plates, as they produce finer results:

Here are the proofs. The prevailing blackish or rather greyish shade comes from the zinc which is greasy in itself and retains the printer's black. The plate is not smooth enough. I feel sure that you have not the same facilities at Pontoise ... in spite of that, you must have something a bit more polished.

Pissarro used aquatint to concentrate on the changing effects of light at different times in a series of remarkable landscapes based on the countryside around Pontoise. The new vigour of his style is marvellously expressed in the virtuosic *Woman on the road*, 1879 (cat. 27), depicting a small peasant figure wandering along a winding country road beneath a dramatic sky with towering clouds. The interest of the scene is in the starkly contrasted sky with its vigorously painted clouds and the dotted effect of the surrounding hills, the shadows deeply bitten and worked to allow the light to gleam through. In *Rainy effect*, 1879 (cat. 28 and 29), Pissarro captured the blurry atmosphere of a sodden landscape by conceiving the plate in aquatint rather than adding it at a later stage over an existing matrix of etched or drypoint lines. The early state included here (cat. 28) consists of a gossamer-soft aquatint foundation that evokes the bare minimum of landscape. In the final state (cat. 29), the forms of the figures, trees and haystack were made more substantial with etching and drypoint work before the final lashing strokes were brilliantly scratched in, without doubt inspired by the rainy scenes depicted in Japanese woodcuts. Extensive aquatint is also the basis of *Horizontal landscape*, 1879 (cat. 37). It has been suggested that the print's unusual elongated format was inspired by a set of horizontal panels illustrating the seasons that were painted by Pissarro in 1872.[19] The unique impression of the second state presented here is annotated in pencil with the word "zinc", and presumably was printed before a second copper plate was made by means of electrotyping.[20]

fig. 26
cat. 41 (detail)

Woman emptying a wheelbarrow
1880
Sterling and Francine Clark
Art Institute, Massachusetts

Pissarro's outstanding achievement is perhaps the highly wrought *Wooded landscape at L'Hermitage, Pontoise*, 1879 (cat. 34, 35 and 36) which offers a glimpse of village houses through a thicket of trees. The idea for the work originated in a proposal by Degas in 1879 to launch a monthly journal of original prints, *Le Jour et la nuit*, with contributions from Degas himself, Mary Cassatt, Félix Bracquemond, Jean-Francois Raffaëlli, Gustave Caillebotte and Pissarro, among others.[21] A letter from Degas to Pissarro exhorts him to "remember that you must make your debut with one or two very, very beautiful plates of your own work. I am also going to get down to it in a day or two."[22] Sadly, the enterprise came to nothing, but nonetheless it did yield some exceptional results which surely must constitute the most original use of intaglio printmaking in the nineteenth century. Degas, Cassatt and Pissarro all prepared plates for the first issue, all employed common techniques abundantly seasoned with aquatint, and all based their prints on earlier works they had done. Degas' contribution (fig. 25) was a reworking of a pastel drawing showing Mary Cassatt at the Louvre, while Cassatt and Pissarro based theirs on oil paintings. All three artists showed their prints in 1880 at the fifth Impressionist exhibition. Among the works exhibited by Pissarro were four states of the *Wooded landscape at L'Hermitage, Pontoise*, displayed together on a yellow mount in a single purple frame.[23]

Pissarro evolved this complex print over six states, loading his copper plate with concoctions of aquatint, softground etching and drypoint until the surface began to approximate in appearance the texture of his paint-clotted canvases. In the unique impression of the first state shown here (cat. 34), Pissarro exploited the wash-like effects of aquatint to lay in the composition broadly and to establish the different areas of tone. The outlines of the houses and hills, along with the figure in the bottom left corner, cut through by one of the tress, were sketched using softground, a variant technique of etching that Pissarro seized from Degas and whose effect resembles a crayon line more than the astringent line of the etching needle.

Next, the screen of trees and sparsely indicated foreground were painted directly on the plate using sugar-lift aquatint.[24] In successive states, acid was brushed directly on the copper plate in small flecks, select areas were burnished back and drypoint accents were added. The final state was printed on Japan paper in an edition of fifty by the printer Salmon, evidently in preparation for its appearance in *Le Jour et la nuit*.

The exploratory nature of Pissarro's printmaking is nowhere more evident than in his practice of making states. Early proof impressions were valued as equally legitimate alternatives to the final state, and Pissarro willingly exhibited them. Clearly, he was conscious of the appeal of presenting different states of a single print as a related sequence of variations on a theme. Of this interesting feature of his prints, Barbara Stern Shapiro has perceptively observed: "Pissarro forces the viewer to examine the entire package, so to speak, to appreciate the integral parts rather than merely focus on the so-called final state … Even Pissarro's vocabulary of annotations – *épreuve d'état, épreuve d'artiste, épreuve définitif, épreuve de choix* – is essentially unhierarchical and gives equal weight to the differing proof impressions."[25]

No print by Pissarro exists in as many states as his almost bucolic vision of rural life, *Woman emptying a wheelbarrow*, 1880 (fig. 26, cat. 41). At the rear of a cottage, a wooden gate opens into a sunny garden plot with a tree in full bloom. A woman with geese at her feet carries heavy pails, while in the foreground a second figure empties the contents of a wheelbarrow onto a compost heap. Early states of this print were executed in drypoint, but the twelfth state demonstrates the elaborate reworking of the plate in the technique Pissarro called *manière grise* – creating tone and texture by abrading the copper plate with metallic brushes, sandpaper and emery sharpened to a point. Pissarro employed it in conjunction with aquatint or etching to produce a new vocabulary of lines, patches and smears. When the plate was inked and printed, the outcome was a picture of the subtlest gradations of tone, imbued with airy luminosity and vibrant texture. The drawback, however, was that the fragile plate could yield only a few good impressions before wearing down, as Pissarro explained to his son: "*Woman emptying a wheelbarrow* is rare, you recall that there are only two proofs, only one of which, the one I am sending you, is good; this is the one in *manière grise.*"[26]

When he wanted to vary his impressions, Pissarro could address the question differently by changing the colours of the printing ink. The four versions of *Twilight with haystacks*, 1879 in this exhibition (cat. 30–33) are repetitions of the same motif, but each single impression has been made into a unique variant with differently weighted emphases of light and atmospheric effect. Some of the rare colour impressions, such as the red-brown and Prussian blue versions (cat. 32 and 33), from Ottawa and Boston respectively, were printed by Degas and are inscribed in pencil by Pissarro with "*imp par E. Degas*". This quintessentially Impressionist print shows two figures making their way home at day's end along a winding road that passes a pair of haystacks and leads to a row of poplars on the horizon. The scene suggests a landscape recalled from memory rather than directly observed and conjures a shimmering vision of rural France in the last glow of daylight. It is a print that justly has been compared with Monet's famous haystack series from 1890–91.[27]

Colour etching and lithography

After the intense months spent working with Degas, Pissarro's printmaking took new directions. He continued making etched views of Pontoise, Osny and Rouen, applying his *manière grise* technique with superb effect (see *Place de la République, Rouen*, 1883, cat. 91). In 1884, he moved to Éragny-sur-Epte, a village near

Gisors, where, in deep rural seclusion, he commenced a series of new experiments in colour etching and lithography.

In this period, the original print was still understood as a black and white medium, although isolated trials in colour etching throughout the 1870s and 1880s were beginning to challenge this idea. The widespread enthusiasm for colour printing witnessed during the 1890s is demonstrated by the highly esteemed work of Mary Cassatt, who, with Pissarro, pioneered the new medium. Cassatt's colour etchings were revealed to the public in 1891 in a small exhibition held with Pissarro at the Durand-Ruel gallery; it had been organised to protest against the exclusion of non-French artists from the recently restructured Société des Peintres-Graveurs Français, which was showing in the gallery's main room. Shortly before their exhibition opened, Pissarro wrote to his son of Cassatt's prints:

It is absolutely necessary, while what I saw yesterday at Miss Cassatt's is still fresh in mind, to tell you about the coloured engravings she is to show at Durand-Ruel's at the same time as I. We open Saturday, the same day as the patriots, who, between the two of us, are going to be furious when they discover right next to their exhibition a show of rare and exquisite works. You remember the effects you strove for at Éragny? Well, Miss Cassatt has realised just such effects, and admirably; the tone matte, fine, delicate, without grubbiness or spots, adorable blue, fresh pinks etc.[28]

In 1894, Pissarro purchased an etching press and began, through trial and error, to perfect the technique himself: "The press I bought from Delâtre has been installed in the large studio; I am waiting for ink to make some prints. We tried to print with oil colour, the effect is astonishing. It gives me the urge to do more etchings."[29] His initiatives resulted in no more than five colour etchings, probably because of the complicated method of multi-plate printing, which meant that colours had to be superimposed separately, one on top of the other, using three or four plates. Nonetheless, Pissarro was evidently pleased with what he achieved, as he wrote to Lucien:

I will send you soon a fine print of my little peasant girls in the grass and a market in black, retouched with tints; I think some excellent things can be made in this way … It has no resemblance to Miss Cassatt, it involves nothing more than retouching with colours, that is all. I have already obtained some fine proofs; it is very difficult to find just the right colours.[30]

Church and farm at Éragny-sur-Epte, 1894/95 (fig. 27, cat. 64) and *The market at Gisors: rue Cappeville*, 1894/95 (cat. 71) – the market Pissarro refers to in his letter – were both carefully worked out in preliminary drawings (cat. 63 and 67, as well as cat. 65, not included in this catalogue) and working proofs before he decided to complete them with colour. In these, his finest achievements in colour etching, Pissarro used four plates, one for each colour: blue, yellow and red, together with a key plate in black or grey for the outlines. Each plate was passed through the press in succession, with great care taken to ensure the paper was perfectly aligned with the plate so that the image did not print out of register. The colours in *The market at Gisors* are more or less contained within the forms, quite unlike the effect Pissarro achieved in *Church and farm at Éragny-sur-Epte*, where colour is used more as a constructive agent to evoke the vibrative quality of an entire village dissolving in golden light.

Finally, mention must be made of Pissarro's late involvement in lithography. Although as a medium it was always closely linked with reproduction, Pissarro's first works, completed in 1874, were purely personal expressions – intimate portraits of his family and rustic scenes – almost certainly done as technical exercises and perhaps inspired by his deep admiration for Daumier. The motivation for his return to the medium in 1894 might be explained by his meeting with the professional lithographic printer, Charles-André Tailliardat, and the requests for illustrations he received from publishers. A small number of Pissarro's late lithographs was produced with publication in mind: *Sower*, 1896, *The vagrants*, 1896 and *The ploughman*, 1901 (cat. 72–74) were destined for the anarchist journal, *Les Temps Nouveaux*, whereas *Bathers in the*

shade of wooded banks, 1894/95 (fig. 28, cat. 78) appeared in 1895 in the album, *L'Estampe Originale*, in an edition of 100. Independently of publication, Pissarro also completed a number of broadly sketched views of Paris' bustling boulevards (see *Place du Havre, Paris*, 1897, cat. 92), which he made from hotel rooms he rented high above the crowds and traffic.

That Pissarro appears not to have placed the same value on his lithographs as his etchings is not surprising, given the hierarchies of the time. He neither made a feature of them in exhibitions, nor did the gift he made in 1900 to the Musée du Luxembourg of his choicest impressions include a single lithograph.[31] In 1894, he wrote to Lucien of a new project:

The weather is very uncertain for the unhappy art of painting; I am content to grind away indoors. I have done a whole series of lithographic drawings in romantic style which seemed to me to have a rather amusing side: bathers, plenty of them, in all sorts of poses, in all sorts of paradises.[32]

Images of bathing women (cat. 78, 79) represent a sudden change in Pissarro's repertoire, but they demonstrate his supreme control when working in lithography and surpass his treatment of that theme in any other medium.[33] Clearly, Pissarro was aware of Cézanne's bathers compositions in oil and watercolour, but the lithographs made after Cézanne's works are feeble by comparison with Pissarro's. In these romanticised visions, he brilliantly reinterpreted the conventional chalk-drawn lithograph and approached the medium as if he were making spontaneous wash drawings, evoking a full range of tones, from the palest grey to the most sombre black. He worked directly on specially prepared zinc plates, although he also made use of the more traditional lithographic stone. He deftly manipulated broad areas of litho-wash, allowing it to blend with the soft touch of crayon while reserving patches of white to suggest glistening bodies blurring with water and foliage. If we are to see in Pissarro's printmaking a return to the vigour and adventure of 1879, surely it is here.

1 Posthumous editions were printed under the aegis of Pissarro's family, who owned the plates. These impressions are regarded to be of inferior quality to those printed under Pissarro's direct supervision. Later editions were printed in 1907 (stamped with a blue monogram), in 1920 (stamped with brown initials *C.P.* in a circle) and 1922, 1923 and 1929 (stamped with grey initials *C.P.*). See Eric Gillis, "Degas et Pissarro: les tirages posthumes. État provisoire de la question", in Nicole Minder, *Degas et Pissarro. Alchimie d'une Rencontre*, Cabinet Cantonal des Estampes, Musée Jenisch, Vevey, 1998, pp. 149–57.

2 *Correspondance de Camille Pissarro*, ed. Janine Bailly-Herzberg, Paris, 1980, vol. 1, p. 141, 23 December 1880; hereafter, *CCP*.

3 Janine Bailly-Herzberg, *Dictionnaire de l'Estampe en France 1830–1950*, Paris, 1985, p. 258 and p. 260. Probably the best and most complete early collection was formed by Samuel P. Avery (1822–1904), who donated it to the New York Public Library in 1900.

4 *CCP*, 1989, vol. 4, p. 55, 8 April 1895.

5 *CCP*, 1986, vol. 2, p. 380, 26 December 1890.

6 *CCP*, 1988, vol. 3, p. 31, 10 February 1891.

7 *CCP*, 1989, vol. 4, p. 26, 28 January 1895.

8 *CCP*, 1980, vol. 1, p. 178, 28 February 1883.

9 Jonathan Mayne (trans. and ed.), *Art in Paris 1845–1862. Salons and Other Exhibitions Reviewed by Charles Baudelaire*, London, 1965, pp. 218–19.

10 Invented in 1850 by the printer Firmin Gillot, this method involved turning a lithographic plate into a relief process. The drawing, executed in a greasy medium, was made on or transferred (by hand or, later, by photography) to a zinc plate and dusted with a resin that acted as a resist. The plate was then etched, leaving the protected areas in relief, producing a relief block.

11 *CCP*, 1980, vol. 1, p. 293, undated, early March 1884.

12 *CCP*, 1988, vol. 3, p. 23, 15 January 1891. Pissarro's stance was quite the opposite of Degas', for example, who was willing to embrace mechanical processes for the reproduction of his drawings. See Michel Melot, *The Impressionist Print*, New Haven and London, 1996, pp. 156–59.

13 Janine Bailly-Herzberg, *L'Eau-Forte de Peintre au Dix-Neuvième Siècle: La Société des Aquafortistes (1862–1867)*, Paris, 1972, vol. I, pp. 266–67.

14 Bailly-Herzberg, *L'Eau-Forte de Peintre au Dix-Neuvième Siècle*, vol. II, p. 159.

15 See Carla Esposito, "Notes sur la collection d'un excentrique: les estampes du Dr. Gachet", in *Nouvelles de l'estampe*, no 126, 1992, pp. 4–8.

16 See Nicole Minder, *Degas et Pissarro. Alchimie d'une Rencontre*, Cabinet Cantonal des Estampes, Musée Jenisch, Vevey, 1998.

17 Barbara Stern Shapiro, "Pissarro as print-maker", in *Camille Pissarro 1830–1903*, Arts Council of Great Britain and Museum of Fine Arts, Boston, 1980, p. 192. See also her *Camille Pissarro. The Impressionist Printmaker*, Museum of Fine Arts, Boston, 1973.

18 Marcel Guerin, *Degas Letters*, Oxford, 1948, no 34, pp. 56–59.

19 Shapiro, "Pissarro as print-maker", pp. 192, 193.

20 Shapiro, "Pissarro as print-maker", pp. 192, 193. Electrotyping was invented in 1839. It involved making an exact duplicate of a block or plate in copper by electrolysis. It was especially popular for reproducing woodblocks and allowed a limitless number of impressions to be printed without fear of damaging the original woodblock. Pissarro almost certainly experimented with this process for different reasons, probably on the advice of Degas, who preferred impressions that had been printed from copper plates, not zinc plates.

21 This project has been documented by the following: Ronald Pickvance, *Degas 1879*, National Gallery of Scotland, Edinburgh, 1979, pp. 76–77; Douglas Druick and Peter Zegers, "Degas and the printed image", in *Edgar Degas: the Painter as Printmaker*, Museum of Fine Arts, Boston, 1984, pp. xxxix–lv; Michel Melot, *The Impressionist Print*, trans. Caroline Beamish, New Haven and London, 1996, pp. 149–55; Barbara Stern Shapiro, "A printmaking encounter", in *The Private Collection of Edgar Degas*, The Metropolitan Museum of Art, New York, 1997, pp. 135–45.

22 Marcel Guerin, *Degas Letters*, Oxford, 1948, no 34, pp. 56–59.

23 J.-K. Huysmans, "L'exposition des indépendents en 1880", in Ruth Berson, *The New Painting. Impressionism 1874–1886. Documentation. Volume I. Reviews*, Fine Arts Museums of San Francisco, 1996, p. 286. The catalogue of the fifth Impressionist exhibition specifies: "A frame. Four states of the landscape intended for the first issue of the publication *Le Jour et la nuit*". See Berson, *Volume II. Exhibited Works*, p. 153.

24 In standard aquatint, the design on the plate has to be negative because the areas the artist paints with stopping-out varnish are those which will remain white. Sugar-lift is a refinement of this and avoids the laborious process of stopping out. The design is brushed directly on the plate with a fluid in which sugar has been dissolved. When dry, the entire plate is then covered with stopping-out varnish and immersed in water which dissolves the sugar, causing it to swell and dislodge the varnish, thereby exposing the bare metal underneath. These areas are then covered with an aquatint ground and bitten in the usual way.

25 Shapiro, "Pissarro as print-maker", p. 203.

26 *CCP*, 1988, vol. 3, p. 38, 1 March 1891.

27 Michel Melot, "La Pratique d'un artiste: Pissarro graveur en 1880", in *Histoire et Critique des Arts, II*, June 1977, pp. 14–38. Other known colour impressions: ultramarine blue and crimson lake, both Ashmolean Museum, Oxford; vermilion, The Art Institute of Chicago; van Dyck brown, Bibliothèque Nationale de France, Paris.

28 *CCP*, 1988, vol. 3, p. 55, 3 April 1891.

29 *CCP*, 1988, vol. 3, p. 416, 3 January 1894.

30 *CCP*, 1989, vol. 4, p. 19, 18 January 1895.

31 Bailly-Herzberg, *Dictionnaire de l'Estampe en France 1830–1950*, p. 260.

32 *CCP*, 1988, vol. 3, p. 445, 19 April 1894.

33 Between 1894 and 1896, Pissarro completed in oil paint a series of bathing nudes. See Shapiro, "Pissarro as print-maker", p. 221.

Between the sky, this land and this water there is necessarily a relationship ...

What interests me less and less in my art is the material side of painting (the lines). The great problem to solve is to bring everything, even the smallest details, into the harmony of the whole ...

Camille Pissarro, 1903

Early works

pp. 72–73: cat. 3 *Landscape*, c.1865 (detail), Indianapolis Museum of Art

Pissarro began his artistic training in the early 1850s in Saint-Thomas, in the Danish Antilles, in the company of the painter Fritz Melbye (1826–96). Right from the start, he showed a partiality for landscape painting. After he arrived in France in 1855, he recommenced his education, enrolling in private classes held by François-Édouard Picot (1786–1868) and Isidore Dagnan (1794–1873), teachers from the École des Beaux-Arts; he also enrolled for classes with a former pupil of Ingres, Henri Lehmann (1814–82). *Landscape with trees*, c.1855 (cat. 1) is probably connected to an academic exercise aiming to produce a "classic" landscape in the manner of Nicolas Poussin and Gaspard Dughet.[1]

Pissarro moved from the academic masters to the "free" ateliers, where he could draw, unsupervised, from a live model. In one of these, the Atelier Suisse, he met Claude Monet, Paul Cézanne and Ludovic Piette, commencing three of the most important artistic friendships of his life. During the latter part of the 1850s, he also received informal tuition from the eminent landscapist, Camille Corot (1796–1875). Corot encouraged him to make studies outdoors, which led Pissarro to explore places accessible through the railway networks around Paris: he painted in Bougival, La Roche-Guyon, La Varenne-St-Hilaire (where he lived from 1863 to 1866) and Pontoise (where he lived from 1866 to 1869, and from 1872 to 1882).

The Marne at La Varenne-St-Hilaire, c.1864 (cat. 4) shows Pissarro's allegiance to the picturesque landscape exemplified by Corot. He inherited Corot's predilection for generalisation and suggestion, avoiding a minute, literal description of the landscape in favour of mood and general effect. Especially striking in this work is the sense of "vacancy", the conspicuous lack of anecdotal subject matter.

This emptiness became a characteristic feature of Pissarro's paintings, but how was it to be explained? In the well-known essay, "City vs. country", in which a demographic shift in France during the Second Empire is described (a depopulation of the countryside, a corresponding crush in the cities), the art historian Robert Herbert discusses the representation of peasant life by numerous artists, including Pissarro.[2] However, Pissarro's contemporaries never made such a literal interpretation of his imagery; rather, they seized on the air of loneliness and melancholy: "The canvases of Pissarro vividly communicate the sensation of space and solitude, from which he extracts an impression of melancholy,"[3] wrote Duret. "It's the modern countryside. We sense that man has passed, rummaging in the ground, carving it up, saddening the horizons,"[4] wrote Zola.

Another strong influence on the young Pissarro was Gustave Courbet's (1819–77) unflinchingly realistic portrayal of rural life. His rendition of the harshness of peasant life and the ruggedness of the landscape was striking and original, but his handling of paint – particularly his use of impasto applied with a palette knife – was equally influential. As Christopher B. Campbell has noted, "Courbet generated paint surfaces of such continuous and experimental variability that one of the myriad roles he played for the succeeding generation of the avant-garde painters was to redirect attention back to the potential inherent in painting's own materiality."[5] This was certainly true of his influence on Pissarro.

Two works in this section have an especially interesting provenance: *Banks of the Marne in winter*, 1866 (cat. 7) was shown in the Salon of 1866, where it was favourably reviewed by Zola and Jean Rousseau (see page 228), and *River landscape with boat*, c.1866 (cat. 6) was possibly given to Zola in gratitude for his eloquent support of Pissarro's work.

1 See "Pissarro and the picturesque" in this catalogue, p. 30, footnote 41.

2 Robert L. Herbert, "City vs. Country: The Rural Image in French Painting from Millet to Gauguin", *Artforum*, February 1970, pp. 44–55.

3 Théodore Duret, "Les Peintres Impressionnistes", in *Critique d'avant-garde*, École nationale supérieure des Beaux-Arts, Paris, 1998, p. 56.

4 Émile Zola, "Les Naturalistes", reprinted in *Mon Salon – Manet – Écrits sur l'art*, Garnier-Flammarion, Paris, 1970, p. 148.

5 Christopher B. Campbell, "Pissarro and the palette knife – Two pictures from 1867", *Apollo*, London, November 1992, p. 311.

1

Landscape with trees, two figures on a road
and mountains in the background
Paysage classique
c.1855

black and white chalk on blue paper, 54 x 69
Yale University Art Gallery, gift of
Joseph F. McCrindle, LL.B., 1948 (1973.132)

2
Road to Port-Marly
Route de Port-Marly
c.1860–67

oil on panel, 22.8 x 35.1
Lent by the Syndics of the Fitzwilliam
Museum, Cambridge (PD58-1958)

3

Landscape
Paysage
c.1865

oil on canvas, 28.4 x 44.1
Indianapolis Museum of Art,
gift of Mrs Joseph E. Cain (75.981)

4

The Marne at La Varenne-St-Hilaire
La Marne à La Varenne-St-Hilaire
c.1864

oil on canvas, 24 x 32
Kunstmuseum Bern, Legat Eugen Loeb, Bern
(G1871)

Sydney only

5

Barge on the Seine
Péniche sur la Seine
c.1863

oil on canvas, 46 x 72
Musée Camille Pissarro, Pontoise
(P1980.3)

6

River landscape with boat
Paysage fluviale avec bâteau
près de Pontoise
c.1866

oil on canvas, 43 x 65
Kunstmuseum Sankt Gallen, Sturzeneggersche
Painting Collection, purchased 1936

7 (overleaf)

Banks of the Marne in winter
Bords de la Marne, en hiver
1866

oil on canvas, 91.8 x 150.2
The Art Institute of Chicago, Mr and Mrs Lewis
Larned Coburn Memorial Fund (1957.306)

Towards Impressionism

pp. 84–85: cat. 11 *Landscape in the vicinity of Louveciennes (Autumn)*, 1870 (detail),

In 1871, during the Franco-Prussian war, Prussian troops were garrisoned in Pissarro's house in Louveciennes for almost four months. They vandalised many of the paintings and drawings Pissarro left behind when he and his family fled from the invading army. This irreparably distorted the record of Pissarro's early career[1], and the gravest losses were related to his production between 1869 and 1870. These years saw the emergence of a style of painting that Pissarro developed in common with Claude Monet, Auguste Renoir and Alfred Sisley: it was the period of the origin of Impressionism. As a consequence, the evidence of Pissarro's role in the invention of Impressionism is incomplete.

Lying to the north-west of Paris, the region of Louveciennes, Voisins, Marly and Bougival is extremely compact. It has been called "the cradle of Impressionism". Pissarro was the first to settle in the area, soon to be joined by Monet (who rented a house in Saint-Michel, the old part of Bougival) and Renoir (whose mother lived in the village of Voisins). They were all enthusiasts of plein-air painting and were fascinated by the possibilities of developing the vivid, vibrant effects of their outdoor sketches and studies into more carefully considered and deliberately wrought *tableaux*. They painted in each other's company and made the results of their experiments available to each other.

At the time, Monet was the most committed and dogmatic *pleinairiste*, whereas Pissarro was more pragmatic, reworking his paintings in the studio whenever he thought it necessary. Because of its size and complexity, *Landscape in the vicinity of Louveciennes (Autumn)*, 1870 (cat. 11) is unlikely to have been painted outdoors in its entirety, if at all. Pissarro exhibited it in the Salon of 1870, so it is one of the earliest works that featured the typical Impressionist palette to be shown to a large public.[2] It evokes the mild climate and gentle light of the Ile-de-France with exquisite fidelity. The composition moves between form and formlessness – an explosion of spring foliage breaks up and reorders the architectural framework. Typically, paintings that were conceived for exhibition in the Salons

were slick and impersonal in handling and large and rhetorical in effect, making an ostentatious feature of subject matter. In contrast, Impressionist paintings tended to be small and intimate, deliberately ordinary in subject, and painted with spontaneity, evidently at speed (see *Effect of fog at Creil*, 1873, cat. 13).

Near Sydenham Hill (cat. 9) was painted in 1871 in London. On the back is inscribed: "To my wife … C. Pissarro". Pissarro married Julie Vellay on 14 June 1871, and this was probably his marriage gift to her. It hung in an honoured place in a succession of Pissarro family homes throughout her lifetime.

The fence, 1872 (cat. 10) is a composition dominated by a single tree. The tree has been struck by lightning and has regrown its branches from a single node. In effect, it functions as a central nervous system for the painting. The lower right branch partly retraces the line of the road that connects the houses. Clouds are ranged along the twigs of the upper branches. The folds of land, the borders of fields and the paths are all rhythmically interrelated, extending and resolving the spread of the tree into the shape of the canvas.

La Sente de Justice, Pontoise, c.1872 (cat. 12) is another extremely novel composition, with the spatial prospect ascending and closing off to the left, and descending and opening out to the right. Pissarro exploits a wild gamut of colours that the scene has suggested to him: palest blue sky, pink-ochre earth, brilliant greens and purplish-brown roofs. The art historian Michael Fried was intrigued by the sidelong, dispersed quality of this composition, likening it to a scene that had been viewed back to front. Pissarro, he wrote, "apprehended his motifs as if from the rear".[3]

---

1 On the destruction of Pissarro's works, see Ralph E. Shikes and Paula Harper, *Pissarro – His Life and Work*, Quartet Books, London, 1980, p. 95. Richard Brettell has questioned whether the assessment of damage and loss was somewhat exaggerated. See Brettell, "Pissarro in Louveciennes", *Apollo*, London, November 1992, p. 317.

2 See pp. 229–31.

3 Michael Fried, *Manet's Modernism*, University of Chicago Press, Chicago, 1998, p. 395.

8

The road near the farm
La route près de la ferme
1871

oil on canvas, 38.1 x 46
Fine Arts Museums of San Francisco,
bequest of Marco F. Hellman (1974.5)

9

Near Sydenham Hill
Près de Sydenham Hill
1871

oil on canvas, 43.5 x 53.5
Kimbell Art Museum, Fort Worth,
Texas (AP71.21)

10

The fence
La barrière
1872

oil on canvas, 37.8 x 45.7
National Gallery of Art, Washington, DC,
collection of Mr and Mrs Paul Mellon
(1985.64.31)

11

Landscape in the vicinity
of Louveciennes (Autumn)
Louveciennes (Automne)
1870

oil on canvas, 88.9 x 115.9
The J. Paul Getty Museum, Los Angeles,
California (82.PA73)

12

La Sente de Justice, Pontoise
c.1872

oil on canvas, 52 x 81
Memphis Brooks Museum of Art, Memphis,
Tennessee, gift of Mr and Mrs Hugo N. Dixon
(53.60)

13

Effect of fog at Creil
Effet de brouillard à Creil
1873

oil on canvas, 38 x 56.5
Private collection, Switzerland

14

Still-life: apples and pears in a round basket
*Nature morte: pommes et poires dans
un panier rond*
1872

oil on canvas, 45.7 x 55.2
Mrs Walter Scheuer; on long-term loan to
Princeton University Art Museum (L1988.62.15)

Sydney only

Impressionism

pp. 96–97: cat. 23 *Piette's house, Montfoucault, in the snow*, 1874 (detail),

Impressionism was "a system of blotches"[1] (*un système des taches*), a critic observed in 1880. This was a good description of Pissarro's paintings: they were made of *taches* (spots, blots, patches) which the artist deliberately neglected to blend or fuse together. He strove for economy and elegance in his notation, using the simplest, sparest means to achieve the most vivid and compelling effect. He analysed the motif and broke it down into constituent *taches*, assembling his picture from these. The viewer is invited to follow the breaking down and building up the image. This is a defining feature of Pissarro's art.

The banks of the Oise near Pontoise, 1873 (cat. 15) is a precious example of his analytic skill. The economy, measure and clarity of the notation are remarkable. Every chip and every speck of the tiny painting are perfectly integrated into the whole, perfectly legible from as many as ten paces away. The subject is not so much the topography as an event of light, the breadth of space, the instant of time. The silvery light is a fugitive phenomenon, like wind in the trees, racing clouds, people on the road, a barge on the stream, possibly a train on the far bank of the river.

In 1872, Pissarro returned to live in Pontoise, a town twenty-six kilometres to the north-west of Paris (he had lived there before, between 1866 and 1869). The semi-rural neighbourhood of L'Hermitage, on the eastern outskirts of Pontoise, provided him with an abundance of motifs for his paintings. *Factory near Pontoise*, 1873 (cat. 16) portrays a distillery on the bank of the Oise (the same factory is glimpsed in cat. 15 and 44). Industrialisation was slow to arrive in France. The sight of tiny brindled cows grazing in front of the factory in cat. 16 underlines just how far Pontoise was from the "dark satanic mills" of the nineteenth century industrial centres.

Pissarro's difference from his closest Impressionist colleagues (Monet, Renoir, Sisley and Cézanne) can be explained through his attachment to the range of subjects outlined by Théodore Duret. "Among the impressionists," wrote Duret, "Pissarro is the one in whom one finds, to the most marked degree, the point of view of the purely naturalist painters. He sees nature by simplifying it, and [he sees it] through its permanent aspects ... He portrays fields that have been ploughed or laden with the harvest, trees in blossom or stripped by the winter, main roads with lopped elm trees and hedges along the sides, rustic pathways which run under leafy trees. He likes the houses of the village which have gardens surrounding them, the farmyards with plough-animals, the ponds where ducks and geese dabble. The humans he introduces into his pictures are rustic peasants and calloused workers."[2]

In addition, Pissarro was eager to explore new problems and possibilities of pictorial composition, which sometimes led to results that were too synthetic and stylised to pass as "naturalist painting" – as in the brilliantly quirky *Hill at L'Hermitage, Pontoise*, 1873 (cat. 18).

Pissarro became renowned for his paintings of small market gardens (cat. 18–22). Two works featuring this motif – *A cowherd on the route du Chou, Pontoise*, 1874 (cat. 22) and *The orchard at Maubuisson, Pontoise*, 1876 (cat. 21) – were shown in the second Impressionist exhibition in 1876 and the fourth Impressionist exhibition in 1879 respectively.

Pissarro participated in all eight of the Impressionist exhibitions (1874–86). He and Claude Monet were principal organisers of the first Impressionist exhibition, but the critical reception of their courageous initiative was a disaster (see pages 233–34). In the latter half of 1874, Pissarro's paintings seemed to reflect the depressing setback he and his colleagues had suffered. *The pond at Ennery* (cat. 24) and a series of winter landscapes he painted at Montfoucault (for example, cat. 23) seem to express his chastened mood.

1 J.-K. Huysmans, "L'Exposition des Indépendants en 1880", reprinted in Denys Riout, *Les écrivains devant l'impressionnisme*, Macula, Paris, 1989, p. 260.

2 Théodore Duret, "Les Peintres impressionnistes" (1878), in *Critique d'avant garde*, École nationale supérieure des Beaux-Arts, Paris, 1998, p. 56.

15

The banks of the Oise near Pontoise
Bords de l'Oise près de Pontoise
1873

oil on canvas, 38.1 x 55.2
Indianapolis Museum of Art,
James E. Roberts Fund (40.252)

16

Factory near Pontoise
Usine près de Pontoise
1873

oil on canvas, 45.7 x 54.6
Museum of Fine Arts, Springfield, Massachusetts,
The James Philip Gray Collection (37.03)

17
The road to Ennery, near Pontoise
La route d'Ennery, près de Pontoise
1874

oil on canvas, 55 x 92
Musée d'Orsay, Paris, donated by Max
and Rosy Kaganovitch, 1973 (RF1973-19)

18

Hill at L'Hermitage, Pontoise
Coteau de L'Hermitage, Pontoise
1873

oil on canvas, 61 x 73
Musée d'Orsay, Paris, *acquis par dation*,
1983 (R.F. 1983-8)

19

Kitchen garden at L'Hermitage, Pontoise
Jardin potager à L'Hermitage, Pontoise
1874

oil on canvas, 54 x 65.1
The National Gallery of Scotland, Edinburgh,
presented by Mrs Isabel M. Traill, 1979 (NG2384)

20
Villa at L'Hermitage, Pontoise
Maison bourgeoise à L'Hermitage, Pontoise
1873

oil on canvas, 50.5 x 65.5
Kunstmuseum Sankt Gallen, Sturzeneggersche
Painting Collection, purchased 1936 (G34)

21

The orchard at Maubuisson, Pontoise
Le verger à Maubuisson, Pontoise
1876.

oil on canvas, 42.5 x 50.2
Private collection, London

Sydney only

22

A cowherd on the route du Chou, Pontoise
Une vachère sur la route du Chou, Pontoise
1874

oil on canvas, 54.9 x 92.1
The Metropolitan Museum of Art,
gift of Edna H. Sachs, 1956 (56.182)

23

Piette's house, Montfoucault, in the snow
La maison de Piette, Montfoucault, effet de neige
1874

oil on canvas, 60 x 73.5
Lent by the Syndics of the Fitzwilliam Museum,
Cambridge (PD10-1966)

24

The pond at Ennery
L'Étang à Ennery
1874

oil on canvas, 53.3 x 64.1
Yale University Art Gallery, collection of
Mr and Mrs Paul Mellon, BA, 1929 (1983.7.14)

The Impressionist print

"Of all the Impressionists, Pissarro was the truest *peintre-graveur* [painter-engraver] and most diligently pursued the art of printmaking,"[1] wrote Barbara Stern Shapiro. His production of prints was marked by the same restlessness and desire for new challenges that characterised his career as a painter. Another leading authority on Impressionist printmaking, Michel Melot, observed that "his is a continuity of breaking-points [...] A chronological reconstruction of his prints cannot rely on a harmonious and continuous development."[2] In fact, this generalisation could apply to Pissarro's entire creative life.

Pissarro began printmaking in the 1860s and his early etchings were rather conventional poetic landscapes, indebted to Camille Corot and the painters of the Barbizon school (see fig. 24, *By the water's edge*, c.1863, page 63). Having set it aside for almost a decade, he resumed etching in Auvers-sur-Oise during the summer of 1873, in the company of his Impressionist colleagues Guillaumin and Cézanne. The incentive to explore this medium again came through an invitation to use the etching press of Dr Paul Gachet. (Gachet exhibited paintings and etchings under the name of Paul van Ryssel.) Pissarro's *Portrait of Paul Cézanne*, 1874 (cat. 100) dates from this period.

Printmaking was a highly socialised activity for Pissarro, and his outstanding achievements of 1879–80 (which are highlighted in this section) owed a great deal to the stimulating companionship of Edgar Degas. This was Pissarro's most fruitful and innovative period of printmaking, resulting in eleven works from 1879 and six from 1880; many of them rank among the greatest achievements of Impressionist graphic art.

In opposition to the traditional emphasis given to line in etching, Pissarro invented an equivalent to the "system of blotches" of his paintings. Thus, he developed an unprecedented range of painterly effects, using unorthodox combinations of etching, aquatint and drypoint and having occasional recourse to unusual tools, such as metal brushes, emery sticks and sandpaper.

In the early state of *Rainy effect*, 1879 (cat. 28), he created a delicate haze of aquatint, on which he built an image – the smoky fuses of a line of trees, a haystack like a fat pear, two peasant women braving the driving rain – in rich, shifting textures. The impression of the sixth state (cat. 29) belonged to Degas, who also owned the earliest state of *Wooded landscape at L'Hermitage, Pontoise*, 1879 (cat. 34). Degas printed many of Pissarro's works himself (including cat. 32 and 33, as Pissarro's inscriptions at the base of the prints testify).

The three states of *Wooded landscape at L'Hermitage, Pontoise* (cat. 34–36) show Pissarro's masterful control of nuance as the print was fine-tuned. The image was based on a painting from the same year (PV444), now in the Kansas City Art Museum. The print exists in six states; the final version was intended for a monthly journal of prints, *Le Jour et la nuit* – a project devised by Degas, which never came to fruition.

The print that had the longest and most complicated gestation in Pissarro's œuvre is *Woman emptying a wheelbarrow*, 1880 (cat. 41): it went through twelve states. Its striking quality of immediacy and freshness belies Pissarro's laborious efforts to perfect this image. His superb orchestration of painterly textures acts like a dappled veil, filtering the snowy whiteness of the paper, and establishes this work as a consummate masterpiece of Impressionist printmaking.

1 Barbara Stern Shapiro, "Pissarro as printmaker", in *Pissarro* (exhibition catalogue), Arts Council of Great Britain, London, 1980, p. 191.

2 Michel Melot, "A rebel's role: concerning the prints of Camille Pissarro", in Christopher Lloyd (ed.), *Studies on Camille Pissarro*, Routledge & Kegan Paul, London, 1986, p. 117.

25

Path in the woods, Pontoise
Chemin sous bois, à Pontoise
1879

aquatint and etching, 16.1 x 21.3
D19, ii of 6 states
The Art Institute of Chicago,
John H. Wrenn Endowment, 1960 (1960.723)

26

Path in the woods, Pontoise
Chemin sous bois, à Pontoise
1879

aquatint and etching, 16.2 x 21.3
D19, v of 6 states
The Art Institute of Chicago,
John H. Wrenn Endowment, 1960 (1960.724)

27
Woman on the road
La femme sur la route
1879

aquatint, softground etching
and etching, 15.6 x 20.9
D18, iv of 4 states
The Art Institute of Chicago,
gift of Gaylord Donnelley (1971.365)

28

Rainy effect
Effet de pluie
1879

aquatint, 16 x 21.4
D24, ii of 6 states
Yale University Art Gallery, Everett
V. Meeks, BA, 1901, and Stephen Carlton
Clark, BA, 1903, Funds (1972.49)

29

Rainy effect
Effet de pluie
1879

aquatint, etching and drypoint, 16 x 21.3
D24, vi of 6 states
Los Angeles County Museum of Art,
Wallis Foundation Fund, in memory
of Hal B. Wallis (AC1996.27.2)

Crépuscule 3e état ?D.23.   vert anglais

31

Twilight with haystacks
Crépuscule avec meules
1879

aquatint, etching and drypoint,
printed in green ink, 10.4 x 18
D23, iii of 3 states
National Gallery of Canada, Ottawa,
purchased 1976 (18724)

30

Twilight with haystacks
Crépuscule avec meules
1879

aquatint, etching and drypoint, 10.4 x 18
D23, iii of 3 states
Museum of Fine Arts, Boston,
Lee M. Friedman Fund (1974.533)

35

Wooded landscape at L'Hermitage, Pontoise
Paysage sous bois à L'Hermitage, Pontoise
1879

softground etching, aquatint and drypoint, 21.6 x 26.7
D16, iv of 5 states; C, v of 6 states
Museum of Fine Arts, Boston,
Lee M. Friedman Fund (1971.268)

32

Twilight with haystacks
Crépuscule avec meules
1879

aquatint, etching and drypoint,
printed in red-brown ink, 10.4 x 18
D23, iii of 3 states
National Gallery of Canada, Ottawa,
purchased 1973 (17292)

33

Twilight with haystacks
Crépuscule avec meules
1879

aquatint, etching and drypoint,
printed in Prussian blue ink, 10.4 x 18
D23, iii of 3 states
Museum of Fine Arts, Boston,
Lee M. Friedman Fund (1983.220)

34

Wooded landscape at L'Hermitage, Pontoise
Paysage sous bois à L'Hermitage, Pontoise
1879

softground etching and aquatint, 21.6 x 26.7
D16 (this state uncatalogued by Delteil); C, i of 6 states
Museum of Fine Arts, Boston,
Lee M. Friedman Fund (1971.267)

36

Wooded landscape at L'Hermitage, Pontoise
Paysage sous bois à L'Hermitage, Pontoise
1879

softground etching, aquatint and drypoint, 21.6 x 26.7
D16, v of 5 states; C, vi of 6 states
Museum of Fine Arts, Boston, Katherine E. Bullard Fund,
in memory of Francis Bullard, Prints and Drawings
Curator's Discretionary Fund, Cornelius C. Vermeule III
and anonymous gifts (1971.176)

37

Horizontal landscape
Paysage en long
1879

aquatint and drypoint, 11.5 x 39.3
D17, ii of 2 states; C, ii of 3 states
The British Museum, London, bequeathed
by Campbell Dodgson (1949-4-11-2594)

38

View of L'Hermitage (Pontoise)
Paysage à L'Hermitage (Pontoise)
1880

etching, 11 x 12.3
D28, ii of 2 states
Bibliothèque Nationale de France

39

Field with mill at Osny
Prairie et moulin à Osny
1885

etching, drypoint and aquatint, 16 x 23.8
D59, vi of 6 states
Bibliothèque Nationale de France

40

The Rondest house at L'Hermitage
La maison Rondest à L'Hermitage
1882

etching and aquatint, 16.5 x 11.2
D35, only state, posthumous printing of 1920
Art Gallery of New South Wales, Sydney,
purchased 2004 (132.2004)

41

Woman emptying a wheelbarrow
Femme vidant une brouette
1880

aquatint and drypoint, 32 x 23.2
D31, xi of 11 states; C, xii of 12 states
Sterling and Francine Clark Art Institute,
Williamstown, Massachusetts (1962.92)

131

Late Impressionism

pp. 132–33: cat. 47 *The highway*, 1880 (detail), The Baltimore Museum of Art

During the late 1870s and the first half of the 1880s, Pissarro worked to improve, diversify and broaden the scope of Impressionism. The following plates show several examples of his daring experiments and bold departures from precedent during this period. Despite the fact that Pissarro had used harmonious, close-valued colours for many years, during the summers of 1875 and 1876 he opted for bright hues and strong tonal contrasts in a series of landscapes painted in Montfoucault (Mayenne). Some of these paintings made a "theme" of yellow, green, blue, white and black. *The large pear tree at Montfoucault*, 1876 (cat. 42) shares its colour scale with other works, including a very well-known painting in the Musée d'Orsay, *The harvest, Montfoucault*, 1876 (PV364).

In the same period, Pissarro experimented with evenly accented, homogeneous brushwork, as in *Resting beneath the trees, Pontoise*, 1878 (cat. 43), where the surface is consolidated into a glittering crust. This painting was exhibited in the seventh Impressionist exhibition in 1882; *The path to Le Chou, Pontoise*, 1878 (cat. 44) appeared in the sixth Impressionist exhibition in 1881 and was commented on by the renowned novelist and critic Joris-Karl Huysmans (see page 237).

In other works – such as *The banks of the Viosne at Osny, grey weather, winter*, 1883 (cat. 45) and *Steep road at Osny*, 1883 (cat. 46) – Pissarro developed a thread-like stroke, creating tapestry-like densities. *Steep road at Osny* was once in the collection of Paul Gauguin and is remarkable for its radically simplified composition and dazzling silver tonality.

The highway, 1880 (cat. 47) makes a fantastic web of blue, green, beige and white. A figure on the right near the foreground has been painted out, yet Pissarro was not too concerned about hiding his revision. The surviving cluster of figures has been transposed into the most daring "equivalences": there is a tiny spot of pure vermilion on the face of the man in the cart, and the horse, the cart and the dark part of the figures are painted violet (although it may

not be possible to see this in reproduction). The sparkle and surprise of this painting help us to identify the qualities prized by Pissarro: "the richness, suppleness, freedom, spontaneity and freshness of sensation of our Impressionist art".[1]

During the initial phases of Impressionism, Pissarro had been closely allied to his fellow landscape painters, Claude Monet, Alfred Sisley and Paul Cézanne. However, following his involvement with Degas in collaborative printmaking ventures, Pissarro became a more committed and prolific figure painter. This was a notable departure from the works that had made his reputation. Pissarro's biographers, Ralph Shikes and Paula Harper, described his figure paintings of the early 1880s as "problematic" and conjectured that they produced a "feeling of uncertainty" which contributed to the lack of success of Pissarro's first one-man exhibition at the Durand-Ruel gallery in May 1883.[2]

Problematic or not, there is no denying the powerful conception of *Peasants resting*, 1881 (cat. 50), shown in the seventh Impressionist exhibition in 1882. The figures are embedded in a dazzling, all-over effervescence of blue, green, blonde and white. Swirling, serpentine rhythms course through the picture. This extraordinary dynamism prefigures the works of Vincent van Gogh (who did not arrive in Paris until 1886).

1 Pissarro, letter to Lucien Pissarro, 6 September 1888, in *Correspondance de Camille Pissarro*, ed. Janine Bailly-Herzberg, Éditions du Valhermeil, Paris, vol. 2, p. 251.

2 Ralph E. Shikes and Paula Harper, *Pissarro – His Life and Work*, Quartet Books, London, 1980, p. 183.

42

The large pear tree at Montfoucault
Le grand poirier à Montfoucault
1876

oil on canvas, 54 x 65
Kunsthaus Zürich, Johanna and
Walter L. Wolf Collection (1984/12)

43

Resting beneath the trees, Pontoise
Dans le bois, le repos, Pontoise
1878

oil on canvas, 65 x 54
Hamburger Kunsthalle/bpk (1090)

44
The path to Le Chou, Pontoise
La Sente du Chou, Pontoise
1878

oil on canvas, 57 x 92
Musée de la Chartreuse,
Douai (2231)

45

The banks of the Viosne at Osny,
grey weather, winter
*Bords de la Viosne à Osny,
temps gris, hiver*
1883

oil on canvas, 65.3 x 54.5
National Gallery of Victoria, Melbourne,
Felton Bequest, 1927 (3466-3)

46
Steep road at Osny
Chemin montant à Osny
1883

oil on canvas, 55.6 x 46.2
Musée des Beaux-Arts de Valenciennes
(P.46.1.406)

47

The highway
La côte de Valhermeil
1880

oil on canvas, 64.2 x 80
The Baltimore Museum of Art, The Cone Collection,
formed by Dr Claribel Cone and Miss Etta Cone
of Baltimore, Maryland (BMA1950.280)

48
Woman washing dishes
La laveuse de vaisselle
1882

oil on canvas, 81.9 x 64.5
Lent by the Syndics of the Fitzwilliam Museum,
Cambridge, bought with the assistance of the
National Art Collections Fund, 1947 (PD.53-1947)

49
The little country maid
La petite bonne de campagne
1882

oil on canvas, 63.5 x 53
TATE, bequeathed by Lucien Pissarro,
the artist's son, 1944 (NO.5575)

50

Peasants resting
Paysannes au repos
1881

oil on canvas, 81 x 65.4
Toledo Museum of Art, purchased
with funds from the Libbey Endowment,
gift of Edward Drummond Libbey (1935.6)

Neo-Impressionism

Monet and Pissarro were deeply impressed by the dazzlingly white snow they saw in one of Turner's paintings during their stay in London, circa 1871. They realised Turner had obtained the effect "not from solid white, but from a quantity of diversely coloured touches, placed side by side, which reconstituted the desired effect at a distance".[1] This was the insight lying behind the Impressionists' technique, a technique consisting of small dabs of juxtaposed colour that blended together in the mind's eye to create a dazzling effect.

Pissarro liked to engage with new problems and new possibilities of pictorial composition. Given his sober, orderly disposition, he had tried to discipline, regulate and unify the more or less random scatter of brushstrokes that typified Impressionism. He was also interested in finding better ways of exploiting the vibrancy of colour to represent effects of light. An example of his love of experimentation is *The artist's palette with a landscape*, c.1878 (cat. 51).

The transition from Impressionism to Neo-Impressionism seemed utterly logical to Pissarro. Wasn't this the way to the more rational, formal, synthetic version of Impressionism he had always desired? The art critic Edmond Duranty described the aims of *Impressionist* painters: "To decompose the solar light into its rays, into its elements, and to recompose its unity through the general harmony of the iridescences that they spread over their canvases"[2] – wasn't this a perfect description of *Neo*-Impressionism?

According to Georges Seurat (1859–91), the inventor of the Neo-Impressionist technique, Pissarro was the first convert and the first to exhibit a Neo-Impressionist picture. In January or February 1886, he showed a little painting in a dealer's window in Paris. *Ploughing at Éragny*, c.1886 (cat. 53) seems to belong to this period, and the Art Gallery of New South Wales' *Peasants' houses, Éragny*, 1887 (cat. 52) shows his rapid mastery of the technique and the tightly integrated pictorial surface it enabled him to achieve.

Many of his subsequent paintings were more premeditated and self-conscious in their design. Close inspection reveals a pencil outline around the figures in *Apple picking at Éragny-sur-Epte*, 1888 (cat. 54), proving that the composition was thought out before Pissarro began painting. The canvas is modest in size – perhaps too small for its monumental conception. The decorative effect of the silhouetted forms relates to Pissarro's study of Japanese art. From close range, the brilliant yellow field becomes a haunting mystery – there is far less yellow constituting it than one supposes – and the jewel-like brilliance of the stepped-up reds in the apples creates a captivating effect.

L'Ile Lacroix, Rouen (effect of fog), 1888 (cat. 57) is one of Pissarro's most abstract compositions – a palpitating expanse of pearl grey, completely devoid of anecdote and "human interest". In 1889, it was included in the sixth exhibition of the modernist society, *Les XX (Les Vingt)*, in Brussels. In a letter to his son Lucien, Pissarro described a frame he had designed[3], and his description seems to tally with the frame of *L'Ile Lacroix*, so this may be the only surviving painting which is framed as he intended.

In a letter written in 1888, Pissarro expressed his misgivings with Neo-Impressionism: "I haven't yet solved the question of dividing the pure tone without harshness [...] How does one obtain the qualities of purity and simplicity of the dot, and the fullness, suppleness, liberty, spontaneity, freshness of sensation of our Impressionist art? [...] The dot is meagre, flimsy, diaphanous, more monotonous than simple, even in the Seurats, especially in the Seurats ... I'm very preoccupied with the question; I'm going to the Louvre to look at certain painters who interest me from that point of view. Isn't it stupid not to have any Turners [in the Louvre]?"[4]

1 Paul Signac, *D'Eugène Delacroix au néo-impressionnisme* (1899), Hermann, Paris, 1978, p. 88.

2 Edmond Duranty, *La Nouvelle peinture – À propos du groupe d'artistes qui expose dans les Galeries Durand-Ruel* (1876), reprinted by L'Échoppe, Paris, 1988, p. 30.

3 Pissarro, letter to Lucien Pissarro, 15 March 1887, in *Correspondance de Camille Pissarro*, ed. Janine Bailly-Herzberg, Éditions du Valhermeil, Paris, 1986, vol. 2, p. 141.

4 Pissarro, letter to Lucien Pissarro, 6 September 1888, ibid., vol. 2, p. 251.

51

The artist's palette with a landscape
Paysans près d'une charrette
c.1878

oil on panel, 24.1 x 34.6
Sterling and Francine Clark Art Institute,
Williamstown, Massachusetts (1955.827)

52

Peasants' houses, Éragny
Maisons de paysans, Éragny
1887

oil on canvas, 59 x 71.7
Art Gallery of New South Wales, Sydney,
purchased 1935 (6326)

53

Ploughing at Éragny
Le labour à Éragny
c.1886

oil on panel, 15.6 x 23.5
Private collection, courtesy of
Barbara Divver Fine Art, New York

54

Apple picking at Éragny-sur-Epte
La cueillette de pommes à Éragny-sur-Epte
1888

oil on canvas, 61 x 74
Dallas Museum of Art, Munger Fund (1955.17.M)

55

L'Ile Lacroix, Rouen
L'Ile Lacroix à Rouen
1887

drypoint and aquatint, 11.6 x 15.7
D69, ii of 2 states
The British Museum, London, bequeathed
by Campbell Dodgson (1949-4-11-2612)

56

View of Rouen: L'Ile Lacroix
Vue de Rouen: L'Ile Lacroix
1887

aquatint, 12.8 x 17.5
LM131, only state
Bibliothèque Nationale de France

57

Lacroix, Rouen (effect of fog)
Lacroix, Rouen (effet de brouillard)

canvas, 46.7 x 55.9
elphia Museum of Art,
G. Johnson Collection, 1917 (1060)

58

Hampton Court Green
1891

oil on canvas, 54.3 x 73
National Gallery of Art, Washington, DC,
Ailsa Mellon Bruce Collection (1970.17.53)

The figure
in the landscape

pp. 158–59: cat. 61 *Springtime, grey weather, Éragny,* 1895 (detail), Art Gallery of Ontario, Toronto

In December 1882, Pissarro and his family moved to the village of Osny, a few kilometres north-west of Pontoise. They moved again in April 1884, settling permanently in the village of Éragny-sur-Epte, near Gisors. This was to be Pissarro's principal residence for the rest of his life. On arrival in Éragny, he wrote to his dealer in Paris: "I haven't been able to restrain myself from painting, so beautiful are the motifs that surround my garden."[1]

A selection of paintings, drawings and prints in this section display the attractions that Éragny held for him. *Springtime, grey weather, Éragny*, 1895 (cat. 61) has a blossoming tree at the centre of the composition. The tree has the flowing limbs and flouncing apparel of a dancer, exulting in its freedom of movement before a backdrop of regimented verticals and a gridlocked fence. The contrast of greens and pinks and the granular texture of the surface are reminiscent of Pissarro's Neo-Impressionist paintings.

The etchings of *Market at Gisors: rue Cappeville*, 1894/95 (cat. 71) and the *Church and farm at Éragny-sur-Epte*, 1894/95 (cat. 64) also take local sights as their subjects – the latter is a view from a barn Pissarro converted into a studio in 1893. These prints were highly esteemed by the pioneering historian of printmaking, Arthur M. Hind: "His [Pissarro's] few colour etchings … are among the best of their kind,"[2] he wrote.

Pissarro liked to show how the figures in his paintings were absorbed into, and seemed to belong to, their surroundings. Nonetheless, most of his drawings portrayed figures in isolation, with the blankness of the paper their only context. He had always been a prolific draughtsman, and his commitment to figure painting in the 1880s resulted in an abundance of drawings. Some were studies for paintings: the Fitzwilliam Museum's *Full-length standing nude of a woman from behind*, c.1896 (cat. 77) depicts a figure that recurred in two paintings – once in a realistic interior (PV936) and once as an idealised Bather (PV938). Nonetheless, the drawing is a delectable work in its own right. Highly coloured, it consists of a continuous

outline of black chalk on pink paper, with the figure coloured in maroon, olive green and flesh-pink pastel. Along with a sense of humble origins, the model's unidealised shape, her awkward shyness and the vulnerability of the nape of her neck are all beautifully characterised.

The drawing in the National Gallery of Victoria (cat. 67) is a study for one of the figures in an ambitious pastel composition, *The poultry market, Gisors*, 1885 (PV1400), now in the Art Institute of Chicago. The model was probably Pissarro's niece, Nini (she also appears in cat. 48 and 49). *Woman carrying a basket* (cat. 70), the drawing from Princeton depicting a field worker, is augmented with incidental studies of her hands, yet it all adds up to a superlative graphic composition, a superb *mise-en-page*. This is a preliminary sketch for a figure in a coloured woodblock print that Pissarro produced in 1893 as part of a series done in collaboration with his son, Lucien (*Weeders*, cat. 68, and *Women herb gathering*, cat. 69).

Pissarro was a friend of Jean Grave, editor of the anarchist journals, *La Revolte* (1887–94) and *Les Temps Nouveaux* (1895–1914). In his memoirs, Grave acknowledged that Pissarro had regularly sent him small sums of money, even when the artist and his family were struggling to make ends meet. In more prosperous times, Pissarro twice paid the printing bills for *Les Temps Nouveaux* and supplied lithographs for reproduction in the newspaper[3] (cat. 72 and 73). *The ploughman*, 1901 (cat. 74) – Pissarro's only colour lithograph – was used for the cover of a brochure (published by Jean Grave) by the famous anarchist writer, Peter Kropotkin.

1 Pissarro, letter to Paul Durand-Ruel, 9 April 1884, in *Correspondance de Camille Pissarro*, ed. Janine Bailly-Herzberg, Presses Universitaires de France, 1980, vol. 1, p. 297.

2 Arthur M. Hind, *A History of Engraving and Etching* (originally published in 1923), Dover Publications, New York, 1963, p. 322.

3 Jean Grave cited in Robert L. and Eugenia W. Herbert, "Artists and Anarchism: Unpublished Letters of Pissarro, Signac and others – I", *Burlington Magazine*, London, November 1960, p. 477.

59
Éragny
1890

graphite and watercolour, 17.7 x 25.3
Lent by the Syndics of the
Fitzwilliam Museum, Cambridge
(PD.10-1982)

60
Landscape
Paysage
undated

pencil, 25 x 29
Private collection, Sydney

61

Springtime, grey weather, Éragny
Printemps, temps gris, Éragny
1895

oil on canvas, 60.3 x 73
Art Gallery of Ontario, Toronto,
purchased 1933 (2111)

62

Washing day at Éragny
La lessive à Éragny
1901

oil on canvas, 33 x 40.5
Queensland Art Gallery, Brisbane,
purchased 1975 (1:1407)

63

Study for Church and farm at Éragny-sur-Epte
Étude pour *L'Église et ferme d'Éragny-sur-Epte*
c.1894

black chalk, 23.8 x 30.7
The Metropolitan Museum of Art,
Harris Brisbane Dick Fund, 1948 (48.10.10)

64

Church and farm at Éragny-sur-Epte
Église et ferme d'Éragny-sur-Epte
1894/95

etching printed in grey, red, yellow and
blue ink on grey-green paper, 15.9 x 24.5
D96, vi of 6 states
National Gallery of Canada, Ottawa,
purchased 1980 (23653)

66

Rear view of a man in a smock
Homme à sarrau, vu de dos
undated

black chalk, 19.7 x 14.9
Princeton University Art Museum,
bequest of Dan Fellows Platt,
Class of 1895 (x1948-472)

67

Study for The poultry market, Gisors
Étude pour *Le marché à la volaille, Gisors*
c.1885

black chalk and pastel, 31 x 24.1
National Gallery of Victoria, Melbourne,
purchased through The Art Foundation
of Victoria, with the assistance of
Mr and Mrs William Jamieson, Members,
1983 (P22-1983)

68

Lucien Pissarro
after a drawing by Camille Pissarro

Weeders
Les sarcleuses
1893

colour woodcut, 17.7 x 11.9
National Gallery of Victoria, Melbourne,
Felton Bequest, 1914 (651-2)

69

Lucien Pissarro
after a drawing by Camille Pissarro

Women herb gathering
Femmes faisant de l'herbe
1893

colour woodcut, 18 x 11.9
National Gallery of Victoria, Melbourne,
Felton Bequest, 1914 (652-2)

70

Woman carrying a basket
Femme portant un panier
undated

black chalk, 20.2 x 16
Princeton University Art Museum,
bequest of Dan Fellows Platt,
Class of 1895 (x1948-463)

71

Market at Gisors: rue Cappeville
Marché de Gisors: rue Cappeville
1894/95

etching printed in grey, red, blue and yellow ink, 20 x 14
D112, vii of 7 states
The Baltimore Museum of Art, purchased with exchange
funds from the bequest of Mabel Garrison Siemonn, in
memory of her husband George Siemonn (BMA 1993.77)

72
Sower
Semeur
1896

lithograph on ochre paper, 20.7 x 26.8
D155, only state
The British Museum, London,
bequeathed by Campbell Dodgson
(1949-4-11-3368)

73
The vagrants
Les trimardeurs
1896

lithograph, 24.8 x 29.4
D154, v of 5 states
National Gallery of Victoria,
Melbourne, purchased 1961 (834-5)

74
The ploughman
La charrue
1901

colour lithograph, 22.5 x 15.2
D194, ii of 2 states
Cincinnati Art Museum,
gift of Herbert Greer French (1940.426)

75

Peasant at a well
Paysanne au puits
1891

etching, 23.2 x 19.4
D101, iii of 3 states, posthumous printing of 1920
National Gallery of Victoria, Melbourne,
gift of James Mollison, 1959 (360-5)

76
Woman kneeling
Femme agenouillée
undated

black chalk, 17.2 x 21.4 (sheet)
Yale University Art Gallery,
gift of John M. Montias,
in memory of his father (1971.65a)

77

Full-length standing nude
of a woman from behind
Étude de femme de dos
c.1896

pastel on pink paper, 47.5 x 24.3
Lent by the Syndics of the Fitzwilliam
Museum, Cambridge (PD.51-1947)

78

Bathers in the shade of wooded banks
Baigneuses à l'ombre des berges boisées
1894/95

lithograph, 15.3 x 21.9
D142, ii of 2 states
Toledo Museum of Art (1912.1182)

79

Line of bathers
Théorie de baigneuses
1894/95

lithograph, 13 x 20
D181, only state
Los Angeles County Museum of Art, gift
of Lewis F. Blumberg and Lynn T. Blumberg,
in honour of the Museum's twenty-fifth
anniversary (M.89.173.16)

80

Back view of bather
Baigneuse, vue de dos
1894/95

etching, drypoint and aquatint, 8.8 x 7.3
D114, v of 5 states,
posthumous printing of 1920
The Baltimore Museum of Art,
Print Fund (BMA 1951.85)

81

Three women bathing
Les trois baigneuses
1894/95

etching and aquatint, 17 x 12.9
D117, ii of 2 states,
posthumous printing of 1923
Art Gallery of New South Wales,
Sydney, purchased 2004 (133.2004)

City views and industrial landscapes

pp. 182–83: cat. 83 *La Place du Théâtre Français*, 1898 (detail), Los Angeles County Museum of Art

"I have always loved the immense streets of Paris, shimmering in the sun, the crowd of all colours, those beautiful linear and aerial perspectives, those eccentric fashions, etc. But how to do it? To install oneself in the middle of the street is impossible in Paris."[1] These lines, written to Pissarro in 1872 by his close friend, the painter Ludovic Piette, were portents of Pissarro's future: two decades later, he would become the pre-eminent painter of modern Paris and of the industrial centres of northern France.

The art historian Richard Brettell has estimated that Pissarro produced more than 300 paintings of the cities of Paris, Rouen, Dieppe and Le Havre during the last decade of his life. "Pissarro painted more cityscapes than any other major Impressionist and, as such, made the most sustained contribution to urban view painting by any great artist since the death of Canaletto in 1768,"[2] he wrote.

A small drypoint etching depicting the railway bridge at Pontoise (cat. 93) reminds us of Pissarro's early attachment to imagery that evoked the Industrial Revolution[3] (compare fig. 2, *Banks of the Oise at Pontoise*, 1867, page 17). *The stone bridge and barges at Rouen* (cat. 94) dates from 1883, the time of his first painting excursion to Rouen. A composition of great subtlety and restraint, it is based on a finely graded scale of greys. During this period, Pissarro chose to exhibit his paintings in white frames, which would have enhanced the delicate tonality. The stillness and "vacancy" of the motif prefigure many Neo-Impressionist compositions (compare *L'Ile Lacroix, Rouen (effect of fog)*, 1888, cat. 57).

Pissarro's paintings of Rouen changed significantly when he returned there to paint in 1896 (cat. 95, 96) and 1898 (cat. 89, 97): the urban landscape was no longer as still and vacant. He responded to the constant traffic of freight and trade, the bustle of conveyance and the explosions of colour and movement on market days and fair days – through which he summoned the eruptive energies of modern times.

In 1888, Pissarro was diagnosed with dacryocystitis, a chronic infection of the tear duct: he needed to keep away from wind and dust. Consequently, most of the paintings from the last fifteen years of his life were views from windows. *Pont Boïeldieu in Rouen, damp weather*, 1896 (cat. 96) was painted from a room in the Hôtel de Paris overlooking the dockside. It is one of Pissarro's most inspired compositions, welding a haphazard flux of elements into a pattern, finding a perfect equilibrium in their midst.

In a letter written in 1897, Pissarro announced: "I have found a room in the Grand Hôtel du Louvre, with a superb view over the avenue de l'Opéra and that corner of the place du Palais-Royal! It is beautiful to paint. Perhaps it is not very aesthetic, but I am delighted to be able to try to do these Paris streets which are customarily called ugly, but which are so silvery, so luminous and so lively and which are so different from the boulevards – it's completely modern!!!"[4] It was from this hotel window that he painted *La Place du Théâtre Français*, 1898 (cat. 83). An argument for his greatness as a pictorial composer could rest on this painting alone. It is a scatter composition *à la japonaise*, retaining the sense of a snapshot as well, and it seems to embrace and assimilate all the chaotic vitality of the modern city's street life. Yet its air of informality and randomness proves to be entirely deceptive. It is a rigorously ordered and integrated universe that Pissarro shows us, with not one speck redundant or out of place.[5]

1 *Mon cher Pissarro – Lettres de Ludovic Piette à Camille Pissarro*, ed. Janine Bailly-Herzberg, Éditions du Valhermeil, Paris, 1985, p. 73.

2 Richard R. Brettell and Joachim Pissarro, *The Impressionist and the City – Pissarro's Series Paintings*, Yale University Press, New Haven, 1993, p. xv.

3 The motif recurs in a painted fan (PV1619).

4 Pissarro, letter to Lucien Pissarro, 15 December 1897, in *Correspondance de Camille Pissarro*, ed. Janine Bailly-Herzberg, Presses Universitaires de France, 1989, vol. 4, p. 418.

5 The American art historian Meyer Schapiro had the inspired idea of pairing this painting with Piet Mondrian's abstract composition, *Broadway Boogie Woogie*. See: Meyer Schapiro, *Modern Art, 19th and 20th Centuries – Selected Papers*, George Braziller, New York, 1978, pp. 252–53.

82

Boulevard Montmartre, morning, cloudy weather
Boulevard Montmartre, matin, temps gris
1897

oil on canvas, 73 x 92
National Gallery of Victoria, Melbourne,
Felton Bequest, 1905 (204-2)

83

La Place du Théâtre Français
1898

oil on canvas, 72.4 x 92.7
Los Angeles County Museum of Art,
Mr and Mrs George Gard De Silva
Collection (M.46.3.2)

84

The Louvre, morning, rainy weather
Le Louvre, matin, temps de pluie
1900

oil on canvas, 66.7 x 81.6
Corcoran Gallery of Art, Washington, DC,
Edward C. and Mary Walker Collection (37.41)

85

Morning, winter sunshine, frost,
the Pont-Neuf, the Seine, the Louvre
*Matin, soleil d'hiver, gelée blanche,
le Pont-Neuf, la Seine, le Louvre*
1901

oil on canvas, 96.5 x 115.9
Honolulu Academy of Arts, gift of Arthur
and Kathryn Murray, 1996 (8439.1)

86

The Louvre under snow
Le Louvre sous la neige
1902

oil on canvas, 65.4 x 87.3
The National Gallery, London,
purchased from Lucien Pissarro, 1932
(NG4671)

87

The Carrousel, autumn morning
Le Carrousel, matin d'automne
1899

oil on canvas, 73.5 x 93
Private collection, Sydney

Sydney only

88

Afternoon sunshine, Pont-Neuf
L'Après-midi, soleil, le Pont-Neuf
1901

oil on canvas, 73 x 92.1
Philadelphia Museum of Art, bequest of
Charlotte Dorrance Wright, 1978 (1978-1-24)

89

Rue de l'Épicerie, Rouen
1898

oil on canvas, 81.3 x 65.1
The Metropolitan Museum of Art,
purchase, Mr and Mrs Richard
J. Bernhard Gift, 1960 (60.5)

90

Fair on a sunny afternoon, Dieppe
La foire à Dieppe, après-midi ensoleillé
1901

oil on canvas, 73.5 x 92.1
Philadelphia Museum of Art, bequest
of Lisa Morris Elkins, 1950 (1950-92-12)

91

Place de la République, Rouen
Place de la République, à Rouen
(avec tramway)
1883

etching, 14.2 x 16.5
D65, ii of 2 states
Bibliothèque Nationale de France

92

Place du Havre, Paris
Place du Havre, à Paris
1897

lithograph, 14.5 x 21.3
D185, ii of 2 states
The British Museum, London, bequeathed
by Campbell Dodgson (1949-4-11-3380)

93

The railway bridge at Pontoise
Le pont du chemin de fer à Pontoise
1882

etching, 13 x 24,8
D37, i of 2 states
Museum of Fine Arts Boston,
Horatio G. Curtis Fund (57.745)

94
The stone bridge and barges at Rouen
Le pont de pierre et les péniches à Rouen
1883

oil on canvas, 54.3 x 65.1
Columbus Museum of Art, Ohio, gift of Howard D.
and Babette L. Sirak, the donors to the campaign for
Enduring Excellence and the Derby Fund (1991.001.053)

95

The stone bridge in Rouen, dull weather
Le pont de pierre à Rouen, temps gris
1896

oil on canvas, 66.1 x 91.5
National Gallery of Canada, Ottawa,
purchased 1923 (2892)

96

Pont Boïeldieu in Rouen, damp weather
Le pont Boïeldieu à Rouen, temps mouillé
1896

oil on canvas, 73.6 x 91.4
Art Gallery of Ontario, Toronto,
gift of Reuben Wells Leonard Estate, 1937 (2415)

97
Sunset, the port of Rouen, steamboats
Coucher de soleil, port de Rouen,
bateaux à vapeur
1898

oil on canvas, 65 x 81.1
National Museums & Galleries of Wales, Cardiff
(NMW A 2492)

Portraits

pp. 204–05: cat. 99 *Minette*, 1872 (detail), Wadsworth Atheneum Museum of Art, Hartford, Connecticut

"The basis of our art is inarguably the French tradition," said Pissarro. "Our masters are Clouet, Nicolas Poussin, Claude Lorrain, the eighteenth century with Chardin and the group of 1830 with Corot."[1] Nowhere are the traditional underpinnings of Pissarro's painting more obvious than in his portraits. The syntax of Ingres' portrait drawings momentarily resurfaces in Pissarro's *Portrait of Paul Cézanne*, 1874 (cat. 100), and hints of the rich, glowing colour and compact arrangements of Chardin hover in *Minette*, 1872 (cat. 99). Pissarro honoured the French tradition in his own fashion: he did not hesitate to contradict and overturn its norms when he felt it was justified.

The nineteenth century critic Jules-Antoine Castagnary discussed portraiture in terms of "a sliding scale of degrees of individuality", John House has remarked.[2] Castagnary thought portraiture was inappropriate for peasants, small children and women – peasants because they belonged to the land where they lived and worked, children because their personalities were underdeveloped, and women because they invested too much vanity and artifice in changing their appearance. As House put it, Castagnary believed that "the only true individual is the successful bourgeois male".[3]

Pissarro's portraits inverted this hierarchy. He never painted a successful bourgeois male. Even though he showed a remarkable aptitude for portraiture, he never accepted a commission. He chose an assortment of women as his subjects, ranging from his grown-up daughter (*Woman sewing*, 1895, cat. 98) to peasant women and domestic servants (*Young farmgirl*, c.1882, cat. 102, and *The young maid*, 1896, cat. 103). He painted and drew his own children (*Minette*). He painted, drew and etched a highly individual but, at the time, conspicuously unsuccessful fellow artist (*Portrait of Paul Cézanne*). On five occasions, he painted himself.

In *Portrait of Eugène Murer*, 1878 (cat. 101), Pissarro painted a pastry cook and unsuccessful *littérateur* whose restaurant and patisserie on the boulevard Voltaire in Paris was lined with Impressionist paintings. Murer owned twenty-five Pissarros, and many works by Sisley, Monet, Renoir, Guillaumin and Cézanne.[4] Renoir had painted a portrait of Murer's sister on an oval format, and Pissarro's painting of his bohemian friend in the guise of a gypsy or bandit was intended to be its companion.[5]

Young farmgirl is an exquisite harmony of blue and pink that features the thread-like stroke characteristic of Pissarro's paintings from 1882 to 1883. Pissarro's decision not to "finish" the figure's hands is proof of his unerring instinct for pictorial wholeness and harmony. The balance of colour and the integration of the texture are perfect: any addition would spoil them.

Self-portrait (cat. 105) was painted in a rented apartment at 28, place Dauphine on the Île de la Cité in Paris (you can see the apartment in a postcard on page 223; cat. 85–89 are views from its windows). Profiled against the light, Pissarro has made his black hat and the line of his shoulder tremulous and crumbly along the edges. Inside and outside, the figure and ground are permeable to each other. The stippling of his white beard takes on an airy delicacy, like a gathering rain cloud. He was an Impressionist to the very end. This was Pissarro's last self-portrait, painted shortly before his death in November 1903.

1 Pissarro quoted by André Chastel, *Introduction à l'histoire de l'art français*, Flammarion, Paris, 1993, p. 55. Chastel wrote an interesting commentary on this statement.

2 John House, "Impressionism and the Modern Portrait", in *Faces of Impressionism* (exhibition catalogue), Rizzoli, New York, 2000, p. 15.

3 ibid.

4 See Ralph E. Shikes and Paula Harper, *Pissarro – His Life and Work*, Quartet Books, London, 1980, pp. 135–136.

5 A. Tabarant, *Pissarro*, trans. J. Lewis May, John Lane, The Bodley Head Limited, London, 1925, p. 43.

98

Woman sewing
Femme qui coude
1895

oil on canvas, 65.4 x 54.4
The Art Institute of Chicago,
gift of Mrs Leigh B. Block (1959.636)

99

Minette
1872

oil on canvas, 45.9 x 35.6
Wadsworth Atheneum Museum of Art,
Hartford, Connecticut, The Ella Gallup
Sumner and Mary Catlin Sumner
Collection Fund (1958.144)

100

Portrait of Paul Cézanne
Portrait de Paul Cézanne
1874

etching, 27 x 21.4
D13, only state
Private collection

101

Portrait of Eugène Murer
Portrait de Eugène Murer
1878

oil on canvas, 64.4 x 54.3 (oval)
Museum of Fine Arts, Springfield, Massachusetts,
The James Philip Gray Collection (52.01)

102

Young farmgirl
Jeune paysanne
c.1882

oil on canvas, 38 x 46
Kunstmuseum Bern,
Legat Eugen Loeb, Bern (G1872)

Sydney only

103

The young maid
La petite bonne
1896

oil on canvas, 61 x 50
The Whitworth Art Gallery,
The University of Manchester,
bequeathed by Dr David
Bensusan-Butt, 1998 (O.1998.2)

104

Self-portrait
Camille Pissarro par lui-même
1890–91

etching, 18.5 x 17.7
D90, i of 2 states
Bibliothèque Nationale de France

105

Self-portrait
*Portrait de Camille Pissarro
par lui-même*
1903

oil on canvas, 41 x 33.3
TATE, presented by Lucien Pissarro,
the artist's son, 1931 (NO.4592)

215

pp. 216–17:
cat. 4 *The Marne at La Varenne-St-Hilaire*,
c.1864 (detail), Kunstmuseum Bern

Pissarro country

This map indicates places where Pissarro painted, lived and visited. It is evident how they tend to cluster around Paris. Even though Paris was not his principal residence, he kept a rented apartment there whenever he could afford it. The railway system gave him access to his chosen painting sites where, over four decades, he witnessed the gradual encroachment of industrialisation and urbanisation.

On the following pages, you will find a selection of postcards which illustrate the principal places of Pissarro's artistic life.

1   Chennevières-sur-Marne
2   La Varenne-St-Hilaire
3   Bougival
4   Louveciennes
5   Marly
6   Maubuisson
7   L'Isle-Adam
8   Pontoise
9   Osny
10 Ennery
11 Auvers-sur-Oise
12 Gisors
13 Éragny-sur-Epte

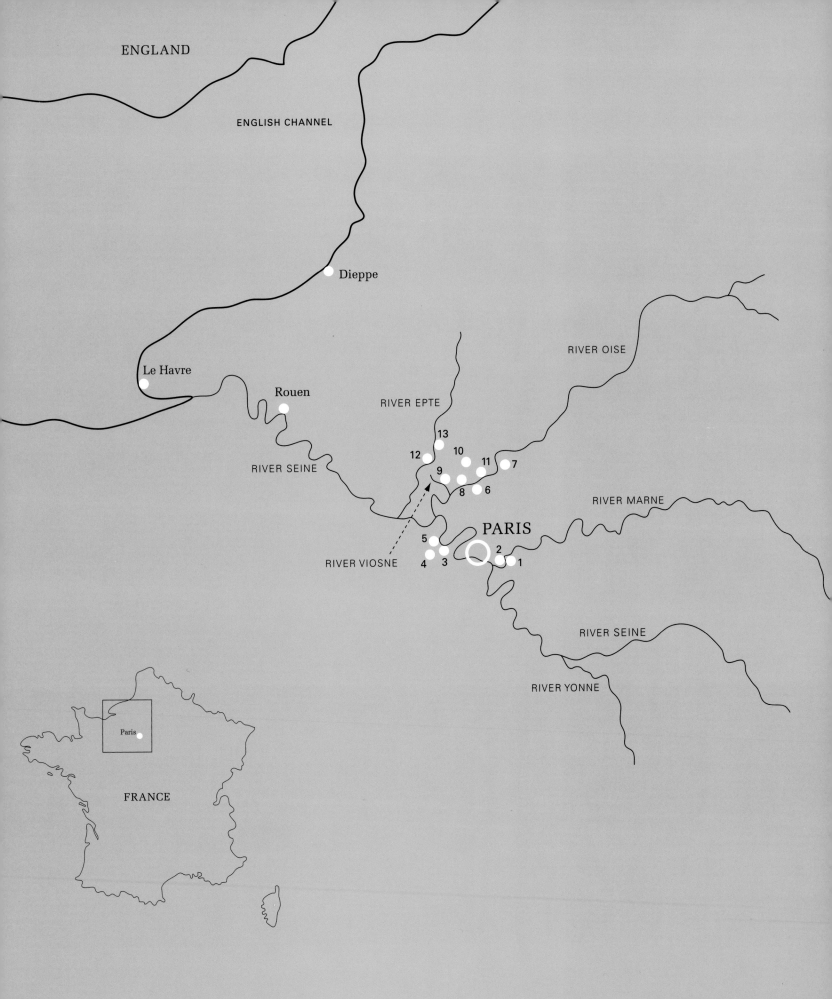

ENGLAND

ENGLISH CHANNEL

● Dieppe

RIVER OISE

Le Havre ●

Rouen ●

RIVER EPTE

13

12   10

RIVER SEINE   9   11   7

8   6

RIVER MARNE

PARIS

5

2

RIVER VIOSNE   4   3   1

RIVER SEINE

RIVER YONNE

Paris

FRANCE

34. Bords de la Seine
à BOUGIVAL

63. BOUGIVAL — La Seine aux Écluses

67. - LA VARENNE-SAINT-HILAIRE (Seine). — Bords de Marne.

Edit. Pilletex

8 - La Varenne-Saint-Hilaire — La Marne et le Coteau de Chennevières

Euc. Legrain, édit.

LOUVECIENNES — Avenue St-Martin et Rue de Versailles

LOUVECIENNES (s.-et-o.) - Rue des Creux. - Chemin de Versailles.

Cafés Mignot, Louveciennes.

L.H. Paris.

top
The Seine at Bougival.

centre
The Marne at La Varenne-St-Hilaire
and Chennevières.

bottom
Country roads at Louveciennes.

top
The countryside near Montfoucault (Mayenne).

centre
L'Hermitage at Pontoise (right), and its vicinity,
showing the distant smokestack of a distillery
on the river Oise.

bottom
The centre of Pontoise (right) and the river Viosne,
which links Pontoise and Osny.

La Moisson

191. — PONTOISE — La rue Victor-Hugo un jour de marché.   ND Ph

Imp. E. Rigault, Pontoise   OSNY. – Les Châtaigniers

OSNY (S.-et-O.) — Lotissement de la Gare

95

2 – Éragny (Oise) - La Ferme et l'Église

J. Bourgeix, lib. édit., Gisors (Eure)

ERAGNY (Oise) — Le lavoir

Phot.-édit., Lamaury, Gisors — Reproduction interdite

top
Staged tableau of a harvest, and market day
in Pontoise.

centre
Rural scenes at Osny, near Pontoise,
showing chestnut trees (left).

bottom
Views of the farm and the church, and the
river Epte at Éragny, very close to Pissarro's house.

212 - ROUEN - Pont Boïeldieu

LE HAVRE — Vue panoramique du Grand Quai
Bateau de Southampton

69   PARIS. — L'Avenue de l'Opéra. — LL.

PARIS. - Boulevard Montmartre.

92. — PARIS. - Le Pont-Neuf.

44. PARIS — Panorama du Louvre   C. M.

top
Industrial landscapes: the Pont Boïeldieu at Rouen,
and dockside at Le Havre.

centre
The boulevards of Paris: the avenue de l'Opéra
photographed from the Hôtel du Louvre, and the boulevard
Montmartre seen from the Grand Hôtel de la Russie.

bottom
The Pont-Neuf looking north (Pissarro's apartment can
be seen to the extreme right), and panorama of the Louvre.

Pissarro and his critics

Terence Maloon and
Claire Durand-Ruel Snollaerts

Nothing is more difficult, when forms of art or modes of thought have achieved success, than to represent the repulsion they first caused.

Théodore Duret[1]

There is no spectacle more edifying than that of a painter, accepted by the critics and the collectors, who makes an extra effort, and who in good faith attempts to re-school himself in his art […] He [Pissarro] never ceased to study and gain in experience. While he thought he was re-educating himself, he was fulfilling a whole lifetime of dedicated study, of knowledge amassed day by day.

Gustave Geffroy[2]

Camille Pissarro (standing, right) and his son, Lucien (centre), with Paul Cézanne (seated), L'Hermitage at Pontoise, c.1884.

There is something utterly touching and beautiful about this artist
[Pissarro]. It is the example of his perpetual renewal.

Gustave Geffroy[3]

How were these paintings, which seem so reasonable, so patient,
so scrupulously studied, and which shine with a tender light,
able to stir up so many critics, so much animadversion, so much
misplaced raillery, when one only had to understand [Pissarro]
and to rank him on the level of the uncontested masters of his art?

Gustave Geffroy[4]

Camille Pissarro's career coincided with a period of extraordinary vitality in the Parisian press. With few exceptions, exhibitions featuring his works were widely noted by journalists, almost always with a couple of sympathetic advocates, even when the critical consensus was most adamantly against Pissarro's work. A great diversity of publications carried art criticism during the latter decades of the nineteenth century: in 1870, for example, there were an estimated 36 daily newspapers in Paris alone; by 1880, this number had grown to about 60.[5] The ubiquity of commentary prompts us to wonder: why was there such widespread interest in contemporary art at the time?

One reason, surely, was that painting was such an unstable and volatile phenomenon in French nineteenth century culture. The fine arts had a peculiar institutional history in France, a history that in many ways was still bound up with the hierarchical values of the ancien régime and the methods of artistic training that had been established in the seventeenth century. However, at the time Pissarro began exhibiting in Paris, contemporary art had become embroiled in a highly publicised furore that raised some disturbing questions about the legitimacy of artistic expression, and about how, and by whom, excellence in art was to be judged.

Given these circumstances, it was obviously in the best interests of Academic artists and conservative critics to minimise the threat to their authority and to emphasise "business as usual". Conversely, it was in the best interests of journalists to fan the atmosphere of controversy and to emphasise the provocation and shock of contemporary painting to attract a large reading public.

In the 1860s, there was still only one way for a young painter to make his mark: to show his works in the Salon. Pissarro had exhibited a painting in the Salon of 1859, and another in the Salon of 1861, but his work was not noted in the press before 1863. From 1863, the Salon became an annual event that took place in the Palais d'Industrie on the Champs-Élysées. The number of works submitted each year was awe-inspiring. In 1861, out of more than 7000 works entered, 4097 were accepted; in 1863, from more than 5000 works submitted, 2217 were retained; in 1874, 2457 were accepted out of 5727 considered. The Salons were extensively reviewed. To be noticed at all in such a crowd was an achievement; to be discussed at length or to feature in the opening remarks of a review was seen as a stroke of great good fortune.

Critics agreed that the Salon was not an ideal place for encountering works of art. "At the Salon it is difficult to distinguish individual sensibility in the mass of disparate canvases hung next to each other. A picture needs isolation or harmonious neighbouring works," observed one critic.[6] The "official" character of the Salons meant that they always

seemed redolent of the dead hand of authority, expressing the conservative bias of the jury. Many writers protested: "For several years, the more or less representative regimes who direct politics extend their deadly influence all the way into our exhibitions, and paralyse the efforts of the jury and administration. In art, as in politics, parliamentarianism kills all spirit of initiative."[7] "It [the Salon] has a makeshift, sad look about it, and blaring cacophonies. If it is visited too many times, it imparts a kind of pervasive sadness to one's thoughts; one has the sense of death."[8]

Could the insensitive hanging of a work conceal an insult to its artist? The critic Louis Leroy thought so when he visited the 1863 Salon des Refusés: "With malicious intent, the administration has, here and there, put some grotesque canvases into prominent positions, towards which the majority of visitors eagerly press."[9] In the Salons, bad paintings could be put "on the line" or good ones could be "skyed" (hung too high to be properly seen), as sometimes happened to Pissarro's works.[10]

Responding to a flood of complaints from rejected artists, in 1863 Emperor Napoléon III established the Salon des Refusés to allow the public to judge the merits of these rejected works themselves. The intensity of response to the Salon des Refusés made it a flashpoint in the history of modern art.

1863–70

In 1863, Pissarro exhibited three works in the Salon des Refusés. Jules-Antoine Castagnary (1830–88), a prominent critic well known in artistic circles as the friend and defender of Gustave Courbet, was not especially impressed by the two paintings singled out as the focus of controversy at the Salon des Refusés – Whistler's *The white girl* and Manet's *Le déjeuner sur l'herbe*[11] – yet he was interested in Pissarro's landscapes. Castagnary began his review with the following dialogue:

"So, there are works of merit [in the Salon des Refusés]?"
"Are you in any doubt of it?"
"Worthy of featuring in the official Salon?"
"With honour."
"In very small quantity, then?"
"What would you say if I told you they were in the majority?"

In the list of twenty-one artists singled out for commendation, Castagnary included Camille Pissarro – the first occasion Pissarro had been noticed by an important Parisian critic.[12] Even more remarkable, in another review of the Salon des Refusés written by Castagnary – this one in the periodical, *L'Artiste* – Pissarro's was the first name cited:

Mr. Pissaro [sic]. – Not finding his name in the previous handbook, I suppose he is a young man. Corot's manner seems to please him. He's a good master, sir, but above all one must be careful not to imitate.[13]

Pissarro was favourably mentioned in two other reviews – by the standards of the day, a very successful debut. These reviews – by the critic and satirist Louis Leroy, and by the columnist Carle Desnoyers, who had adopted the pseudonym of Le Capitaine Pompilius[14] – explained the gulf that had opened up between the official Salon and the Salon des Refusés.

It was caused, they said, by the ideas of elevated style and elevated subject matter held by the Academicians of the École des Beaux-Arts. The problem was that the majority of contemporary painters rejected or were indifferent to those ideas. In fact, Castagnary had proclaimed that a "new school" was revealed by the Salon des Refusés, a school he called "naturalist":

Look at these landscapes, these interiors, these still-lifes, spread out in such number across the gallery walls. It is those diverse genres which have been especially emphasised. Why?

Still-life is an insignificant genre, the members of the Institute, custodians of "great art", reply. Maybe so, but doesn't painting incorporate all the objects of the visible world? Can you prevent a painter from seeking all the possible applications of his craft, treating sometimes a figure, sometimes a landscape, sometimes a still-life according to his whim or mood of the day?[15]

The Academicians were offended by the lack of idealisation and what they saw as a lack of style in the type of painting that Castagnary called naturalist. Le Capitaine Pompilius explained to his readers that, as far as landscape paintings were concerned, "… the Academy only admits classical landscape [*le paysage composé*] into the domain of art – that is to say: nature rectified, according to the principles of a school. A landscape imitated from nature is only a student exercise, according to [the Academy's] views."[16]

Pissarro submitted several paintings of distinction to the Salons of 1864 and 1865, but nothing could prepare us for the critical reception of his painting, *Banks of the Marne in winter*, 1866 (cat. 7), in the Salon of 1866.

Pissarro had no doubt already met the young *littérateur* Émile Zola (1840–1902) through their mutual friend, Paul Cézanne. Pissarro must have felt a good deal of admiration for Zola's outspokenness as the art critic for the newspaper, *L'Événement*. Yet, as it turned out, Zola's preoccupation with Courbet, Manet and the "new school" of naturalism proved too much for the newspaper's readers, and his editors saw fit to relieve him of his post. In his valedictory review for *L'Événement*, Zola praised Pissarro's *Banks of the Marne in winter* in the 1866 Salon:

Mr Pissarro is an unknown of whom, no doubt, nobody will speak. It is my duty to shake his hand heartily before I leave. Thank you, sir, your landscape refreshed me for a good half-hour in the great desert of the Salon. I know that you were admitted into it very grudgingly, and I sincerely congratulate you for that. However, you ought to know that you will please no-one, that they will find your picture too bare, too black. Also, why the devil did you make the signal blunder of painting solidly and studying nature honestly!

Look, you choose wintertime; you have a simple stretch of avenue, then a hillside behind, empty fields all the way to the horizon – nothing at all to feast the eyes upon; an austere and grave painting; an extreme care for truth and accuracy; a harsh, strong wilfulness. You are a clumsy fellow, sir – you are an artist that I like.[17]

Pissarro's *River landscape with boat*, c.1866 (cat. 6) – now in the Kunstmuseum Sankt Gallen in Switzerland – once belonged in Émile Zola's collection. Its bareness, moody tonality and rugged, palette-knifed impasto are similar to the qualities in *Banks of the Marne in winter* that Zola praised in his review. In fact, the two works were probably painted at about the same time.[18] Given Zola's uncertain future and shaky finances in 1866 and the years that immediately followed, it is unlikely he had the means to buy *River landscape with boat*; possibly, it came to him as a gift from Pissarro, grateful for his spirited defence.

Jean Rousseau (1829–91), secure in the employ of a middle-of-the-road, bourgeois publication, was a critic of more conservative inclination than Zola, yet he too found *Banks of the Marne in winter* arresting, and alerted his readers to its originality:

I would like to mention a canvas signed by a still unknown name, Mr Pissaro [sic]. This canvas is obliquely cut by a long road planted with trees, all upright and completely naïve in character. At the end of the road is a square white house, most honest and banal in aspect. To the right stretches a plain which is as bare and flat as a hand, and which exerts no more claim to character. Above all this is a heavy grey sky. Nothing, assuredly, could be more vulgar than this site, yet I defy you to pass by it without noticing. It becomes original because of the abrupt energy of the execution, which tends to underline the ugliness instead of trying to disguise it. One sees that Mr Pissaro [sic] is not banal through an inability to be picturesque. On the contrary, he exercises a robust, exuberant talent to bring out the vulgarities of the contemporary world, like a satirical poet – more eloquent when the truth is more brutal and less arranged.[19]

The relatively large canvases Pissarro sent to the Salons would have been unwieldy objects to move from place to place, and difficult to shield from gusts of wind, eddies of dust or sudden showers when the paintings were wet. It is unlikely that

fig. 30
Honoré Daumier
*Amélie, it is not seemly for you
to pause so long*
1855
from *L'Exposition Universelle*
lithograph, 18.3 x 23.5
Art Gallery of New South Wales,
gift of Mrs Hanne Fairfax, 1967

these canvases were painted entirely outdoors; some may never have been painted outdoors at all. Throughout his working life, Pissarro retained a pragmatic attitude to the advantages of reworking and transposing in the studio the studies he had begun *sur le motif*. He had a talent for making things look candid and fresh, and for keeping paintings looking that way.

In fact, the sense of truth in Pissarro's pictures was so novel and compelling for his contemporaries that virtually all the critics of the period assumed he was a thorough-going *pleinairiste* whose works were painted outdoors from start to finish. His paintings in the Salons were always said to be faithful copies of nature rather than creative transpositions or formal inventions. A cliché came into circulation around this time: that Pissarro didn't compose his pictures at all; they were arbitrarily chosen "slices of life". The critic Léon Billot was probably the earliest to make this assertion in print:

Mr. Pissarro takes his palette and departs for the fields. He stops anywhere, sets up his easel and does the first thing that comes to him: *La Côte du Jallais, near Pontoise* [PV55], for example … It is very simply painted; the details executed in groups have the effect of having been studied one at a time. There is obviously a great talent for painting here. Unfortunately the subject is missing.[20]

On the occasion of the Salon of 1868, Zola wrote an entire article about Pissarro, whom he presented to his readers as "maybe the least known" representative of naturalism. If contemporary painters had really learnt how to paint, Zola argued, "they would have the lush, solid craft [*le métier gras et solide*] of the masters; the subject [of their works] would matter little."[21] According to Zola, Pissarro followed the genuine tradition of the old masters, and his originality was "profoundly human":

It resides in the very temperament of the painter, consisting of exactitude and gravity. Never have paintings seemed to me to have such magisterial breadth. You hear in them the deep voices of the earth; you intimate in them the powerful life of trees. The austerity of the horizons, the disdain for commotion and the complete absence of any irrelevant flavour give the ensemble a kind of epic grandeur.[22]

Pissarro submitted *Landscape in the vicinity of Louveciennes* (*Autumn*), 1870 (cat. 11) to the Salon two years later. Nobody could plausibly claim (as Léon Billot had done in 1868) that any details in the painting had been studied independently, "one at a time". In fact, everything merges and melds with seeming effortlessness into an evocation of broad, silvery light. "Pizzarro [sic] … sees the totality of a place, its tender harmony, and […] seizes its atmospheric aspect," the art critic Jean Ravenel reported.[23]

Landscape in the vicinity of Louveciennes (*Autumn*) was one of Pissarro's earliest Salon paintings to feature the reformed palette we have come to associate with Impressionism. He had all but eliminated black and earth colours (umbers, ochres, siennas); they occurred only in contrasting accents, in the face and cap of the boy, in the peasant woman's shoes, in details of their clothing. Instead, Pissarro relied on blends of prismatic colour (blues, reds, yellows, greens) to create a range of subtly differentiated greys, fawns, greens and off-whites. As well as obtaining an enveloping effect of pearly luminosity and transparency, he reduced the contrasts of light and dark to a minimum – the conventional role played by tonal contrasts in establishing the "design" of a landscape was deftly sidestepped and underplayed.

Did Pissarro's rejection or neglect of chiaroscuro mean that he didn't design, didn't compose his painting at all? Was *Landscape in the vicinity of Louveciennes* (*Autumn*) just a formless sprawl, an unreconstructed slice of life? On the contrary: not a speck in it is out of place. There is not a superfluous jot in a whole vast explosion of tiny flecks and dapples interspersed with a crush of geometrical fragments. The painting represents the most complicated clash of nature and the man-made environment, yet Pissarro has integrated and synthesised it all into a composition of faultless lucidity. Nonetheless,

Vue prise dans un atelier, quelques jours avant l'ouverture de l'exposition.

fig. 31
Honoré Daumier
*Scene in a studio, a few days before
the opening of an exhibition*
1855
from *L'Exposition Universelle*
lithograph, 19.5 x 25.1
Art Gallery of New South Wales,
gift of Mrs Hanne Fairfax, 1967

it is important to recognise that, in 1870, even the critics most favourably disposed towards Pissarro's paintings were oblivious to his subtlety and resourcefulness as a composer.

Consider the curious case of Théodore Duret. In a landmark essay, "Les Peintres français en 1867", Duret declared his partisanship of naturalism, claiming that "the masters who constitute the group of naturalist painters have known how to satisfy entirely all [the] conditions of originality".[24] As might be expected, Duret took up Pissarro's cause by incorporating him into the camp of realists and naturalists, making him a pretext for further reflections on the phenomenon of originality in art. However, when reviewing the Salon of 1870, Duret gave rise to some rather strange, confusing assertions about Pissarro which had a tenacious afterlife, fuelling a common misunderstanding and prejudice that persisted right up to the time of Pissarro's death in 1903. Duret wrote:

Pissarro is, in certain respects, a realist. Never does he compose a picture – and never does he arrange nature in a landscape […] He imprints the least of his canvases with a feeling of life.[25]

An unsigned obituary notice that appeared in London's *Daily Chronicle* in November 1903, evidently written by Elizabeth Pennell, one of Whistler's biographers, put a damning construction on the remarks Duret had made more than thirty years before:

As far back as 1870 M. Duret, in writing of him among the other men of the 'Avant-Garde', pointed out the weakness of all when he declared that Pissarro would never 'compose' a picture nor, in a landscape, arrange what nature had not troubled to arrange for him.[26]

Duret sometimes played up to the philistine scoffers and the enemies of the naturalists and Impressionists, indulging in witticisms such as this: "He [Pissarro] often comes to paint insignificant sites, where nature herself makes so little of a picture, that he paints a landscape without making a picture."[27] Usually, however, he would jest with the philistines as a prelude to turning the tables on them:

Imagine that he [Pissarro] demeaned himself by painting cabbages and lettuces, and even, I believe, artichokes. Yes, when painting the houses of certain villages, he [also] painted the cottage gardens attached to them, and in the gardens were cabbages and lettuces which, along with all the rest, he reproduced on the canvas. Yet, for the partisans of "great art" there is something degrading in this fact, prejudicial to the dignity of painting, something that reveals the vulgar tastes of the artist, a complete forgetfulness of the ideal, a complete lack of higher aspirations, and so on and so forth.[28]

Art should not isolate itself from life, and it cannot be understood separately from a fresh, personal feeling. If art is thus acknowledged, it embraces all the manifestations of life, all that nature contains. Nothing is noble or low in itself. The artist, according to his aspirations and fancy, has the right to let his eyes wander over all parts of the visible world in order to reproduce them on canvas.[29]

Théodore Duret (1838–1927) was a man of independent means, heir to a family business manufacturing cognac and other spirits. He was a regular presence at the artists' and writers' gatherings at the Café Guerbois in the Batignolles district of Paris, where he first met Pissarro in 1866.[30] A journalist, art collector and critic, Duret was the first to publish a substantial essay on Impressionism (1878). He had the unique distinction of having his portrait painted by both Manet (1868) and Whistler (1883). Over time, he acquired four, possibly more, paintings by Pissarro[31], and for many years he was the stoical recipient of a great many anguished, importuning letters from Pissarro, who had by then become caught up in a chronic struggle to provide for his family and keep his stricken career afloat.[32] But, in 1873, Pissarro's worst years lay far ahead. For the moment, his prospects seemed full of hope, and Duret, as his friend and public defender, was keeping a loyal, protective eye on his interests.

1870–74

The imposing size and the relative formality of the composition Pissarro submitted to the Salon of 1870 – *Landscape in the vicinity of Louveciennes (Autumn)* – disguised how much and how swiftly his painting was changing at the time. Claude Monet and Auguste Renoir, whom Pissarro had known for many years, moved into the area of Louveciennes, where Pissarro had rented a house in 1869.[33] The three painters had an important interest in common: they were all fascinated by the artistic possibilities of plein-air painting and by the challenge of boosting the fresh, novel qualities obtained in their outdoor landscape studies into more ambitious, intelligently considered and precisely wrought *tableaux*. Many of their technical experiments were conducted during painting excursions together, and the results were made available to all. Given the stimulation and moral support offered by this brilliant peer group, Pissarro, Monet and Renoir were probably able to alter their painterly language, fine-tune their means of representation and extend their formal speculation in much more pointed and daring ways than they might have done individually. Their circle widened to include Alfred Sisley, Paul Cézanne, Armand Guillaumin and others, preparing the way for the emergence of the Impressionist group.

Their paintings grew to look increasingly different from and oppositional to the typical fare of the Salons. Their paintings also grew to look very much alike. Today, one of the revelatory exhibits in the Musée d'Orsay in Paris consists of a painting by Pissarro, a painting by Monet and a painting by Sisley, all dated 1872, treated like a triptych and braced together in a single frame. To all intents and purposes, the constitution of the palette, the tonality, the brushwork, the type of composition and the overall mood of the three paintings are identical. During the early 1870s, Pissarro, Monet and Sisley developed a communal style and a collective artistic identity – much as the Neo-Impressionists did around 1886–90, or as Picasso and Braque did during their phase of Analytic Cubism in 1910–11. The art historian John Rewald tells the amusing story of Monet and Renoir painting a duck pond side by side in 1873. Forty years later, when one of the canvases (unsigned) came onto the market, initially neither artist was able to say who had painted it.[34]

The period of the gestation of Impressionism was rudely interrupted by the outbreak of the Franco-Prussian war in July 1870. Pissarro and his family fled from their house in Louveciennes, eventually ending up in London. Pissarro recounted:

In 1870 I found myself in London with Monet, and we met Daubigny and Bonvin. Monet and I were very enthusiastic over the London landscapes. Monet worked in the parks, whilst I, living in Lower Norwood, at that time a charming suburb, studied the effects of fog, snow and springtime [see *Near Sydenham Hill*, 1871, cat. 9]. We worked from Nature, and later on Monet painted in London some superb studies of mist. We also visited the museums. The watercolours and paintings of Turner and Constable, the canvases of Old Crome, have certainly had influence upon us. We admired Gainsborough, Lawrence, Reynolds, &c., but we were struck chiefly by the landscape-painters, who shared more in our aim with regard to "plein air," light, and fugitive effects [...] We had the idea of sending our studies to the exhibition of the Royal Academy. Naturally we were rejected.[35]

In London, Daubigny introduced Pissarro to a young French art dealer, Paul Durand-Ruel (1831–1922), who was setting up a gallery at 168, New Bond Street. From 1870 to 1875, Durand-Ruel regularly included two or three of Pissarro's paintings in mixed exhibitions of French landscapes or in surveys of nineteenth century French paintings. Some of Durand-Ruel's exhibitions were politely, even favourably reviewed by the British press, but before 1874 there was no mention of Pissarro's contributions to any of them, nor was there any mention of the paintings by Manet, Monet, Degas and Renoir which Durand-Ruel also exhibited.[36]

Pissarro's fraternising with Monet in London seems to have been limited to visits to museums and exhibitions, and possibly a few social calls.[37] On his return to France in 1871, Monet moved up-river from Bougival to Argenteuil, and Pissarro decamped from Louveciennes to Pontoise in 1872. Although they now lived further apart, they kept in touch through frequenting the same cafes and artists' dinners in Paris and through their links to the Durand-Ruel gallery. Once he had set up house in Pontoise, Pissarro was able to satisfy his need for the stimulating companionship of other artists and remained involved in what Richard Brettell has called "interactive and non-hierarchical education" through his contact with Paul Cézanne, a younger, less experienced painter who had moved to Auvers-sur-Oise in order to be close to Pissarro.[38]

As well as taking works on consignment, Durand-Ruel bought paintings directly from his gallery artists, thus ensuring Pissarro a steady income.[39] Collectors also bought works directly from Pissarro's studio, with the result that he was now earning rather well. When one of his principal collectors, Ernest Hoschedé, was forced to liquidate some assets in 1874, Pissarro's paintings obtained surprisingly high prices at auction. But neither Pissarro nor Monet exhibited in the Salons of 1872 and 1873. Were they too angry and disillusioned with the jury to enter their works? That was probably true in Monet's case, although there is good reason to suppose that one of Pissarro's largest paintings, *Louveciennes*, 1871 (PV123), was intended for the Salon but was rejected.[40]

Pissarro's and Monet's feelings of dissatisfaction and defiance in regard to the Salon were shared by many of their peers. On 27 December 1873, a charter written by Pissarro was accepted by an assembly of artists, and a new artists' cooperative group, the Société Anonyme des Peintres, Sculpteurs, Graveurs, etc[41], was constituted. Its implicit aim was to break the monopoly of the Salon and to provide alternative opportunities for the member artists to exhibit their work.

The Society's first exhibition opened on 15 April 1874 in the photographer Nadar's former studio at 35, boulevard des Capucines. There were 165 works by thirty artists, most of whose names are remembered mainly because of the accident of their involvement in this historic event. It was first exhibition in Paris to feature gaslight, and the first whose opening times favoured the working people of Paris: there were two sessions – 10am to 6pm, and 8pm to 10pm.

The critic Philippe Burty (1830–90) was tangentially involved with the Société Anonyme through his recruitment of one of the artists – at Degas' behest, he had persuaded the engraver Félix Bracquemond to join. Burty published the fullest account of the raison d'être of the exhibition, probably drafted in consultation with one or more of the organisers:

The chief object of these gentlemen [...] was to present their paintings almost under the same conditions as in a studio, that is, in a good light, isolated from one another, in smaller numbers than in official exhibitions […] without the neighbourhood of other works either too bright or too dull. They feel that their style of painting, whether by simplicity of design, or by simplicity of tone, or by simplicity of composition, looks like a challenge or a caricature when placed side by side with works conceived under the pre-occupation of mannered design, artificial tones, or subjects intended to provoke laughter or emotion by the most vulgar artifices […] They renounce success, medals, decorations, and even the esteem of their fellows, to pursue a purely artistic end. They depend on elements of interest strictly aesthetic, and not social or human – lightness of colouring, boldness of masses, blunt naturalness of impression.[42]

Burty was no mere publicist or proselytiser for the Impressionists, demonstrating his independence by making his reservations and criticisms of their works explicit. However, his allusions to "bright painting" (la peinture claire) and his remarks about unconventional techniques, compositions and subjects were relevant only to a minority of the exhibitors. They definitely did pertain to Renoir, Monet, Sisley, Pissarro and Cézanne, whom Burty called "the extreme realists"[43] of the exhibition.

Another writer, Marc de Montifaud (in fact, Marie-Émilie Chartroule, that rarest of

creatures in the nineteenth century, a female art critic) noted that at the centre of the exhibition were "six or seven daredevils, recognisable by their works, all incomplete, which give off an imperious sense of truth".[44] Her review went on to identify Monet, Degas, Renoir, Cézanne, Sisley, Rouart, Pissarro and Guillaumin as those daredevils. Étienne Carjat also discerned a core group:

Monet, Pissaro [sic], Cézanne, Sisley and Guillaumin form a group apart in this exhibition, of which they are the principal initiators.

What they seem to be looking for, above all, is THE IMPRESSION, a word expressly invented for the needs of their cause.

Impression: maybe. But is it the impression of form, effect or colour? Form! All of them disdain it, since it requires a laborious and patient study that doesn't suit their temperament. Effect! Very little of it, all things considered. Which leaves colour – that is where we have expectations of these new iconoclasts.

Well, even their colour fails them. It is false, heavy and common. What the masters call a tone, they call a blotch [une tache]. For them it is no longer a matter of broken colours merging into a desired harmony, but of flat, multicoloured strokes juxtaposed at random, in all the crudeness of the palette.[45]

A more even-tempered and respectful review by Armand Silvestre emphasised the artists'

merits, but it made some qualifications that the readers of the time would have regarded as no less damning:

At their head are three artists […] who have the unarguable merit of being single-minded in their goals. This doggedness results in a feature all three have in common, which is to make the technical side of their painting conspicuous from the start. At first we can barely discern the difference between Mr Monet and Mr Sisley's pictures, and the latter from Mr Pissaro's [sic]. A little study soon teaches you that Mr Monet is the most skilful and daring; Mr Sisley is the most harmonious and timid; Mr Pissaro, who is essentially the inventor of this painting, is the most real and the most naïve […] One thing is certain, and that is that in no respect do these three landscapists resemble any previous master …

Their paintings have a singularly pleasant tonality. A pale light inundates them, and all is gaiety, clarity, holidays in spring, golden evenings, or apple trees in bloom. Their lightly painted canvases of modest dimensions open windows […] They [the artists] do not, in general, choose their sites with the preoccupation of the old landscapists – on the contrary – and that is where I find that they have an utterly graceless affectation. If [this affectation] is motivated by the philosophical idea that all things in nature are equally beautiful, it is artistically untrue. It may tend to reveal the extent of their interpretative means, but it only emphasises the workman-like aspect of their manner.[46]

One of the most famous and, in many respects, one of the most revealing reviews of the first Impressionist exhibition was written for laughs in the satirical magazine, *Le Charivari*. The writer purported to have seen the exhibition in the company of a doddering, splenetic old painter, "Mr Joseph Vincent, landscapist, pupil of Bertin, medal-winner, honoured by several governments". Louis Leroy described how he had led Mr Vincent to Pissarro's painting, *Ploughed field* (PV203, *Hoarfrost* in the catalogue):

Seeing this remarkable landscape, the gentleman thought that there was something wrong with his glasses. He cleaned them carefully, then put them back on his nose.

"By Michalon!" he exclaimed, "what's that?"
"You see … a white frost over deeply ploughed furrows."
"Those furrows? That frost? They're palette-scrapings uniformly applied to a dirty canvas. It has neither head nor tail, top nor bottom, foreground nor background."
"Perhaps, but the impression is there."
"Well, odd sort of impression."[47]

Théodore Duret had already attempted to dissuade Pissarro from his involvement with the Société Anonyme, to prevent him from taking a step he was sure would damage his career. He wrote to Pissarro in December 1873, urging him to persist in sending paintings to the Salon and warning him of the danger of throwing in his lot with a group whom, he could predict, would surely be called rabblerousers and no-hopers. Duret's way of urging Pissarro to conform to the Salon's conventions and criteria could only have served to offend him and to exacerbate his stubbornness: "I urge you to select pictures that have a subject, something resembling a composition, pictures that are not too freshly painted, that are already a bit finished."[48]

That Duret's advice was sensible and well intended could not be doubted. However:

I maintain my opinion that rustic nature, with animals, is what suits your talent best. You haven't Sisley's decorative feeling nor Monet's fantastic eye, but you have what they have not, an intuitive and profound feeling for nature and a powerful brush, with the result that a beautiful picture by you is something absolutely definitive. If I had a piece of advice to give you, I would say: don't think of Monet or Sisley, don't think of what they are doing, go your own way, your path of rustic nature.[49]

Yet the injunction "not to think of Monet and Sisley, not to think of what they are doing" would have struck a discordant note with Pissarro, who is on record as refuting those who "believed that one invented painting in a fell swoop and that one was original when one resembled nobody".[50] He recognised that art was a social language, not the solipsistic fancy of an isolated individual. His political leanings towards anarchism and socialism did not inhibit him from being a "joiner". In addition, he might have connected Duret's recommendations with *possessive individualism*, an ideology he could have equated with the bourgeois class, with the bourgeois "enemy".[51]

Pissarro's conviction that he and the Impressionists were ultimately on the right path, that they would eventually be vindicated, offered little consolation in the circumstances. He and his fellow artists had been ferociously trounced. The scandal of their exhibition proved a catastrophic setback for sales. Formerly considered a rising painter of exceptional promise, Pissarro was now seen as part of a nihilist, hooligan fringe in contemporary art. To make matters worse, his most marketable quality – his individuality – was now thrown into question.

1876–83

It would seem that Pissarro dutifully read, pondered and responded to criticisms of his work. His works painted in Ennery and Montfoucault during the late autumn and winter of 1874 appear to have taken to heart some of the pejorative remarks made about the first Impressionist exhibition. Paintings of this time (for example, *The pond at Ennery*, 1874, cat. 24) drew away from Impressionism's chromatic flamboyance and established a new kind of muted, close-valued tonalism – a tonalism that is elusive and unpredictable because it is so precisely and exquisitely coloured. The slapdash brushstroke, the delinquent *tache*, the hyperkinetic flurries that were universally reprimanded were held in abeyance. Well-chosen picturesque motifs and well-constructed compositions refuted the accusations that the Impressionists didn't care about such things. Far from pursuing ephemeral effects (as he had done in *Effect of fog at Creil*, 1873, cat. 13), Pissarro emphasised the solidity and permanency of these motifs.[52] In October 1873, he even indicated to Duret that he had taken his advice to heart and was beginning to act on it, announcing an imminent departure for Montfoucault, Brittany, "to study the figures and animals of the real countryside".[53]

The second Impressionist exhibition was held two years later. Several of Pissarro's Montfoucault paintings were included in a selection of works portraying a variety of seasonal effects – fog, wintry bleakness, a heavy snowfall at L'Hermitage, a glimmering pond overhung by trees and backed by yellow wheatfields at harvest time, a profusion of midsummer verdure along the route du Chou (see *A cowherd on the route du Chou, Pontoise*, 1874, cat. 22). The exhibition was held in the Durand-Ruel gallery, at 11, rue Le Peletier, Paris. The number of participating artists was reduced to a more coherent group of nineteen, with each shown in greater depth – Pissarro exhibited twelve works.

Although Philippe Burty thought this exhibition was better received by the public than its predecessor, the press notices were more plentiful – and, in general, more savage.[54] There was an article signed *Gène-Mur* (Eugène Murer) which was overtly friendly towards Pissarro, spelling his name correctly and calling him "*le chef de la nouvelle école*".[55] Another review, signed by Alexandre Pothey, called attention to his landscapes "whose uneven terrains are so well constructed"[56], but finesse of construction was what most reviewers did not see. The Impressionists were said to be:

… a mob of hoaxers, pretentious and impotent simpletons ["*épeleurs*"] who treat composition with scorn because they don't know how to compose, affecting to despise study because they don't like to work […] and advocating a system destructive of art because they will never be true artists.[57]

It is the school of impaired sight applied to painting. Objects are seen as if through a prism which decomposes light into its primitive colours. Red, yellow and blue spread everywhere, running riot in their native crudity, devouring with relentless ferocity skies, plots of land, trees, flowers, beaches, paths, fields, white frosts, shepherds, peasants, railway bridges and Japanese women. They make an insane wreckage of all these elements.[58]

According to Charles Bigot, the exhibition was a "collection of flashy sketches"[59], with only the works by Marcellin Desboutin spared from his censure:

For an artist a sketch is not a picture [*un tableau*]; it is the first impression of things, where the principal planes are indicated by lively, jerky, often brutal touches. For the artist, that is where the work of the hand and eye really begins. It involves rendering each detail faithfully and in modelled relief, without the general tone weakening, without the unity of effect disappearing. The new school suppresses the tableau, dispenses with work, and offers the sketch for public admiration.[60]

There were another six Impressionist exhibitions held between 1876 and 1883. The third, in 1877, had special significance because it was the principal occasion when Cézanne exhibited with the group. Even discounting Cézanne's presence, it must have been the most dazzling of all the Impressionist exhibitions. Monet showed a superb selection of landscapes and urban scenes; Renoir demonstrated his prowess as a portraitist and showed two of his best loved figure compositions; Degas also showed some of his most celebrated figure compositions; and Caillebotte showed large realist paintings of modern Paris. Was Pissarro's contribution overshadowed by the impact of his fellow exhibitors? Interestingly enough, he chose this occasion to display the range of results of his technical experiments over the previous several years. No doubt as a gesture of solidarity towards Cézanne, his paintings emphasised the cultivation of texture, the exploration of material density and the unity of image, surface and substance. Needless to say, the few critics who noticed this were either nonplussed or disapproving.

The 1876 exhibition had drawn a long tirade in *Le Figaro* from Albert Wolff, the most widely read and most influential art critic in Paris at the time. Wolff described the Impressionists as "five or six lunatics with a woman amongst them, a group of unfortunate creatures stricken with the madness of ambition".[61] Following this sensational outburst, they were now widely regarded as marginal, disreputable and defenceless publicity seekers. They became easy targets for tyros of the press exercising their "Parisian wit":

Messrs. Pissaro [sic] and Cézanne, who have their partisans, form a school apart, even two schools, within the school. I acknowledge their qualities of drawing and even of arrangement – but as for colour, that's a different matter which I decline to dwell upon. I admit to not finding them very comprehensible and would be wary of analysing the merit that others attribute to them. Nevertheless, I ought to say that the interior of a rotten Dutch cheese by Mr Cézanne, catalogued under the title of *impression from nature*, seemed exceptionally successful.[62]

Pissaro [sic] is the most independent of the group. He's bordering on sixty. In his face are only two eyes of a caressing brown, and a long multicoloured beard which he uses to wipe his brushes.[63]

In a dictionary of artists, Pissarro was listed as: "Pissarro (Camille) – impressionist market gardener, speciality cabbages."[64]

The novelist Joris-Karl Huysmans waded into the fray in 1880, alleging that the Impressionists were suffering from a defect of vision that was an acknowledged symptom of a mental illness – namely, hysteria. Pissarro, like the others, was said to suffer from a "mania for blue", which Huysmans dubbed "indigomania".[65] By the following year, Huysmans had become rather more considerate of the Impressionists, whose eyesight, it seemed to him, had mysteriously improved, although his interest in their paintings was little more than a

fig. 32
Camille Pissarro
Artist in the studio
1890
from *Turpitudes Sociales*
pen and ink (facsimile edition)
Private collection, Sydney

pretext for letting loose a gymnastics of mixed metaphors and a riot of colour, rehearsing the "decadent" effects that would soon make his prose style famous:

From closeby, [Pissarro's] *Sente du Chou* [cat. 44] is a masonry-work, a bizarre rough thrumming, a hodgepodge of tones of all sorts covering the canvas with lilac, Naples yellow, rose madder and green; from a distance it is the air circulating, the sky unlimited, nature quivering, water vaporising, the sun irradiating, the earth fermenting and fuming![66]

The redoubtable Henry Havard thought Impressionism was really not French at all: "As a Frenchman, I like clarity, precision, frankness, which are thequalities [sic] of our my [sic] nation [...] I confess that I find nothing in Impressionism recalling that simplicity and ancient logic, that French clarity and elegance".[67] The following year, oozing malicious satisfaction, Havard noted:

Let us recognise [...] that Impressionism is dying. The sacred phalange isn't recruiting any more. Mr Degas remains without disciples and Mr Pissaro [sic] isn't creating pupils. Even better, the old pontiffs are deserting: Mr Monnet [sic] has gone over to the enemy; this year he is showing at the Salon [...] They are no longer even attracting curiosity. The insatiable public wants something new, and they have ceased to be that.[68]

Indeed, the exhibition in 1880 was really "no longer an impressionist show", as John Rewald has flatly stated.[69] With Monet, Sisley and Renoir absent, the character and the orientation of the exhibition was now very different. Émile Cardon reported his awareness of fewer "pure impressionists" and more "faux intransigeants"[70] (false intransigents). Pissarro sprang the surprise of showing mostly figure paintings and examples of his recent etchings, which added to the preponderance of figures over landscapes, studio works over *pleinairisme*.

The fifth Impressionist exhibition showed faltering energy and no clear sense of direction; Monet and Renoir had left the group to try their chances in the Salon. Their works were accepted but badly placed, while Sisley's submission had been rejected by the jury. Nonetheless, the public had begun to get used to Impressionism. An influence had even started to affect the "official" painters. Émile Zola chose this vulnerable moment to express his disillusionment with the Impressionists:

The great tragedy is that no artist of this group has powerfully and definitively fulfilled the method they all share, scattered in their works. The method is there, infinitely widespread, but nowhere in any of their works does one find it implemented by a master. They are all precursors, the man of genius isn't born [...] They leave off working too soon,

too often; they are too easily satisfied; they show themselves incomplete, illogical, exaggerated, powerless.[71]

Zola did not mention, and the poisonous Henry Havard was obviously not aware, that Pissarro was currently "creating" one of his outstanding "pupils": Paul Gauguin. Gauguin was a successful young Parisian banker who had taken to spending his Sundays painting in Pontoise with Pissarro. Between 1879 and 1883, Gauguin spent his summer holidays in Pontoise or in Osny as a full-time painter under Pissarro's benevolent eye. During the years 1879–80, Pissarro had also became involved in a process of "interactive, non-hierarchical education" with no less a personage than Edgar Degas.[72] The two men's collaborative ventures in printmaking bore extraordinary results for both (see *Rainy effect*, cat. 29, *Twilight with haystacks*, cat. 32 and 33).

In addition, in 1879–80 the economic depression that had commenced in 1873 began to lift. A new climate of speculation favoured the art market. Durand-Ruel started to prosper again, with the benefits quickly felt by Pissarro. Pissarro could now afford to hire models to pose for his paintings.[73] All of these factors contributed to his reorientation from his earlier alliance with the Impressionist landscape painters to his much closer ties with the figure painters – Degas, Renoir, Morisot, Cassatt, Caillebotte. The juxtaposition of Pissarro's new figure paintings with his landscapes seems to have made critics fully aware, for the first time, of the artificial, confected character of the landscapes, which they no longer treated as if they were naturalistic "slices of life".

Mr Pissarro takes us far from this communicative gaiety [of Berthe Morisot's works]. He paints painfully in bright tones; he saddens the spring and the flowers, he weighs down the air. His technique is pasty, woolly, tormented; his figures of a melancholic character are treated with the same means as the trees, the grass, the walls and houses. However, a considerable stylishness, a forcefulness in the rendition of some well-made landscapes redeems this heaviness. Mr Pissarro distantly recalls Millet, but he recalls him in blue.[74]

Mr Pissarro is devoted to a heavy, thick painting without any transparency and without the least care for truth. One would say his paint was distemper whitewash.[75]

In 1883, Pissarro was given a one-man show at Durand-Ruel's gallery, hard on the heels of similar shows held during the immediately preceding months by Monet and Renoir. Pissarro exhibited seventy works, the majority consisting of recent figure paintings and a selection of earlier works, including paintings going back to 1870–72. Visitors to the exhibition were said to be "numerous", and reviewers began to treat him as an historical character:

Mr Pissarro is one of the fathers of impressionism, an inflexible character. He has the reputation of not ceding an inch of his territory. However, enter number 9, boulevard de la Madeleine and you will be stupefied ... not to be stupefied at all.[76]

It had become apparent that the public was growing accustomed to Impressionism. However, people's non-stupefication could also mean their indifference and boredom. Several reviewers were disturbed by Pissarro's "indifferent" treatment of his motifs, and they argued that indifference and boredom were appropriate responses to his works:

[These paintings] proceed by following a ready-made formula, which repudiates every intervention of the intellect and every flight of fancy. The difference of the seasons, of the land, of the hours of the day [...] are never seen in Mr Pissarro's paintings. Of drawing, composition, study, the least said, as the painter does not even seem to suspect they exist. So the fifty canvases of Mr Pissarro are completely neutral.[77]

System has placed a heavy hand on the artist. [In his recent productions,] the painter seems dominated by an implacable pre-conception. Why does he adopt a hatched, woolly technique of wearying uniformity, with little tufts that give his painting the appearance of plush, or seem to have come out of a strainer?[78]

[His latest works are] jerky, ragged, and the colour is thickly deployed in narrow, straight touches, broken-up by pockmarks which give the canvas the aspect of a geographical map in relief. [His execution] commandeers skies, lands, foliage, water, rocks without a care for verisimilitude. In practice, this is the absolute negation of the goal of painting – which is to give illusion and to endow each object, beyond its form and colour, with the rough or polished, opaque or transparent, light or solid aspects it has in nature.[79]

The critics were still unable to acknowledge Pissarro's brushstrokes and textures as a legitimate source of interest. They thought the medium and the surface of a painting were things that should be discounted, that ought not to be noticed. Paintings were meant to be looked *through*, not at. Consequently, the emphatic materiality of Pissarro's paintings was taken as the proof of his superficiality and clumsiness, as the evidence of his artistic failure. An untransmuted medium and an overobtrusive style had stifled the image, it was thought. These gorgeously cultivated textures and exquisitely humming tones were not acceptable proof of the paintings' intrinsic quality. They were seen as a negation or a subversion of painting's legitimate goals – a defeat of content, variety and lifelikeness.

1883–98

Political beliefs were considered more or less a private matter in the press of nineteenth century France, and it was rare for the Impressionists to be attacked in any overt way for their politics, even when this was a journalist's obvious intention – for example, when the group was called a "phalange", likened to a "mob", called "intransigents", and so on. Despite the political differences within the Impressionist group, and despite the apoliticism of the majority, Pissarro was convinced that their art was political and even revolutionary at its root. The proof of this was the incomprehension and hysterical overreaction of the bourgeois press and public to the Impressionist exhibitions, the deadly enmity of the Academicians and the Salon jury, the decades of gruelling poverty and struggle to which Pissarro and his fellow Impressionists had been subjected. "Our ideas, impregnated with anarchist philosophy, rub off on our work," Pissarro wrote to his son, Lucien, in 1891.[80]

In a left-wing Parisian newspaper, the columnist Trublot – famous for adopting the accent and idiom of the *titi parisien* in his journalistic patter – wrote: "In painting, [the Impressionists] stand for just what the Naturalists are in literature and the Socialists are in politics. In it is what is most logical and advanced, what represents the future. They have the common privilege of scaring the timid and the pedants, disturbing the tranquillity of those holding the pan-handle – the pan-handle of officialdom."[81]

When an exhibition of the Impressionists organised by Durand-Ruel in New York took place in 1886, it was not the New York equivalents of the *titi parisien* but the conservative critics who saw red when they saw Impressionism. Then, as now, American journalists did not consider an individual's political and religious beliefs to be a private matter: they were all too ready to guess aloud what those beliefs and affiliations might be, and to use their guesses as the basis of a character assassination:

The Paris Impressionists [...] are pressing a deliberate campaign against legitimate art [...] The landscapes after [Monet], Pissaro [sic] and Guillaumin, are simply insolent in the crudity and rudeness of their work [...] Impressionists, at least this group, are simply expositors of the social and moral cultus of Zola-ism, of sensualism and voluptuousness, or sheer atheism. And while the dreariness and unspirituality of their landscapes is thus accounted for, we have a canon which interprets their figure and genre subjects. One of the foremost, Degas, is nothing more than a peeping Tom.[82]

The tiresome old roof of our New York art exhibitions has been broken up very efficiently by the sudden appearance of this ultra-modern group of French painters in our midst [...] Communism incarnate, with the red flag and the Phrygian cap of lawless violence boldly displayed, is the art of the French Impressionist.[83]

fig. 33
Camille Pissarro
Anarchie
1890
from *Turpitudes Sociales*
pen and ink (facsimile edition)
Private collection, Sydney

A book published in 1898 baldly asserted that "Impressionism in art is something parallel to socialism in politics".[84] Published in French, the author was an American, Frederick Arthur Bridgman (1847–1928), who had attended the Atelier Suisse in Paris (he enrolled there the year after Cézanne left) and had studied at the École des Beaux-Arts under Gérôme. As a painter, Bridgman was a well-known Orientalist. In his preliminary remarks, he set forth his beliefs: Art should be dedicated to the ideal of a refined, noble culture, a culture predicated on "Taste, Beauty and Choice".[85] The Impressionists had intentionally sabotaged this ideal by imposing a kind of levelling on pictorial form, rendering everything equal, everything equivalent:

Hatching in a single direction, with strokes as thick as a thumb, destroys all the lines of the drawing, trees, animals, old men's trousers, halos of saints, all become constituted of the same material, a crude embroidery made with coarse wool.[86]

Bridgman believed that art should uphold an ideal of clarity, stability and certainty, which the Impressionists had desecrated through the ill-considered haste of their improvisations:

The depiction of the so-called impressionists does not go beyond a glance at a landscape seen from the window of a railway carriage driven at 80 kilometres an hour, and as for

technique or craft, it is carefully avoided. It is as if one looked at nature while blinking one's eyes, in such a manner that objects were left only as confusedly mixed marks.[87]

The successors of Manet [...] have most often taken up and exaggerated his grave faults, his coarseness and his vulgarity.[88]

In the autumn of 1885, Pissarro, who by now was aged fifty-five, was introduced by Armand Guillaumin to the neophyte Impressionist painter, Paul Signac. Signac was a friend of Georges Seurat, then aged twenty-six, who was in the process of conducting experiments with "optical mixture", making paintings by juxtaposing small points of unmixed colour, which tended to blend energetically in the mind's eye. Signac introduced Pissarro to Seurat in October. The venerable Impressionist was impressed by Seurat's talent and intelligence, and was very much taken with the results of his experiments. Due to the almost mechanical regularity with which Seurat had applied his stipples of colour, the texture of his paintings seemed remarkably even and unified. Pissarro no doubt recognised aims he had in common with Seurat's painting, as the art historian Martha Ward has explained. She noted that Pissarro's "conversion" to Neo-Impressionism was consistent with the "sporadic regularization" of technique in paintings he had made during the early 1880s (for example, see *The banks of the*

Viosne at Osny, grey weather, winter, 1883, cat. 45, *Steep road at Osny*, 1883, cat. 46 and *The artist's palette with a landscape*, c.1878, cat. 51).[89]

With typical inquisitiveness and enthusiasm, Pissarro began to try out Seurat's technique (see *Ploughing at Éragny*, c.1886, cat. 53) and found that he liked the results. However, his Impressionist colleagues, dealer Paul Durand-Ruel and the small group of committed, serious collectors of his paintings were all taken aback by what they saw as Pissarro's inconstancy, his susceptibility to new influences, even his betrayal of the Impressionist cause. "Many people will argue with Mr Pissarro about his new method," Octave Mirbeau wrote – as indeed they did.[90] Needless to say, the senior critics took a dim view of it, although there was some support from the younger ones. Once again, just as Pissarro had begun to obtain a modicum of success, his change of direction sent his career into a tailspin.

Mr Pessaro [sic] sets himself the task of imitating the newcomers and puts himself at the tail of Mr Seurat. Here you see him as a pointillist and an exclusive violettist.[91]

Mr Pissarro is the intransigent of the intransigents. Whatever his qualities may be, I refuse to believe the sincerity of his impression translated through a method of stippling which removes the fresh and free appearance that a work should always preserve.[92]

How did it come about that the artist, in the prime of his talent, saw fit to interrupt his labours and grew infatuated with a method? Obviously he thought the method could produce superior results to those he had previously obtained – a better distribution of tone, a more intense light [...] Unfortunately this method takes an excessive place and prevents one from seeing the picture [...] True colouration disappears. The mixture occurs neither in the observer's eye, nor on the canvas. Drawing is also affected by these deteriorations. And it is not light that is produced in the end, but lividity.[93]

Mr Pissaro [sic], as one knows, is a seeker, a conscientious and indefatigable seeker who, at the age when others rest on their achievements, wanted to continue in pursuit of a "differently", a "further" and a "better", letting drop the formulas that enabled him to create so many masterful works already. The canvases on show all belong to this last manner. The colours are infinitely divided. Their combination operates in the retina, not on the canvas. Whatever the inconveniences of this rather complicated procedure, it is undeniable that Mr Pissaro [sic] has obtained prodigious effects of shimmering and vibration.[94]

The eighth and last Impressionist exhibition was held in 1886 at 1, rue Laffitte. Monet, Renoir and Sisley again refused to participate, partly if not wholly because they disapproved of Pissarro's new friends.

The Neo-Impressionists' paintings were exhibited together in one room, with the huge canvas of Seurat's *Un dimanche à la Grande Jatte* dominating, and Pissarro's large, complex *Apple picking* (PV695) also given pride of place. By exhibiting his works in the context of the group, Pissarro once again challenged the creed of "possessive individualism" (see footnote 51), emphasising the collective, communal ownership of his artistic idiom – with all due recognition of Seurat as the inventor of the "dot". With the possible exception of Seurat, all of the Neo-Impressionists were – like Pissarro – adherents of the political philosophy of Anarchism, so it would have been natural for them to question the assumptions of intellectual property in the practice of painting.[95] Neo-Impressionism was a visual language and field of investigation that they shared, not a patented technique belonging to a single individual. The critics tended to distinguish Pissarro from the others by recourse to a cliché, implying that Pissarro's subject matter was limited and was, indeed, stereotyped. A dozen times over, he was characterised as "a painter of sunlit fields and peasants".

The longest, most engaged and partisan review of the exhibition was written by Félix Fénéon (1861–1944), an anarchist intellectual and a brilliant young man of letters. Looking back over more than a decade, Fénéon saw that there was an analogy between the transpersonal, communal, "embedded" visual language of the Impressionists and the contextual, embedded way they represented the world:

[They] had seen the objects in solidarity, each amongst the others, without chromatic autonomy, participating in the luminous mores [*les mœurs lumineuses*] of their neighbours. Traditional painting had considered [objects] ideally isolated and lit by a poor, artificial day.[96]

To expand on Fénéon's inferences: the Impressionists' paintings showed the vital interrelatedness of people, places, objects and phenomena. In a sense, theirs was a social vision – a vision of community, a vision of democracy. Unlike "traditional painting", their works conveyed a sense of the lived dimension of time, with an imagery that was explicitly time-bound. They chose to paint motifs that were transitory, that were in flux, that were modern, and they registered the changing world in a pictorial language that was also subject to change – Impressionism had now become Neo-Impressionism.

It was obvious to all who saw them that Pissarro's new paintings were laboriously and patiently made, owing little to spontaneity and improvisation, making only an intermittent evocation of naturalism and *pleinairisme*. In his new paintings, as in the older ones, the viewer was invited to witness the painter's creative process, seeing the motif broken down into fragmentary elements and seeing the composition built up from these; seeing the work of analysis and seeing the work of synthesis. Pissarro was generally not a colourist in the sense that Monet and Renoir were. Monet and Renoir created swelling symphonies of colour, whereas Pissarro usually converted his colours into tones – iridescent, coruscating, intensely alive aggregates of tone. Pissarro was essentially a tonal painter, like his master, Corot, and he shared this exigent, refined feeling for tone with the other Neo-Impressionists.[97]

It was within the context of the Neo-Impressionist group exhibitions that Pissarro began to achieve wider recognition as the dedicated, highly inventive pictorial composer he had always been. Octave Mirbeau lauded his "eye habituated to the grand syntheses of colour and design"[98]; Gustave Kahn praised "such luminous landscapes, so free of lines, so harmonic and synthetic".[99] However, the laborious method soon began to lose its appeal for Pissarro, who complained that he missed the "fullness, suppleness, liberty, spontaneity, freshness of sensation of our Impressionist art".[100] And so, by 1890, he had rehabilitated a robust, rapid, scintillatingly energetic touch, painting in an idiom that owed less and less to Neo-Impressionism (see *Hampton Court Green*, 1891, cat. 58 and *Springtime*,

Éragny, 1895, cat. 61). In fact, Pissarro was painting better, more confidently and more prolifically than ever, and everything now started to come right for him.

However, the retrospective and survey exhibitions of Pissarro's paintings held at this time were still a stumbling block for critics. How could they explain, how could they condone a lifetime of restless experimentation? Pissarro's unusual comportment could not be ascribed to his eclecticism, his lack of seriousness or weakness of character. Before the advent of the generation of Matisse, Picasso and Derain, painters of the first rank who frequently changed their style and the formal and technical parameters of their art were virtually unknown. Arsène Alexandre broached the thorny question in his review of Pissarro's show at the Boussod et Valadon gallery in 1890:

Mr Pissarro is a very good painter [...] but it seems that he lacks a certain originality which makes the true masters [...] Involuntarily he imitates Cézanne, then Monet; then in a more recent epoch he turned himself into a pointillist ... Beside some very stippled, even too stippled canvases, he shows others – market scenes handled in a different, quasi-classic manner. All of this betrays a certain hesitation, a difficulty of settling down, of deciding, and Mr Pissarro has experienced this hesitation all his life.[101]

Alexandre had further thoughts on Pissarro's supposed indecisiveness, concluding that his comments were untrue and/or unjust, perhaps jolted by Pissarro's protestations about what he had written. Two years later, he was able to make amends on the occasion of an exhibition of sixty works at the Durand-Ruel gallery:

One sees fully, then, what one took to be [Pissarro's] hesitation was perfect awareness, what one judged as the influence of so and so, was only the wish to endow his thought with the best and most complete means of expression.[102]

The 1892 exhibition at the Durand-Ruel gallery coincided with a moment when the tide of critical opinion turned decisively in Pissarro's favour. The reversal of attitudes is best demonstrated by a review written by a young painter associated with the Nabi group, Félix Vallotton, which was published in a Swiss newspaper:

[Pissarro] has the supreme honour of being considered a young person by young people. There is no literature in his canvases, no suggestive or engaging titles, not even subjects in the strict sense of the word. This art has to pass absolutely through the eye to be experienced and understood.

The banks of the Oise, an absolute masterpiece, [...] will remain the perfect example of the painter's manner twenty years ago. The harmony of the ensemble results from a predilection for large, flat tonalities, and I don't believe that anyone has ever exceeded the power of the impression made by this great white sky and these riverbanks eaten away by light.[103]

The very things that Pissarro had been berated for were suddenly adduced as the shining virtues of his art and as the quintessence of his modernity. There was now a consensus in his favour, supported by a younger generation of painters and by the best Parisian critics of the time: Théodore Duret, Gustave Geffroy, Arsène Alexandre, Octave Mirbeau, Félix Fenéon, Georges Lecomte, Thadée Natanson …

An almost unanimous acclamation of Pissarro's modernity, exceptional creativity and perennial youthfulness thundered out in response to his exhibition of paintings of contemporary Paris held in June 1898 at the Durand-Ruel gallery. This success was all the more remarkable because it was concurrent with a exhibition of Monet at the Galerie Georges Petit. The critics and collectors were thrilled:

It is especially interesting that Mr Pissarro's recent works [...] have never been more vibrant, stronger or younger. Above all they are decorative and harmonious interpretations of Paris, views of the boulevards, the avenue de l'Opéra, the rue Saint-Honoré, at varied hours, day and night, or in different weathers.[104]

[Pissarro's goal was] to contemplate and surprise the gestures, among the tumultuous, colourful hours, of the throngs in our streets [...] He had easy access to the best advice from the Japanese painters.[105]

After so many canvases, so many masterpieces where the intimate and radiant poem of nature is written in such splendid harmonies, here the old master wanted to present the stone of the cities, the swarming life of the streets, the grey of the façades, the skies complicated by smoke and dust [...] And how audacious is the architecture of these pictures, how boldly and unexpectedly they are built! The pedestrians, the vehicles, the animals, the stars seen from on high are seized in their true shape and in their true light [...] At the same time, all of this balances, harmonises and makes very decorative compositions.[106]

They are portraits always under revision, because the physiognomies they portray are infinitely changeable: dream of Paris yesterday while you look at Paris today.[107]

"Decidedly it's a success, my exhibition," Pissarro wrote to his son Lucien, noting the consistently favourable reviews it had received.[108] At the age of sixty-seven, Pissarro had at last well and truly arrived.

1 Théodore Duret, *Histoire des Peintres Impressionnistes (4th edition)*, Librairie Floury, Paris, 1939, p. 7.

2 Gustave Geffroy, "Camille Pissarro" (written 15 February 1890, to accompany Pissarro's exhibition at the Galerie Boussod et Valadon), in *La Vie Artistique*, vol. 1, E. Dentu, Paris, 1892, p. 40.

3 Gustave Geffroy, "Pissarro" (written 10 June 1898, to accompany Pissarro's exhibition at Galeries Durand-Ruel), in *La Vie Artistique*, vol. 6, H. Floury, Paris, 1900, pp. 181–82.

4 Gustave Geffroy, introduction to *Catalogue de la Collection de Madame Vve C. Pissarro*, Galerie Nunès & Fiquet, Paris, 20 May–20 June 1921, n.p.

5 Ruth Berson, "Introduction", *The New Painting – Impressionism 1874–1866. Documentation*, 2 vols, Fine Arts Museums of San Francisco, San Francisco, 1996, vol. 1, p. xi.

6 Gustave Geffroy, "Sisley", *Les Cahiers d'Aujourd'hui*, nos 13–14, Paris, 1923, p. 12.

7 Claude Vignon: "Variétés – Salon de 1852", *Le Public*, Paris, 12 April–2 June 1852; reprinted in Jean-Paul Bouillon, Nicole Dubreuil-Blondin, Antoinette Ehrard, Constance Naubert-Riser (eds): *La Promenade du Critique Influent*, Hazan, Paris, 1990, p. 30.

8 Jules-Antoine Castagnary, "Salon des Refusés", *L'Artiste*, vol. II, 1863, p. 95. Gérard Monnier has described very well the compromised, unsustainable role of the Salon: "The Salon in the nineteenth century was […] the place where the artist established his relationship with power – administrative power, academic power and economic power. Stemming from its status as a shop-window, demonstrative of the realities operating elsewhere, the Salon was transformed into a place of confrontation and competition, a place of sales, a place determining the extent of the social existence of the artist." – Gérard Monnier: *L'art et ses institutions en France*, Gallimard, Paris, 1995, p. 129.

9 Louis Leroy, "Salon de 1863 – IX Les Refusés", *Le Charivari*, Paris, 20 May 1863, p. 2.

10 For example, in the Salon of 1868: "*La Côte du Jallais* and *l'Ermitage* [sic] by Pissarro, which this year they have again placed very high, but not high enough to prevent art lovers from following the solid qualities that distinguish them." – Jules-Antoine Castagnary: "Salon de 1868: le paysage", *Le Siècle*, 29 May 1868, p. 1.

11 Jules-Antoine Castagnary on Whistler and Manet – see "Salon de 1863 – III. Les Refusés", *Le Courrier du Dimanche*, 17 May 1863, Paris, p. 5.

12 ibid., p. 4.

13 Jules-Antoine Castagnary: "Salon des Refusés – XIV", *L'Artiste*, vol. II, 1 December 1863, p. 95. Pissarro had identified himself in the catalogue as "pupil of Corot".

14 Louis Leroy, art. cit. (footnote 9), p. 3. Le Capitaine Pompilius, untitled article, *Le Petit Journal*, 5 June 1863, p. 1. For the identification of Le Capitaine Pompilius as Carle Desnoyers, see Michael Fried: *Manet's Modernism*, University of Chicago Press, London, 1996, p. 528, n. 126.

15 Castagnary, art. cit. (footnote 8), p. 4. Castagnary's recognition of a "new school" revealed by the Salon des Refusés was corroborated in a contemporaneous review by Théophile Thoré (writing under the nom de plume of William Bürger) – see "Salon de 1863", in *Salons de W. Bürger, 1861–1868*, 2 vols, Paris 1870, vol. 1, p. 414.

16 Le Capitaine Pompilius: art. cit. (footnote 14).

17 Émile Zola: "Adieux d'un critique d'art" (originally in *L'Événement*, 20 May 1866), *Mon Salon – Manet – Écrits sur l'art*, Garnier-Flammarion, Paris, 1970, p. 87.

18 Until recently, this painting has been titled *River landscape with boat near Pontoise* and has been dated circa 1872. Its similarities to The Art Institute of Chicago's painting and the circumstances of Pissarro's relationship with Zola would appear to favour its redating to circa 1866 and require a reconsideration of its presumed location: might it have been painted in or near La Varenne-St-Hilaire rather than Pontoise?

19 Jean Rousseau, "Le Salon de 1866 – IV", *L'Univers Illustré*, Paris, 14 July 1866, pp. 447–48.

20 Léon Billot, "Exposition des Beaux-Arts", *Le Journal du Havre*, Le Havre, 25 September 1868.

21 Émile Zola, "Les Naturalistes", *L'Événement illustré*, Paris, 19 May 1868, reprinted in *Mon Salon – Manet – Écrits sur l'art*, op. cit. (footnote 17), p. 146.

22 ibid., p. 147.

23 Jean Ravenel, "Préface au Salon de 1870", *Revue Internationale de l'art et de la curiosité*, vol. III, no 4, Paris, 15 April 1870, p. 323.

24 Théodore Duret, *Les Peintres français en 1867*, Dentu, Paris, 1867, reprinted in Jean-Paul Bouillon, et al, *La Promenade du critique influent*, op. cit. (footnote 7), p. 160.

25 Théodore Duret, "Le salon – Les naturalistes", *L'électeur libre*, Paris, 12 May 1870, p. 61.

26 Anonymous, "Pissarro", *Daily Chronicle*, London, 16 November 1903, p. 3, reprinted in Kate Flint (ed.): *Impressionists in England – The Critical Reception*,

Routledge & Kegan Paul, London, 1984, p. 348.

27 Théodore Duret, "Le salon – Les naturalistes", art. cit. (footnote 25).

28 Théodore Duret, "Les Peintres impressionnistes" (originally published in 1878 as a pamphlet by Librairie parisienne H. Heymann et J. Perois, Paris), reprinted in *Critique d'avant-garde*, École nationale supérieure des Beaux-Arts, Paris, 1998, pp. 56–57.

29 ibid., p. 57.

30 Meeting of Duret and Pissarro in 1866 – see the letter of condolence written by Théodore Duret to Lucien Pissarro on the occasion of Pissarro's death, 13 November 1903, in Anne Thorold (ed.), *Artists, Writers, Politics – Camille Pissarro and His Friends*, Ashmolean Museum, Oxford, 1980, p. 36.

31 On Duret's Pissarros, see Anne Distel, "Some Pissarro collectors in 1874", in Christopher Lloyd (ed.), *Studies on Camille Pissarro*, Routledge & Kegan Paul, London, 1986, p. 73, n. 28.

32 Pissarro's importuning letters to Duret, see vol. 1 of Janine Bailly-Herzberg (ed.): *Correspondance de Camille Pissarro, 5 vols, 1980–1996*, vol. 1, Presses Universitaires de France, Paris; vols 2–5, Éditions du Valhermeil, Saint-Ouen-l'Aumône.

33 Renoir was living in Voisins, a hamlet very close to Pissarro's house, and Monet in the neighbouring town of Bougival. On Pissarro's paintings of the period, see Richard R. Brettell, "Pissarro in Louveciennes", *Apollo*, London, November 1992, pp. 315–19.

34 John Rewald, *The History of Impressionism*, 4th revised edition, Secker and Warburg, London, 1980, pp. 284–85.

35 Letter from Pissarro to Wynford Dewhurst, November 1902, quoted in Wynford Dewhurst, *Impressionist Painting*, George Newnes Limited, London, 1904, pp. 31–32.

36 On the earliest exhibitions in London to feature works by the Impressionists, see Kate Flint (ed.): op. cit. (footnote 26), pp. 356–60, and for the earliest press notices to mention the Impressionists, ibid., pp. 34–35. See also John House: "New Material on Monet and Pissarro in London in 1870–71", *Burlington Magazine*, vol. CXX, no 907, London, October 1978, p. 638.

37 Cf. Georges Lecomte: "Their eyes had been educated by Turner who, long ago, limited himself only to the colours of the prism […] Returning to Paris, Messrs Pissarro and Monet made themselves exegetes of the new technique." – "Camille Pissarro", *Les Hommes*

d'Aujourd'hui, vol. 8, no 366, 1890, n.p. The importance of this encounter with Turner for the invention of Impressionism's characteristic "vibrato" of juxtaposed colours was emphasised by Paul Signac in D'Eugène Delacroix au néo-impressionnisme, Hermann, Paris, 1978, pp. 88–89. It should be remembered that both Signac and Henri Matisse made "Turner pilgrimages" to London, in 1898, probably at Pissarro's urging.

38 "enseignement interactif et non-hiérarchique" – see Richard Brettell, "Cézanne/Pissarro: élève/élève", in Françoise Cachin, Henri Loyrette and Stéphane Guégan (eds), Cézanne aujourd'hui, Réunion des musées nationaux, Paris, 1997, p. 30. On Pissarro and Cézanne, see Joachim Pissarro (ed.): Pioneering Modern Painting: Cézanne and Pissarro 1865–1885, Museum of Modern Art, New York, 2005.

39 On Pissarro and the other Impressionists' income from Durand-Ruel before the first Impressionist exhibition in 1874, see Ralph E. Shikes and Paula Harper, Pissarro – His Life and Work, Quartet Books, London, 1980, p. 105.

40 Richard Brettell, art. cit. (footnote 38), p. 32.

41 The Société Anonyme was one of the first but not the first artists' cooperative in Paris – see Gérard Monnier, op. cit. (footnote 8), pp. 176–177. On the founding charter, see Rewald, op. cit. (footnote 34), pp. 312–13.

42 Philippe Burty, "The Paris Exhibitions: Les Impressionnistes", The Academy, London, 30 May 1874, p. 616.

43 ibid.

44 Marc de Montifaud, "Exposition du boulevard des Capucines", L'Artiste, Paris, 1 May 1875, p. 307.

45 Étienne Carjat: "L'Exposition du boulevard des Capucines", La Patriote française, Paris, 27 April 1874, p. 3.

46 Armand Silvestre, "Physiologie du refusé – L'Exposition des revoltés", L'Opinion nationale, Paris, 22 April 1874, p. 2.

47 Louis Leroy, "L'Exposition des impressionnistes", Le Charivari, Paris, 25 April 1874, p. 79.

48 ibid.

49 Théodore Duret, letter to Pissarro, December 1873, quoted in Shikes and Harper, op. cit. (footnote 39), p. 107.

50 Pissarro, letter to Lucien Pissarro, 22 November 1895, in Bailly-Herzberg, op. cit. (footnote 32), vol. 4, p. 121. Cf. Théophile Thoré's famous statement, "To be a master is to resemble no one."

51 The emergence of possessive individualism in seventeenth century England is analysed in C.B. Macpherson, The Political Theory of Possessive Individualism – Hobbes to Locke, Oxford University Press, Oxford, 1964.

52 "He sees nature by simplifying it, and [he sees it] through its permanent aspects." – Théodore Duret, "Les Peintres impressionnistes", op. cit. (footnote 28), p. 56.

53 "pour étudier les figures et les animaux de la vraie campagne" – Pissarro, letter to Théodore Duret, 20 October 1874, in Bailly-Herzberg, op. cit. (footnote 32), vol. 1, p. 95.

54 Philippe Burty, "Fine Art – The Exhibition of the 'Intransigeants'", The Academy, London, 15 April 1876, p. 363.

55 Gène-Mur, "Revue artistique: les impressionnistes", La correspondance française, Paris, 16 April 1876, p. 1.

56 Alex. Pothey, "Chronique", La Presse, Paris, 31 March 1876, p. 3.

57 Simon Boubée, "Beaux-Arts – Exposition des Impressionnistes chez Durand-Ruel", La Gazette de France, 5 April 1876, p. 2.

58 D. d'Olby, "Salon de 1876", Le Pays, Paris, 10 April 1876, p. 3. This review of the Salon included a digression about the "Exposition des Intransigéants chez M. Durand-Ruel, rue Le Peletier, 11".

59 "une collection d'ébauches tapageuses" – Charles Bigot, "Causerie artistique", La Revue politique et littéraire, 2nd series, X, Paris, 8 April 1876, p. 352.

60 ibid., p. 351.

61 Albert Wolff, "Le Calendrier parisien", Le Figaro, Paris, 3 April 1876, p. 1.

62 Jacques, "Menus Propos – Exposition Impressionniste", L'Homme libre, Paris, 12 April 1877, p. 2.

63 Alexandre Hepp: "Impressionnisme", Le Voltaire, Paris, 3 March 1882, p. 1. Although he looked older, Pissarro was only 51 when the article was published.

64 E. Giraud (ed.), Petit Bottin des lettres et des arts, Paris, 1886, p. 109.

65 Joris-Karl Huysmans, "L'Exposition des indépendants en 1880", L'Art moderne, Charpentier, Paris, 1883, reprinted in Riout (ed.), Les écrivains devant l'impressionnisme, Macula, Paris, 1989, pp. 255–56.

66 Joris-Karl Huysmans, "L'Exposition des indépendants en 1881" from L'Art moderne, in Rioult, op. cit. (footnote 64), p. 283. Huysmans' famous "decadent" novel, A Rebours, was published in 1884. If the painting

referred to is The path to Le Chou, Pontoise, 1878 (cat. 44), as seems likely, the description of colours is surprisingly inexact.

67 "Français, j'aime la clarté, la précision, la franchise, qui sont lesqualités [sic] de notre ma [sic] nation" – Henry Havard, "L'Exposition des artistes indépendants", Le Siècle, Paris, 27 April 1879, p. 3.

68 Henry Havard, "L'Exposition des artistes indépendants", Le Siècle, Paris, 2 April 1880, p. 2.

69 John Rewald, op. cit. (footnote 34), p. 439.

70 Émile Cardon, "Choses d'art: L'Exposition des impressionnistes", Le Soleil, Paris, 5 April 1880, p. 3.

71 Émile Zola, "Le Naturalisme au Salon" (originally published as a series of four articles in Le Voltaire, Paris, 18–22 June 1880), reprinted in Le Bon Combat, Hermann, Paris, 1974, pp. 214–15.

72 Pissarro confided the opinion to his son Lucien that "Degas is certainly the greatest artist of our epoch" – Pissarro, letter to Lucien Pissarro, 9 May 1883, in Bailly-Herzberg (ed.), op. cit. (footnote 32), vol. 1, p. 204.

73 See Shikes and Harper, op. cit. (footnote 39), pp. 154–155 passim. Unfortunately, the boom was very short-lived: in 1882, a crash of the Banque de France sent Durand-Ruel and his artists back to their accustomed struggle.

74 Charles Ephrussi, "Exposition des Artistes Indépendants", Gazette des Beaux-Arts, Paris, 1 May 1880, reprinted in Denys Riout (ed.), Les écrivains devant l'impressionnisme, Macula, Paris, 1989, p. 236.

75 "badigeon à la colle" – Élie de Mont, "Cinquième Exposition des impressionnistes, 10, rue des Pyramides", La Civilisation, Paris, 20 April 1880, p. 2.

76 Edmond Jacques, "Exposition de M. Pissarro", L'Intransigéant, Paris, 14 May 1883, p. 2.

77 G. Dargenty, "Exposition des œuvres de M. Pissarro", Courrier de l'art, Paris, 31 May 1883, p. 255.

78 Frédéric Henriot, "Exposition – Des œuvres de C. Pissaro [sic]", Le Journal des Arts, Paris, 25 May 1883, p. 2.

79 A. Hustin, "Exposition de Pissarro", Moniteur des Arts, Paris, 11 May 1883, p. 166. Pissarro responded to or may have anticipated the criticisms of his densely charged surfaces. See the letter to Lucien Pissarro, 4 May 1883, in Bailly-Herzberg, op. cit. (footnote 32), vol. 1, p. 202.

80 Pissarro, letter to Lucien Pissarro, 13 April 1891, in Bailly-Herzberg, op. cit (footnote 32), vol. 3, p. 63.

81 Trublot (Paul Alexis), "A Minuit – Les impressionnistes", *Le cri du peuple*, Paris, 1 May 1885, p. 3.

82 Anonymous, *The Churchman*, New York, 12 June 1886. For a survey of the American responses to Impressionism, see H. Huth, "Impressionism comes to America", *Gazette de Beaux-Arts*, Paris, April 1946, pp. 226–52.

83 Anonymous, "The Impressionists", *Art Age*, New York, March 1886.

84 F.-A. Bridgman, *L'Anarchie dans l'art*, Société Française d'Editions d'Art, L.-Henry May, Paris, 1898, p. 5. Bridgman's master, Gérôme, set the tone for this polemic in 1894, when the Caillebotte collection of 65 paintings was offered to the French state. Gérôme made a public protest: "I do not know these gentlemen, and of this bequest I know only the title ... Does it not contain paintings by M. Monet, by M. Pissarro and others? For the government to accept such filth, there would have to be a great moral slackening" – quoted by John Rewald, *The History of Impressionism*, op. cit. (footnote 34), p. 570.

85 ibid., p. 3.

86 ibid., p. 81.

87 ibid., pp. 11–12.

88 ibid., p. 70.

89 "It was Pissarro's sporadic regularization of his technique in the early 1880s – his desire to make a surface plain and accessible; his application of color contrasts through reference to artisanal making; his avoidance of dazzling improvization – that made the uniformity, simplicity and anonymity of the 'point' of color attractive" – Martha Ward, *Pissarro, Neo-Impressionism and the Spaces of the Avant-Garde*, The University of Chicago Press, Chicago, 1996, p. 75. Richard Brettell has also discerned the use of complementary colours (a typical feature of Neo-Impressionist painting) in some of Pissarro's works of the 1870s. See Richard Brettell, "Camille Pissarro: A revision", in *Camille Pissarro* (exhibition catalogue), Arts Council of Great Britain and Museum of Fine Arts, Boston, 1980, p. 30.

90 Octave Mirbeau, "L'Exposition Internationale de la rue de Sèze (II)", originally in Gil Blas, Paris, 14 May 1887, reprinted in *Combats esthétiques*, Séguier, Paris, 2 vols, 1993, vol. 1, p. 333.

91 Ernest Hoschedé, "Exposition des 33. Peinture et sculpture", *L'Événement*, Paris, 2 January 1888, p. 2.

92 Alexandre Georget, "Exposition International de Peinture", *L'Echo de Paris*, 17 May 1886.

93 Gustave Geffroy, "Salon de 1887", *La Justice*, Paris, 13 June 1887, p. 1. The quote occurs in "Hors du Salon – rue de Sèze et rue Laffitte", a section of the article concerning the exhibitions.

94 G. Alb. A. (G.-A. Aurier), "Beaux-Arts", *Mercure de France*, Paris, April 1890, p. 143.

95 See Martha Ward, op. cit. (footnote 89); Robert L. and Eugenia W. Herbert, "Artists and Anarchism: Unpublished Letters of Pissarro, Signac and others – I", *Burlington Magazine*, London, November 1960, pp. 473–82.

96 Félix Fénéon, "La VIIIe exposition impressionniste", originally in *La Vogue*, 13–20 June 1886, reprinted in *Au-delà de l'impressionnisme*, Hermann, Paris, 1966, p. 64.

97 Their concern for tone was too subtle for some reviewers. For example, Firmin Javel upbraided Pissarro for his "voluntary ignorance of tonal values" – see Firmin Javel, "Les Impressionnistes", *L'Événement, Paris*, 16 May 1886, p. 1.

98 Octave Mirbeau, art. cit. (footnote 89), p. 334.

99 Gustave Kahn, "L'Art français à l'Exposition", *La Vogue*, Paris, August 1889, p. 132. Pissarro wrote a letter of thanks to Mirbeau ("Thank you, Sir, you have dressed the wounds and eased the torments") – Pissarro, letter to Octave Mirbeau, 21 May 1887, in Bailly-Herzberg, op. cit. (footnote 32), vol. 2, pp. 170–71.

100 Pissarro, letter to Lucien Pissarro, 6 September 1888, in Bailly-Herzberg, op. cit. (footnote 32), vol. 2, p. 251.

101 Arsène Alexandre, "Camille Pissarro", *Le Figaro*, Paris, 28 February 1890, p. 2. For an excellent essay treating the questions raised by Pissarro's stylistic shifts, see John House, "Camille Pissarro's idea of unity", in Christopher Lloyd (ed.), *Studies on Camille Pissarro*, Routledge & Kegan Paul, London, 1986, pp. 15–34.

102 Arsène Alexandre, "Chroniques d'Aujourd'hui – Camille Pissarro", *Le Figaro*, Paris, 3 February 1892, p. 2.

103 Félix Vallotton, "L'Exposition Pissarro", *Gazette de Lausanne*, Lausanne, 7 March 1892, p. 2. The admiration was mutual: Pissarro saw Vallotton's work and admired his talent. See letter to Lucien Pissarro, 15 July 1893, in Bailly-Herzberg, op. cit. (footnote 32), vol. 3, p. 347.

104 C. d'H., "Exposition Camille Pissarro", *Moniteur des Arts*, 3 June 1898, p. 559.

105 André Fontainas, "Art Moderne", *Mercure de France*, July 1898, p. 280. Pissarro, like all the Impressionists, fell under the influence of Japanese art in the 1870s.

106 Georges Lecomte, "Quelques Syndiqués – Camille Pissarro", *Les Droits d'Homme*, Paris, 4 June 1898, p. 1.

107 Gustave Geffroy, "Camille Pissarro", *Le Journal*, Paris, 25 June 1898, p. 2.

108 Pissarro, letter to Lucien Pissarro, 6 June 1898, in Bailly-Herzberg (ed.), op. cit. (footnote 32), vol. 4, pp. 487–88.

List of works

• exhibited in Sydney only

Early works

1
**Landscape with trees,
two figures on a road
and mountains in the
background**
Paysage classique
c.1855
black and white chalk on
blue paper, 54 x 69
Inscribed in black chalk,
lower right: *C.P.*
Yale University Art Gallery,
gift of Joseph F. McCrindle,
LL.B., 1948 (1973.132)

2
Road to Port-Marly
Route de Port-Marly
c.1860–67
oil on panel, 22.8 x 35.1
PDRS68
Lent by the Syndics of
the Fitzwilliam Museum,
Cambridge (PD58-1958)

3
Landscape
Paysage
c.1865
oil on canvas, 28.4 x 44.1
PDRS87
Indianapolis Museum of Art,
gift of Mrs Joseph E. Cain
(75.981)

4 •
**The Marne at
La Varenne-St-Hilaire**
*La Marne à
La Varenne-St-Hilaire*
c.1864
oil on canvas, 24 x 32
PV42, PDRS94
Kunstmuseum Bern, Legat
Eugen Loeb, Bern (G1871)

5
Barge on the Seine
Péniche sur la Seine
c.1863
oil on canvas, 46 x 72
PDRS79
Musée Camille Pissarro,
Pontoise (P1980.3)

6
River landscape with boat
*Paysage fluviale avec
bâteau près de Pontoise*
c.1866
oil on canvas, 43 x 65
PDRS108
Kunstmuseum Sankt Gallen,
Sturzeneggersche Painting
Collection, purchased 1936

7
**Banks of the Marne
in winter**
*Bords de la Marne,
en hiver*
1866
oil on canvas, 91.8 x 150.2
PV47, PDRS107
The Art Institute of Chicago,
Mr and Mrs Lewis Larned
Coburn Memorial Fund
(1957.306)

> Work is a marvellous regulator of moral
> and physical health. All the sadness,
> all the bitterness, all the grief – I forget,
> I overlook these in the joy of working.

Camille Pissarro

Towards Impressionism

8
The road near the farm
La route près de la ferme
1871

oil on canvas, 38.1 x 46
PDRS 197

Fine Arts Museums of
San Francisco, bequest of
Marco F. Hellman (1974.5)

9
Near Sydenham Hill
Près de Sydenham Hill
1871

oil on canvas, 43.5 x 53.5
PV115, PDRS 190

Kimbell Art Museum,
Fort Worth, Texas (AP71.21)

10
The fence
La barrière
1872

oil on canvas, 37.8 x 45.7
PV135, PDRS 231

National Gallery of Art,
Washington, DC, collection
of Mr and Mrs Paul Mellon
(1985.64.31)

11
**Landscape in the vicinity
of Louveciennes (Autumn)**
Louveciennes (Automne)
1870

oil on canvas, 88.9 x 115.9
PV87, PDRS 157

The J. Paul Getty Museum,
Los Angeles, California
(82.PA73)

12
**La Sente de Justice,
Pontoise**
c.1872

oil on canvas, 52 x 81
PDRS 264

Memphis Brooks Museum
of Art, Memphis, Tennessee,
gift of Mr and Mrs Hugo
N. Dixon (53.60)

13
Effect of fog at Creil
Effet de brouillard à Creil
1873

oil on canvas, 38 x 56.5
PDRS 324

Private collection, Switzerland

14 •
**Still-life: apples and
pears in a round basket**
*Nature morte:
pommes et poires
dans un panier rond*
1872

oil on canvas, 45.7 x 55.2
PV194, PDRS 269

Mrs Walter Scheuer; on
long-term loan to Princeton
University Art Museum
(L1988.62.15)

Impressionism

15
**The banks of the Oise
near Pontoise**
*Bords de l'Oise près
de Pontoise*
1873

oil on canvas, 38.1 x 55.2
PV222, PDRS 303

Indianapolis Museum
of Art, James E. Roberts Fund
(40.252)

16
Factory near Pontoise
Usine près de Pontoise
1873

oil on canvas, 45.7 x 54.6
PV215, PDRS 208

Museum of Fine Arts,
Springfield, Massachusetts,
The James Philip Gray
Collection (37.03)

17
**The road to Ennery,
near Pontoise**
*La route d'Ennery,
près de Pontoise*
1874

oil on canvas, 55 x 92
PV254, PDRS 349

Musée d'Orsay, Paris,
donated by Max and Rosy
Kaganovitch, 1973 (RF1973-19)

18
**Hill at L'Hermitage,
Pontoise**
*Coteau de L'Hermitage,
Pontoise*
1873

oil on canvas, 61 x 73
PV209, PDRS 291

Musée d'Orsay, Paris, *acquis
par dation*, 1983 (R.F. 1983-8)

19
**Kitchen garden at
L'Hermitage, Pontoise**
*Jardin potager à
L'Hermitage, Pontoise*
1874

oil on canvas, 54 x 65.1
PV267, PDRS 342

The National Gallery
of Scotland, Edinburgh,
presented by Mrs Isabel
M. Traill, 1979 (NG2384)

20
**Villa at L'Hermitage,
Pontoise**
*Maison bourgeoise à
L'Hermitage, Pontoise*
1873

oil on canvas, 50.5 x 65.5
PV227, PDRS 308

Kunstmuseum Sankt Gallen,
Sturzeneggersche Painting
Collection (G34), purchased
1936

21 •
**The orchard at
Maubuisson, Pontoise**
*Le verger à
Maubuisson, Pontoise*
1876

oil on canvas, 42.5 x 50.2
PV345, PDRS 442

Private collection, London

22
**A cowherd on the route
du Chou, Pontoise**
*Une vachère sur la route
du Chou, Pontoise*
1874

oil on canvas, 54.9 x 92.1
PV260, PDRS 354

The Metropolitan Museum
of Art, gift of Edna H. Sachs,
1956 (56.182)

23
**Piette's house,
Montfoucault, in the snow**
*La maison de Piette,
Montfoucault, effet
de neige*
1874

oil on canvas, 60 x 73.5
PV285, PDRS 386

Lent by the Syndics of
the Fitzwilliam Museum,
Cambridge (PD10-1966)

24
The pond at Ennery
L'Étang à Ennery
1874

oil on canvas, 53.3 x 64.1
PV259, PDRS 340

Yale University Art Gallery,
collection of Mr and Mrs Paul
Mellon, BA, 1929 (1983.7.14)

The Impressionist print

25
Path in the woods, Pontoise
Chemin sous bois, à Pontoise
1879

aquatint and etching
16.1 x 21.3
D19, ii of 6 states
Inscribed in pencil, lower left:
2e état

The Art Institute of Chicago, John H. Wrenn Endowment, 1960 (1960.723)

26
Path in the woods, Pontoise
Chemin sous bois, à Pontoise
1879

aquatint and etching
16.2 x 21.3
D19, v of 6 states
Inscribed in pencil, lower left:
5e état

The Art Institute of Chicago, John H. Wrenn Endowment, 1960 (1960.724)

27
Woman on the road
La femme sur la route
1879

aquatint, softground etching and etching, 15.6 x 20.9
D18, iv of 4 states
Inscribed in pencil, not by the artist, lower right: *no 3*

The Art Institute of Chicago, gift of Gaylord Donnelley (1971.365)

28
Rainy effect
Effet de pluie
1879

aquatint, 16 x 21.4
D24, ii of 6 states
Stamp of Marcel Louis Guérin, lower left (L.1872b)

Yale University Art Gallery, Everett V. Meeks, BA, 1901, and Stephen Carlton Clark, BA, 1903, Funds (1972.49)

29
Rainy effect
Effet de pluie
1879

aquatint, etching and drypoint
16 x 21.3
D24, vi of 6 states
Studio stamp of Edgar Degas, lower right (L.657). Inscribed in blue crayon, lower right: *1893*, and in pencil, lower right: *wo*, and lower left: *94857*

Los Angeles County Museum of Art, Wallis Foundation Fund, in memory of Hal B. Wallis (AC1996.27.2)

30
Twilight with haystacks
Crépuscule avec meules
1879

aquatint, etching and drypoint, 10.4 x 18
D23, iii of 3 states
Inscribed in black chalk and pencil, lower left: *no 3 / Epreuve d'artiste / crépuscule (cuivre)*, and lower right: *C. Pissarro / imp. par Salmon*

Museum of Fine Arts, Boston, Lee M. Friedman Fund (1974.533)

31
Twilight with haystacks
Crépuscule avec meules
1879

aquatint, etching and drypoint, printed in green ink, 10.4 x 18
D23, iii of 3 states
Inscribed in pencil, not by the artist, lower left: *Crépuscule 3e état D.23*, and lower right: *vert anglais*

National Gallery of Canada, Ottawa, purchased 1976 (18724)

32
Twilight with haystacks
Crépuscule avec meules
1879

aquatint, etching and drypoint, printed in red-brown ink
10.4 x 18
D23, iii of 3 states
Inscribed in black chalk, lower left: *no 9 / Epreuve d'artiste / cuivre – crépuscule*, and lower right: *C. Pissarro / imp par E. Degas*, and in pencil, in another hand: *brun rouge*

National Gallery of Canada, Ottawa, purchased 1973 (17292)

33
Twilight with haystacks
Crépuscule avec meules
1879

aquatint, etching and drypoint, printed in Prussian blue ink
10.4 x 18
D23, iii of 3 states
Inscribed in black chalk, lower left: *no 11 / Epreuve d'artiste / (cuivre) Crépuscule*, and lower right: *C. Pissarro / imp par E. Degas*, and in pencil, in another hand: *Bleu de Prusse*

Museum of Fine Arts, Boston, Lee M. Friedman Fund (1983.220)

34
Wooded landscape at L'Hermitage, Pontoise
Paysage sous bois à L'Hermitage, Pontoise
1879

softground etching and aquatint, 21.6 x 26.7
D16 (this state uncatalogued by Delteil); C, i of 6 states
Studio stamp of Edgar Degas, verso (L.657)

Museum of Fine Arts, Boston, Lee M. Friedman Fund (1971.267)

35
Wooded landscape at L'Hermitage, Pontoise
Paysage sous bois à L'Hermitage, Pontoise
1879

softground etching, aquatint and drypoint, 21.6 x 26.7
D16, iv of 5 states;
C, v of 6 states
Inscribed in black chalk, lower left: *no 1–4e état / Paysage sous bois à l'hermitage / Pontoise*, and lower right: *C. Pissarro*

Museum of Fine Arts, Boston, Lee M. Friedman Fund (1971.268)

36
Wooded landscape at L'Hermitage, Pontoise
Paysage sous bois à L'Hermitage, Pontoise
1879

softground etching, aquatint and drypoint, 21.6 x 26.7
D16, v of 5 states;
C, vi of 6 states
Inscribed in pencil, lower right: *C. Pissarro*

Museum of Fine Arts, Boston, Katherine E. Bullard Fund, in memory of Francis Bullard, Prints and Drawings Curator's Discretionary Fund, Cornelius C. Vermeule III and anonymous gifts (1971.176)

37
Horizontal landscape
Paysage en long
1879

aquatint and drypoint
11.5 x 39.3
D17, ii of 2 states;
C, ii of 3 states
Inscribed in black chalk, lower left: *no 1–2e état / paysage en long / Zinc*

The British Museum, London, bequeathed by Campbell Dodgson (1949-4-11-2594)

38
View of L'Hermitage (Pontoise)
Paysage à L'Hermitage (Pontoise)
1880

etching, 11 x 12.3
D28, ii of 2 states
Inscribed in pen and ink, lower left: *no 4 / Epreuve d'artiste / paysage à l'hermitage / cuivre manière grise*, and lower right: *C. Pissarro*

Bibliothèque Nationale de France

39

Field with mill at Osny
Prairie et moulin à Osny
1885

etching, drypoint and aquatint
16 x 23.8
D59, vi of 6 states

Inscribed in black chalk, lower
left: *no 4 / Epreuve d'artiste
(extra) / Prairie et moulin à Osny
(Pontoise)*, and lower right:
C. Pissarro

Bibliothèque Nationale
de France

40

**The Rondest house
at L'Hermitage**
*La maison Rondest
à L'Hermitage*
1882

etching and aquatint
16.5 x 11.2
D35, only state, posthumous
printing of 1920

Inscribed in pencil, lower right:
1/50

Art Gallery of New South
Wales, Sydney, purchased
2004 (132.2004)

41

**Woman emptying a
wheelbarrow**
Femme vidant une brouette
1880

aquatint and drypoint
32 x 23.2
D31, xi of 11 states;
C, xii of 12 states

Inscribed in black chalk, lower
left: *11e état no 1 / femme vidant
une brouette / aquateinte (belle
épreuve)*

Sterling and Francine Clark
Art Institute, Williamstown,
Massachusetts (1962.92)

Late Impressionism

42

**The large pear tree
at Montfoucault**
*Le grand poirier
à Montfoucault*
1876

oil on canvas, 54 x 65
PV360, PDRS462

Kunsthaus Zürich, Johanna
and Walter L. Wolf Collection
(1984/12)

43

**Resting beneath
the trees, Pontoise**
*Dans le bois, le repos,
Pontoise*
1878

oil on canvas, 65 x 54
PV466, PDRS573

Hamburger Kunsthalle/bpk
(1090)

44

**The path to Le Chou,
Pontoise**
*La Sente du Chou,
Pontoise*
1878

oil on canvas, 57 x 92
PV452, PDRS542

Musée de la Chartreuse,
Douai (2231)

45

**The banks of the Viosne
at Osny, grey weather,
winter**
*Bords de la Viosne à Osny,
temps gris, hiver*
1883

oil on canvas, 65.3 x 54.5
PV586, PDRS704

National Gallery of Victoria,
Melbourne, Felton Bequest,
1927 (3466-3)

46

Steep road at Osny
Chemin montant à Osny
1883

oil on canvas, 55.6 x 46.2
PV585, PDRS700

Musée des Beaux-Arts de
Valenciennes (P.46.1.406)

47

The highway
La côte de Valhermeil
1880

oil on canvas, 64.2 x 80
PV510, PDRS623

The Baltimore Museum of Art,
The Cone Collection, formed
by Dr Claribel Cone and Miss
Etta Cone of Baltimore,
Maryland (BMA1950.280)

48

Woman washing dishes
La laveuse de vaisselle
1882

oil on canvas, 81.9 x 64.5
PV579, PDRS685

Lent by the Syndics of the
Fitzwilliam Museum,
Cambridge, bought with the
assistance of the National
Art Collections Fund, 1947
(PD.53-1947)

49

The little country maid
*La petite bonne de
campagne*
1882

oil on canvas, 63.5 x 53
PV575, PDRS681

TATE, bequeathed by Lucien
Pissarro, the artist's son, 1944
(NO.5575)

50

Peasants resting
Paysannes au repos
1881

oil on canvas, 81 x 65.4
PV542, PDRS655

Toledo Museum of Art,
purchased with funds from
the Libbey Endowment, gift
of Edward Drummond Libbey
(1935.6)

Neo-Impressionism

51

**The artist's palette
with a landscape**
*Paysans près
d'une charrette*
c.1878

oil on panel, 24.1 x 34.6
PV454, PDRS562

Sterling and Francine Clark
Art Institute, Williamstown,
Massachusetts (1955.827)

52

Peasants' houses, Éragny
*Maisons de paysans,
Éragny*
1887

oil on canvas, 59 x 71.7
PV710, PDRS844

Art Gallery of New South
Wales, Sydney, purchased
1935 (6326)

53

Ploughing at Éragny
Le labour à Éragny
c.1886

oil on panel, 15.6 x 23.5
PV706, PDRS837

Private collection, courtesy
of Barbara Divver Fine Art,
New York

54

**Apple picking at
Éragny-sur-Epte**
*La cueillette de pommes
à Éragny-sur-Epte*
1888

oil on canvas, 61 x 74
PV726, PDRS850

Dallas Museum of Art,
Munger Fund (1955.17.M)

55
L'Ile Lacroix, Rouen
L'Ile Lacroix à Rouen
1887

drypoint and aquatint
11.6 x 15.7
D69, ii of 2 states

Inscribed in black chalk, lower
left: *no 4 / belle épreuve / L'Ile
Lacroix à Rouen*, and lower right:
C. Pissarro

The British Museum, London,
bequeathed by Campbell
Dodgson (1949-4-11-2612)

56
View of Rouen:
L'Ile Lacroix
Vue de Rouen:
L'Ile Lacroix
1887

aquatint, 12.8 x 17.5
LM131, only state

Bibliothèque Nationale
de France

57
L'Ile Lacroix, Rouen
(effect of fog)
L'Ile Lacroix, Rouen
(effet de brouillard)
1888

oil on canvas, 46.7 x 55.9
PV719, PDRS855

Philadelphia Museum of Art,
John G. Johnson Collection,
1917 (1060)

58
Hampton Court Green
1891

oil on canvas, 54.3 x 73
PV746, PDRS887

National Gallery of Art,
Washington, DC, Ailsa Mellon
Bruce Collection (1970.17.53)

The figure in the
landscape

59
Éragny
1890

graphite and watercolour
17.7 x 25.3

Inscribed in pencil, lower left:
4 mars 1890 – Éragny, and in
black chalk, lower right: *C. Pissarro*

Lent by the Syndics of
the Fitzwilliam Museum,
Cambridge (PD.10-1982)

60
Landscape
Paysage
undated

pencil, 25 x 29

Inscribed in pencil, lower centre:
violet. Studio stamp of Camille
Pissarro, lower right (L.613e)

Private collection, Sydney

61
Springtime, grey weather,
Éragny
Printemps, temps gris,
Éragny
1895

oil on canvas, 60.3 x 73
PV912, PDRS1075

Art Gallery of Ontario, Toronto,
purchased 1933 (2111)

62
Washing day at Éragny
La lessive à Éragny
1901

oil on canvas, 33 x 40.5
PV1058, PDRS1216

Queensland Art Gallery,
Brisbane, purchased 1975
(1:1407)

63
Study for **Church and**
farm at Éragny-sur-Epte
Étude pour *Église et*
ferme d'Éragny-sur-Epte
c.1894

black chalk, 23.8 x 30.7

Studio stamp of Camille Pissarro,
lower right (L.613a)

The Metropolitan Museum of
Art, Harris Brisbane Dick Fund,
1948 (48.10.10)

64
Church and farm
at Éragny-sur-Epte
Église et ferme
d'Éragny-sur-Epte
1894/95

etching printed in grey, red,
yellow and blue ink on
grey-green paper, 15.9 x 24.5
D96, vi of 6 states

Inscribed in pencil, lower left:
*5e état – no 2 / Église et ferme
d'Éragny*, lower right: *Rouge no.
2 / Traits ardoise*, and lower left,
not by the artist: *L.D.96*

National Gallery of Canada,
Ottawa, purchased 1980
(23653)

65
Not illustrated in catalogue
Church and farm
at Éragny-sur-Epte
Église et ferme
d'Éragny-sur-Epte
c.1895

pen and ink, 31.8 x 24.8

Wadsworth Atheneum
Museum of Art, Hartford,
Connecticut, bequest of
Albert R. Olsen (200.24.8)

66
Rear view of a man
in a smock
Homme à sarrau,
vu de dos
undated

black chalk, 19.7 x 14.9

Signed in black chalk,
lower right: *C.P.*

Princeton University Art
Museum, bequest of Dan
Fellows Platt, Class of 1895
(x1948-472)

67
Study for **The poultry**
market, Gisors
Étude pour *Le marché*
à la volaille, Gisors
c.1885

black chalk and pastel
31 x 24.1

Studio stamp of Camille Pissarro,
lower left (L.613a)

National Gallery of Victoria,
Melbourne, purchased through
The Art Foundation of
Victoria, with the assistance of
Mr and Mrs William Jamieson,
Members, 1983 (P22-1983)

68
Lucien Pissarro
after a drawing by
Camille Pissarro
Weeders
Les sarcleuses
1893

colour woodcut, 17.7 x 11.9

National Gallery of Victoria
Melbourne, Felton Bequest,
1914 (651-2)

69
Lucien Pissarro
after a drawing by
Camille Pissarro
Women herb gathering
Femmes faisant de l'herbe
1893

colour woodcut, 18 x 11.9

National Gallery of Victoria,
Melbourne, Felton Bequest,
1914 (652-2)

70
Woman carrying a basket
Femme portant un panier
undated

black chalk, 20.2 x 16

Signed in black chalk,
lower right: *C.P.*

Princeton University Art
Museum, bequest of Dan
Fellows Platt, Class of 1895
(x1948-463)

71
Market at Gisors:
rue Cappeville
Marché de Gisors:
rue Cappeville
1894/95

etching printed in grey, red,
blue and yellow ink, 20 x 14
D112, vii of 7 states

Inscribed in pencil, lower left:
*no 9 ep d'art / Marché de Gisors
(rue Cappeville)*, and lower right:
C. Pissarro

The Baltimore Museum of Art,
purchased with exchange
funds from the bequest of
Mabel Garrison Siemonn, in
memory of her husband George
Siemonn (BMA 1993.77)

72
Sower
Semeur
1896
lithograph on ochre paper
20.7 x 26.8
D155, only state
Inscribed in pencil, lower left:
1er état no 2 / Semeur

The British Museum, London,
bequeathed by Campbell
Dodgson (1949-4-11-3368)

73
The vagrants
Les trimardeurs
1896
lithograph, 24.8 x 29.4
D154, v of 5 states
National Gallery of Victoria,
Melbourne, purchased 1961
(834-5)

74
The ploughman
La charrue
1901
colour lithograph, 22.5 x 15.2
D194, ii of 2 states
Cincinnati Art Museum,
gift of Herbert Greer French
(1940.426)

75
Peasant at a well
Paysanne au puits
1891
etching, 23.2 x 19.4
D101, iii of 3 states,
posthumous printing of 1920
Inscribed in pencil, lower right:
8/50

National Gallery of Victoria,
Melbourne, gift of James
Mollison, 1959 (360-5)

76
Woman kneeling
Femme agenouillée
undated
black chalk, 17.2 x 21.4 (sheet)
Inscribed in black chalk,
lower right: *C.P.*

Yale University Art Gallery,
gift of John M. Montias,
in memory of his father
(1971.65a)

77
**Full-length standing nude
of a woman from behind**
Étude de femme de dos
c.1896
pastel on pink paper
47.5 x 24.3
Studio stamp of Camille Pissarro,
lower right (L.613a)

Lent by the Syndics of
the Fitzwilliam Museum,
Cambridge (PD.51-1947)

78
**Bathers in the shade
of wooded banks**
*Baigneuses à l'ombre
des berges boisées*
1894/95
lithograph, 15.3 x 21.9
D142, ii of 2 states
Inscribed in pencil, lower right:
C. Pissarro

Toledo Museum of Art
(1912.1182)

79
Line of bathers
Théorie de baigneuses
1894/95
lithograph, 13 x 20
D181, only state
Los Angeles County Museum
of Art, gift of Lewis F. Blumberg
and Lynn T. Blumberg, in
honour of the Museum's
twenty-fifth anniversary
(M.89.173.16)

80
Back view of bather
Baigneuse, vue de dos
1894/95
etching, drypoint and aquatint
8.8 x 7.3
D114, v of 5 states,
posthumous printing of 1920
Inscribed in pencil, lower right:
37/50, and lower centre:
*Baigneuse vue de dos, D.114,
IVLIX*

The Baltimore Museum of Art,
Print Fund (BMA 1951.85)

81
Three women bathing
Les trois baigneuses
1894/95
etching and aquatint, 17 x 12.9
D117, ii of 2 states,
posthumous printing of 1923
Inscribed in pencil, lower right:
15/20

Art Gallery of New South
Wales, Sydney, purchased 2004
(133.2004)

City views
and industrial
landscapes

82
**Boulevard Montmartre,
morning, cloudy weather**
*Boulevard Montmartre,
matin, temps gris*
1897
oil on canvas, 73 x 92
PV992, PDRS1166
National Gallery of Victoria,
Melbourne, Felton Bequest,
1905 (204-2)

83
**La Place du Théâtre
Français**
1898
oil on canvas, 72.4 x 92.7
PV1031, PDRS1208
Los Angeles County Museum
of Art, Mr and Mrs George
Gard De Silva Collection
(M.46.3.2)

84
**The Louvre, morning,
rainy weather**
*Le Louvre, matin, temps
de pluie*
1900
oil on canvas, 66.7 x 81.6
PV1157, PDRS1346
Corcoran Gallery of Art,
Washington, DC, Edward C.
and Mary Walker Collection
(37.41)

85
**Morning, winter sunshine,
frost, the Pont-Neuf,
the Seine, the Louvre**
*Matin, soleil d'hiver, gelée
blanche, le Pont-Neuf,
la Seine, le Louvre*
1901
oil on canvas, 96.5 x 115.9
PV1162, PDRS1353
Honolulu Academy of Arts,
gift of Arthur and Kathryn
Murray, 1996 (8439.1)

86
The Louvre under snow
Le Louvre sous la neige
1902
oil on canvas, 65.4 x 87.3
PV1215, PDRS1408
The National Gallery, London,
purchased from Lucien
Pissarro, 1932 (NG4671)

87 •
**The Carrousel,
autumn morning**
*Le Carrousel,
matin d'automne*
1899
oil on canvas, 73.5 x 93
PV1110, PDRS1255
Private collection, Sydney

88
**Afternoon sunshine,
Pont-Neuf**
*L'Après-midi, soleil,
le Pont-Neuf*
1901
oil on canvas, 73 x 92.1
PV1181, PDRS1351
Philadelphia Museum of Art,
bequest of Charlotte Dorrance
Wright, 1978 (1978-1-24)

89
Rue de l'Epicerie, Rouen
1898

oil on canvas, 81.3 x 65.1
PV1036, PDRS1221

The Metropolitan Museum
of Art, purchase, Mr and Mrs
Richard J. Bernhard Gift, 1960
(60.5)

90
**Fair on a sunny
afternoon, Dieppe**
*La foire à Dieppe,
après-midi ensoleillé*
1901

oil on canvas, 73.5 x 92.1
PV1200, PDRS1416

Philadelphia Museum of Art,
bequest of Lisa Morris Elkins,
1950 (1950-92-12)

91
**Place de la République,
Rouen**
*Place de la République,
à Rouen (avec tramway)*
1883

etching, 14.2 x 16.5
D65, ii of 2 states

Inscribed in pencil, lower left:
*no 5 ep d'art / épreuve de choix
/ tirées à 5 ep,* and lower right:
*C. Pissarro / place de la
République / à Rouen (C)*

Bibliothèque Nationale
de France

92
Place du Havre, Paris
Place du Havre, à Paris
1897

lithograph, 14.5 x 21.3
D185, ii of 2 states

Inscribed in purple pencil, lower
left: *Ep defi no 8,* lower centre:
Place du Havre à Paris / sur Z,
and lower right: *C. Pissarro*

The British Museum, London,
bequeathed by Campbell
Dodgson (1949-4-11-3380)

93
**The railway bridge at
Pontoise**
*Le pont du chemin de fer
à Pontoise*
1882

etching, 13 x 24.8
D37, i of 2 states

Inscribed in black chalk, lower left:
*No. 1 – 1er état / Le pont du
chemin de fer à Pontoise / Zinc*

Museum of Fine Arts, Boston,
Horatio G. Curtis Fund (57.745)

94
**The stone bridge and
barges at Rouen**
*Le pont de pierre et
les péniches à Rouen*
1883

oil on canvas, 54.3 x 65.1
PV605, PDRS728

Columbus Museum of Art,
Ohio, gift of Howard D. and
Babette L. Sirak, the donors
to the campaign for Enduring
Excellence and the Derby Fund
(1991.001.053)

95
**The stone bridge in
Rouen, dull weather**
*Le pont de pierre à Rouen,
temps gris*
1896

oil on canvas, 66.1 x 91.5
PV963, PDRS1124

National Gallery of Canada,
Ottawa, purchased 1923 (2892)

96
**Pont Boïeldieu in Rouen,
damp weather**
*Le pont Boïeldieu à Rouen,
temps mouillé*
1896

oil on canvas, 73.6 x 91.4
PV948, PDRS1116

Art Gallery of Ontario, Toronto,
gift of Reuben Wells Leonard
Estate, 1937 (2415)

97
**Sunset, the port of Rouen,
steamboats**
*Coucher de soleil, port de
Rouen, bateaux à vapeur*
1898

oil on canvas, 65 x 81.1
PV1039, PDRS1236

National Museums & Galleries
of Wales (NMW A 2492)

Portraits

98
Woman sewing
Femme qui coude
1895

oil on canvas, 65.4 x 54.4
PV934, PDRS1098

The Art Institute of Chicago,
gift of Mrs Leigh B. Block
(1959.636)

99
Minette
1872

oil on canvas, 45.9 x 35.6
PV197, PDRS282

Wadsworth Atheneum Museum
of Art, Hartford, Connecticut,
The Ella Gallup Sumner and
Mary Catlin Sumner
Collection Fund (1958.144)

100
Portrait of Paul Cézanne
Portrait de Paul Cézanne
1874

etching, 27 x 21.4
D13, only state

Inscribed in pencil, lower left:
à mon ami Luce / C. Pissarro

Private collection

101
Portrait of Eugène Murer
Portrait de Eugène Murer
1878

oil on canvas
64.4 x 54.3 (oval)
PV469, PDRS528

Museum of Fine Arts,
Springfield, Massachusetts,
The James Philip Gray
Collection (52.01)

102 •
Young farmgirl
Jeune paysanne
c.1882

oil on canvas, 38 x 46
PV566, PDRS689

Kunstmuseum Bern, Legat
Eugen Loeb, Bern (G1872)

103
The young maid
La petite bonne
1896

oil on canvas, 61 x 50
PV943, PDRS1111

The Whitworth Art Gallery,
The University of Manchester,
bequeathed by Dr David
Bensusan-Butt, 1998 (O.1998.2)

104
Self-portrait
*Camille Pissarro
par lui-même*
1890–91

etching, 18.5 x 17.7
D90, i of 2 states

Bibliothèque Nationale
de France

105
Self-portrait
*Portrait de Camille
Pissarro par lui-même*
1903

oil on canvas, 41 x 33.3
PV1316, PDRS1528

TATE, presented by Lucien
Pissarro, the artist's son, 1931
(NO.4592)

Contributors

Terence Maloon, curator of *Camille Pissarro*, is senior curator of Special Exhibitions at the Art Gallery of New South Wales, where he has worked in Public Programmes and curated exhibitions for the past eighteen years. He was art critic for the *Sydney Morning Herald* between 1982 and 1987. His recent exhibitions for the Art Gallery of New South Wales include *Picasso: The last decades* (2002), *Drawing the figure – Michelangelo to Matisse* (co-curated with Peter Raissis, 1999) and *Classic Cézanne* (1998). He also curated monographic exhibitions of Tony Tuckson (1989) and Alan Mitelman (1995) for the Museum of Modern Art, Heide, Melbourne, and of Virginia Coventry (2004) for the Drill Hall Gallery, Canberra. A recent major publication was *Sweet Reason: The art of Charles Pollock* (2003), Ball State University, Indiana, USA.

Richard Shiff is Effie Marie Cain Regents Chair in Art at The University of Texas at Austin, where he directs the Centre for the Study of Modernism. He is the author of *Cézanne and the End of Impressionism* (1984) and contributed essays for the Art Gallery of New South Wales exhibitions, *Classic Cézanne* (1998) and *Picasso: The last decades* (2002). His historical and theoretical studies extend from the early nineteenth century to the present. In 2005, he contributed an essay on painting, photography and film for the exhibition, *Impressionnisme et naissance du cinématographe*, held in Lyons, and a study of Gauguin's Impressionism for the exhibition, *Gauguin/van Gogh*, held in Brescia.

Joachim Pissarro, curator in the Department of Painting and Sculpture at The Museum of Modern Art, New York, has taught at Yale University and Hunter College, City University of New York. He is the co-author of the catalogue raisonné of Pissarro's paintings, to be published in late 2005 by the Wildenstein Institute. He has written several books on Impressionism, including a monograph, published in 1993, of Camille Pissarro, his great-grandfather. His latest book, *Altering Egos: Cézanne and Pissarro; Johns and Rauschenberg*, was published in 2005 by Cambridge University Press. He has curated or co-curated the exhibitions, *The Impressionist and the City: Pissarro's Series Paintings* (Dallas, Philadelphia, London, 1993), *Rouen, les Cathédrales de Monet* (Rouen, 1994), *Camille Pissarro, Impressionist Innovator* (Jerusalem, 1995), *Jasper Johns: New Paintings and Works on Paper* (SFMOMA and Yale University Art Gallery, 1999) and *Pioneering Modern Painting: Cézanne and Pissarro 1865–1885* (MoMA, New York, 2005).

Claire Durand-Ruel Snollaerts, an art historian and great-great-granddaughter of the famous art dealer, Paul Durand-Ruel, has devoted the past ten years to researching and writing the Pissarro critical catalogue at the Wildenstein Institute. She has contributed to various exhibitions: *Albert André* (Paris, 1990), *The Impressionist and the City: Pissarro's Series Paintings* (Dallas, Philadelphia, London, 1993), *From Manet to Gauguin, Masterpieces from Swiss Private Collections* (London, 1995), *Les Femmes Impressionnistes: Morisot, Cassatt, Gonzalès* (Tokyo, Hiroshima, Osaka, Hakodate, 1995) and *Camille Pissarro* (Ferrara, 1998).

Peter Raissis is curator of European Prints, Drawings and Watercolours at the Art Gallery of New South Wales. He was co-curator of *Drawing the figure – Michelangelo to Matisse* (1999) and *Albertina: Old Master Drawings from Vienna* (2002). Exhibitions he has organised include *Victorian Watercolours* (2000) and *Whistler to Freud: Etching in Great Britain* (2001). He is co-author, with Richard Beresford, of *The James Fairfax Collection of Old Master Paintings, Drawings and Prints* (2003).

Acknowledgments

My first thanks to the Director of the Art Gallery of New South Wales, Edmund Capon, for assigning me this project three years ago and for his enthusiasm and practical support throughout the process of bringing it to fruition. The exhibition, its associated publications and public programmes are the outcome of very close teamwork. My number one working partner, Erica Drew, managed the exhibition, the production of the catalogue, and much else besides, with her typical one hundred per cent commitment and extraordinary range of diplomatic and organisational skills. Special thanks also to Anne Flanagan, General Manager of Exhibitions, and to Lisa Franey – Anne for having put our team together and for being such a staunch and considerate colleague, and Lisa for remaining unfazed by the flood of correspondence that passed through her hands and for generously volunteering to look after my office while I was travelling overseas.

Peter Raissis, curator of European Prints, Drawings and Watercolours, oversaw the selection of Pissarro's prints and drawings; I thank him for yet another enjoyable and rewarding collaboration. The logistics of transporting an exhibition of this size to Australia and back, with so many lenders from such far-flung places, involved a huge job of work for Charlotte Davy, her predecessor Anna Hayes, Charlotte Cox and Amanda Green, who all proved more than equal to the task. Wayne Chandler from the NSW Treasury Managed Fund looked after the indemnification, and Terry Fahey of Global Specialised Services arranged the conveyance – many thanks to both.

The catalogue and all the graphics associated with the exhibition were designed with customary flair by Analiese Cairis. Anna Macdonald was our meticulous copy editor. Both worked tirelessly to obtain the best results and I congratulate them on their great achievement. Graham Maslen from Spitting Image made an enormous effort, fine-tuning the scans and colour separations. Edwina Brennan took charge of our transparency requests and successfully elicited more than 150 images from more than 50 sources.

As good luck would have it, over the past ten years our expert consultants for the exhibition, Joachim Pissarro and Claire Durand-Ruel Snollaerts, were working together on the new Camille Pissarro catalogue raisonné for the Wildenstein Institute in Paris. They were extremely generous in sharing their very comprehensive, up-to-date knowledge of Pissarro's œuvre and have played a large role in the success of this exhibition. I have a longstanding debt to Richard Shiff, not only for contributing an essay that rounds out so well the content of the catalogue and completes the sense of Pissarro's moral character we have attempted to convey in this exhibition, but for his profound and inspired study of modern art, from which I have gained, and continue to gain, many valuable insights.

The splendour of this exhibition reflects the generosity of our lenders. Our deepest thanks to:

The Art Institute of Chicago James Wood, Douglas Druick, Gloria Groom, Jay Clark, Martha Tedeschi, Suzanne Folds McCullagh • *Philadelphia Museum of Art* Joseph Rishel, Innis Shoemaker, Tom Zarobell, Jennifer Thompson • *Los Angeles County Museum of Art* Andrea Rich, J. Patrice Marandel, Kevin Salatino, Michele Ahern • *The Metropolitan Museum of Art, New York* Philippe de Montebello, Susan Stein, Patrice Mattia • *Kimbell Art Museum, Fort Worth, Texas* Timothy Potts, Malcolm Warner, Liz Johnson, Patty Decoster • *The J. Paul Getty Museum, Los Angeles* Deborah Gribbin, Scott Schaefer, Charlotte Eyerman, Debby Lepp • *Yale University Art Gallery, Connecticut* Jock Reynolds, Suzanne Boorsch, Susan Greenberg, Lisa Hodermarsky, Lynne Addison • *National Gallery of Art, Washington, DC* Earl A. Powell, Philip Conisbee • *Museum of Fine Arts, Boston* Malcolm Rogers, Sue Reed, Patrick Murphy, Nicole Myers, Kate Silverman, Kim Pashko • *Museum of Fine Arts, Springfield, Massachusetts* Heather Haskell • *Fine Arts Museums of San Francisco* Harry S. Parker, Lynn Orr, Marion Stewart • *Dallas Museum of Art* John R. Lane, Dorothy Kosinski • *Memphis Brooks Museum of Art, Tennessee* Kaywin Feldman, Marilyn Masler, Kip Peterson • *Cincinnati Art Museum* Timothy Rub, Kristin Spangenberg, Betsy Wieseman, Rebecca Posage • *Columbus Museum of Art* Nannette V. Maciejunes, Annegreth Nill, Elizabeth Hopkin • *Indianapolis Museum of Art* Anthony G. Hirschel, Ellen W. Lee, Rebekah Marshall • *Toledo Museum of Art, Ohio* Don Bacigalupi, Lawrence W. Nichols, Tom Loeffler, Patricia Whitesides • *Corcoran Gallery of Art, Washington, DC* David C. Levy, Laura Coyle, Terri Anderson • *The Baltimore Museum of Art* Doreen Bolger, Katherine Rothkopf • *Princeton University Art Museum* Susan M. Taylor, Laura Giles, Maureen McCormick • *Wadsworth Atheneum Museum of Art, Connecticut* Willard Holmes, Eric Zafran, Eugene Gaddis, Mary Herbert-Busick • *Sterling and Francine Clark Art Institute, Massachusetts* Michael Conforti, Richard Rand, Jim Ganz, Phyllis Michaelson • *Honolulu Academy of Arts* Stephen Little, Jennifer Saville, Sanna Deutsch • *Fitzwilliam Museum, Cambridge* Duncan Robinson, Jane Monro and Jane Sargent • *TATE, London* Nicholas Serota, Matthew Gale • *The British Museum, London* Anthony Griffiths, Donato Esposito • *The National Gallery, London* Charles Saumarez Smith, David Jaffé, Christopher Riopelle • *The National Gallery of Scotland, Edinburgh* Timothy Clifford • *National Museums & Galleries of Wales, Cardiff* Michael Tooby, Tim Egan • *The Whitworth Art Gallery, The University of Manchester* Graham Allen •

pp. 258–59:

Camille Pissarro (in white jacket) with his family, Éragny, c.1885.
top: sons Georges and Félix
centre: Pissarro with sons Lucien and Ludovic-Rodolphe
seated, from left: the housemaid, Mme Pissarro holding Paul-Émile, daughter Jeanne and Mme Pissarro's niece, Nini

Camille Pissarro at L'Hermitage, Pontoise, c.1875.

Art Gallery of Ontario, Toronto Matthew Teitelbaum, Michael Parke-Taylor, Marcie Lawrence • *National Gallery of Canada, Ottawa* Pierre Théberge • *Musée d'Orsay, Paris* Serge Lemoine, Caroline Mathieu • *Bibliothèque Nationale de France, Paris* Jean-Noël Jeannenay, Claude Bouret • *Musée de la Chartreuse, Douai* Françoise Baligand • *Musée des Beaux-Arts de Valenciennes* Patrick Ramade • *Musées de Pontoise* Christophe Duvivier • *Hamburger Kunsthalle* Uwe M. Schneede, Jens E. Howoldt • *Kunsthaus Zürich* Christoph Becker, Karin Marti • *Kunstmuseum Bern* Matthias Frehner, Judith Durrer • *Kunstmuseum Sankt Gallen* Roland Wäspe, Konrad Bitterli, Alexandra Hänni • *National Gallery of Victoria, Melbourne* Gerard Vaughan, Tony Ellwood, Ted Gott, Alisa Bunbury, Cathy Leahy, Janine Bofill, Tarragh Cunningham, Nicole Monteiro • *Queensland Art Gallery, Brisbane* Doug Hall

To the private lenders who prefer to remain anonymous, our heartfelt thanks – and also to Ully Wille, Akky van Ogtrop, Michael and Judy Gleeson-White, and Joy Glass for their assistance in procuring loans.

Many friends and colleagues helped to alleviate the rigours of travel with the pleasure of their company: Patrice Marandel in Los Angeles; Richard Shiff in Dallas; John and Annasue Wilson in Cincinnati; Gloria Groom in Chicago; Joe Rishel in Philadelphia; Philip Conisbee and Virginia Spate in Washington; Jo Smail and Julien Davis in Baltimore; Michael Parke-Taylor and Barbara Butts in Toronto; Yve-Alain Bois in Cambridge; Gary and Jennifer Wragg, Basil Beattie and John House in London; Sylvia Winter, Patricia Calmès, Hélène Klein, Annette and Alain Bourrut Lacouture, Pierre and Chantal Georgel in Paris; Walter and Maria Feilchenfeldt, Ully and Marie Wille in Zurich. Colleagues in museums who were especially welcoming and hospitable include: Mariantonia Reinhard and Madeleine Gerber at the Museum Oskar Reinhardt am Römerholz; Sylvain Bellenger at the Cleveland Museum of Art; Colin Harrison at the Ashmolean Museum; Eva Maria Preiswerk at the Museum Langmatt Sidney und Jenny Brown; and Eik Khang at the Walters Art Museum. I also must thank Mme Questiaux, Jean Rondot and Sylvain Palfroy in Pontoise; and Alain Mothe – who knows more about Pissarro's painting sites than anyone – for the unforgettable Sunday when he showed me motifs he had discovered in the vicinity of L'Hermitage.

Colleagues in the Conservation department at the Art Gallery of New South Wales, Alan Lloyd, Paula Dredge and Rose Peel, have worked above and beyond the call of duty, as have Claire Martin in Publicity, Belinda Hanrahan, Leith Douglas, Kylie Wingrave and Margaux Simms in Sponsorship and Marketing, Jane Winter from the Art Gallery of New South Wales Foundation, Jenni Carter and Diana Panuccio in the Photography department, and Cara Hickman and Jo Hein in Graphics. The installation crew, painters, workshop and building maintenance staff are the Gallery's quiet achievers. They make an important contribution to the aesthetics and the atmosphere of the Gallery, and I am grateful for their association.

For their moral and practical support, I thank my friends Sylvia Winter, Jean Lévy, Aida Tomescu, Michiel Dolk, Lucia Cascone and Virginia Coventry; my colleagues in Public Programs at the Art Gallery of New South Wales, especially Brian Ladd, Ursula Prunster, Jonathan Cooper, Tristan Sharp, Shoena White and George Alexander; my colleagues in the Art Gallery Society of NSW, especially Judith White, Craig Brush and Jill Sykes; and in the Alliance Française de Sydney, especially Joël Hakim and Valérie Nicolas.

For their help in obtaining research material, I am indebted to Susan Schmocker, Kay Truelove and Robyn Louey in the Art Gallery of New South Wales library; Marie-Christine Maufus and Alexia de Buffévent at the Wildenstein Institute; and the library staffs of the Bibliothèque Fornay and the Musée des Arts-Décoratifs in Paris. Jackie di Diana, Roma Khubchandani and Mathilde Girard volunteered to find, photocopy and transcribe many articles and documents – my thanks for their help.

Robert Herbert curated "Social Turpitudes", a superb programme of films based on Zola and Maupassant stories, to accompany the exhibition, and Marshall McGuire has organised two beautiful concerts. We are delighted to have their contributions. On behalf of Robert Herbert, I would like to thank Chantal Girondin and Sidney Peyrolles, Cultural Attaché of the French Embassy, for their help in locating and transporting films. We gratefully acknowledge the sponsorship of the film programme by the French Embassy in Australia. Our thanks also to Maryanne Leigh from Acoustiguide; Michele Watts from Celestial Harmonies; and SBS television for subtitling and screening Sylvain Palfroy's film, *L'ami Pissarro*.

We are delighted to have the participation of Richard Brettell, John House, Richard Shiff, Joachim Pissarro, Virginia Spate, Ted Gott and Roger Benjamin in the Camille Pissarro conference scheduled to take place at the Art Gallery of New South Wales on 19 November 2005. We thank the Humanities Research Centre of the Australian National University for its contribution to funding this event, and acknowledge Caroline Turner and Ian Donaldson for their support.

Terence Maloon

Curator, *Camille Pissarro*
Sydney, September 2005

Pissarro.